Elements of
Nonequilibrium Statistical Mechanics

Elements of
Nonequilibrium Statistical Mechanics

V. Balakrishnan
Department of Physics
Indian Institute of Technology Madras

Ane Books Pvt. Ltd.
New Delhi ♦ Chennai

Elements of Nonequilibrium Statistical Mechanics

V. Balakrishnan

Price: ₹ 795/-
Pages: xxii + 314

Published by

Ane Books Pvt. Ltd.

4821, Parwana Bhawan, 1st Floor, 24 Ansari Road, Darya Ganj,
New Delhi - 110 002, Tel.: +91(011) 23276843-44, Fax: +91(011) 23276863
e-mail: kapoor@anebooks.com, Website: www.anebooks.com

Branch Office:

- Avantika Niwas, 1st Floor, 19 Doraiswamy Road, T. Nagar,
  Chennai - 600 017, Tel.: +91(044) 28141554, 28141209
  e-mail: anebookschennai@gmail.com, rathinam@anebooks.com

ISBN : 978-93-8015-625-5

Printed at : Milan Enterprises, New Delhi.

To my mother

Preface

Equilibrium statistical mechanics is one of the most beautiful mansions in the vast estate of physics. It is a magnificent edifice, built upon the solid foundations laid by several of the creators of modern physics, including Boltzmann, Maxwell, Gibbs and Einstein. Its underpinnings involve deep concepts regarding the behavior of dynamical systems. It encompasses, and goes well beyond, the subject of thermodynamics. Its domain of applicability is quite extensive, and includes both classical and quantum mechanical systems. Equilibrium statistical mechanics rightfully enjoys a place in the standard curriculum in physics at the university level.

In general, however, systems that are subject to time-dependent phenomena are not in a state of thermal equilibrium. Out-of-equilibrium states are ubiquitous. Their study is therefore of great importance. The class of such states is very large and very diverse. An extension of equilibrium statistical mechanics, such as linear response theory, is possible when a system is 'slightly' out of equilibrium. The role of fluctuations is well understood in this situation. It is this feature that is exploited in linear response theory. But there is, as yet, no overarching theory of nonequilibrium statistical mechanics. In part, this is because of the sheer diversity of the phenomena involved.

A powerful approach to nonequilibrium phenomena and the associated fluctuations is via modeling using random processes, both discrete and continuous. Stochastic differential equations and the related master equations for probability densities play a leading role in this approach. This is because they readily and accurately model a remarkably wide variety of physical situations. The Langevin equation and the Fokker-Planck equation are archetypes of the most useful kinds of stochastic differential equations and master equations, respectively. There is also a fundamental, and most useful, correspondence between the two. In this book, I have used the above as the means to introduce the student to the elements of nonequilibrium statistical mechanics.

The focus is on the simplest physical problem in this regard—which is also the original context in which the Langevin and Fokker-Planck equations were introduced. This is the problem of a particle moving in a fluid in thermal equilibrium, the mass of the particle being much larger than that of a molecule of the fluid. More significantly, *it serves as a convenient medium to study and understand many aspects of nonequilibrium statistical physics, fluctuations and noise.* A glance at the contents of the main chapters, the detailed exercises and the appendixes will show how this strategy has been implemented. In an extended prologue, I have elaborated on this theme and on related matters such as the emphasis, the style, and the overall approach adopted in this book. I would like to stress here that it would be useful and helpful to read this prologue before going on to the main text.

Over and above its intrinsic importance in contemporary physics, nonequilibrium statistical mechanics is applicable to a host of current topics, including nanomaterials and biomolecules, to name a pair of prominent examples. Its relevance is on the increase, justifying its inclusion in the curriculum in its own right.

In writing a book of this sort, the need to strike a balance between pedagogical simplicity and technical accuracy arises continually. One has to curb the almost spontaneous tendency of the book to become a specialized monograph. And yet the need to empower the student to perform calculations independently must be kept in mind. I believe this book can be read not only by graduate (that is, master's level and doctoral) students, but also by senior undergraduate students with some background in thermal physics (or thermodynamics and statistical mechanics). It is my hope that this book will serve as a stepping stone for the student to go on to the quantitative study of the fascinatingly diverse world of nonequilibrium phenomena.

A webpage for this book will be available at the following URL:
http://www.physics.iitm.ac.in/%7Evbalki/eonesm.html

Acknowledgments

I thank my former student, C. Sudheesh, and my colleague, Suresh Govindarajan, for their invaluable help in generating all the figures in this book. I am very grateful to both of them for patiently teaching me some of the nuances of what is, to me at any rate, the arcane and often quixotic set of abbreviations known as UNIX commands. I undertook the 'writing' of this book, from title page to index, directly in LaTeX, and found the experience instructive. Suresh provided sustenance and help whenever the task of straightening out the directories, subdirectories and files seemed to be too daunting to face.

I thank my colleagues in the Department of Physics at IIT Madras, in particular, P. C. Deshmukh, Suresh Govindarajan, Neelima Gupte, S. Lakshmi Bala, Arul Lakshminarayan, Rajesh Narayanan, M. V. Satyanarayana, P. B. Sunil Kumar and Prasanta Tripathy, for providing a congenial work atmosphere and for many interesting discussions on a variety of topics over the years. I should also like to thank specially the considerable number of students of IIT Madras who have enriched my life by their interaction with me for nearly three decades now. Lakshmi Bala and Suresh read the manuscript and made helpful suggestions. Needless to say, any errors that remain are my own.

Parts of this book were written during visits in recent years to the International Centre for Theoretical Physics, Trieste, Italy, and Hasselt University, Diepenbeek, Belgium. I am grateful to my colleagues in these institutions for the warm hospitality I enjoyed during these visits. This book would not have been possible without the affectionate support, constant encouragement, kind advice and unstinting help (both logistic and technical) given by my wife Radha. I am thankful to her more than words can express.

Contents

Prologue

Thermodynamics[1] is an essential part of any undergraduate curriculum in science. This is as it should be: after all, in broad terms, thermodynamics is the 'science of energy'. In somewhat more precise terms, it is the study of the behavior of the *average* values, or *mean* values, of physical quantities pertaining to a system in thermodynamic equilibrium (often abbreviated as **thermal equilibrium**). Familiar examples of such quantities are the internal energy U (the average value of the energy of a system), the pressure P (the average value of the force per unit area exerted by the molecules of a fluid), and so on. Thermodynamics in its pristine form deals almost exclusively with systems in thermal equilibrium. Strictly speaking, time-varying phenomena do not fall within the purview of thermodynamics[2].

Underlying thermodynamics is the subject of **equilibrium statistical mechanics** (ESM). This is the statistical mechanics that forms part of the standard physics curriculum at both undergraduate and master's levels. Indeed, the laws of thermodynamics can be *derived* from equilibrium statistical mechanics. But ESM does more. It helps us understand not only the behavior of average values, but also that of the *scatter* or *dispersion* about the average values, or the *deviations* from the mean values. In other words, ESM enables us to deal with the fluctuations of physical quantities about their mean values, again in states of thermal equilibrium. The extent to which a physical observable is dispersed around its mean value is also of direct physical significance. For instance, the variance of the energy of a system in thermal equilibrium with a heat bath at constant temperature is directly related to its specific heat at constant volume. While quantities such as specific heats are input parameters that must be put in by hand in thermodynamics, they can actually be calculated

---

1 Boldface type will be used for important technical terms on their first (and sometimes, their only) appearance.

2 In this sense, 'thermodynamics' is a misnomer, although the term is now well established in use. 'Thermostatics' might have been a more accurate label. 'Thermostatistics' has also been proposed. Several authors prefer **thermal physics** to encompass thermodynamics along with some statistical physics, kinetic theory, and related matters.

from first principles in ESM. But ESM, too, is essentially applicable only to systems in thermal equilibrium. It does not deal with *time-dependent* phenomena in general. All the **ensembles** usually considered in ESM—the microcanonical, canonical and grand canonical ensembles—pertain to systems in thermal equilibrium.

In real life, however, time-dependent phenomena are the rule rather than the exception. Except under carefully controlled conditions, most physical, chemical and biological systems are **open systems**—that is, they can exchange both energy and matter with other systems with which they are in contact, or with which they interact. In general, they are not in thermal equilibrium. **Transport phenomena**—the transport of energy and matter from one system to another— are of fundamental importance in all such systems. Processes such as convection, advection, diffusion, heat conduction, and all kinds of reactions, are ubiquitous. Thermal equilibrium is often either an idealization or else an asymptotic state in these circumstances. The time-variation of the statistical distributions of systems must be dealt with.

In order to handle such situations, we must go beyond ESM, to **nonequilibrium statistical mechanics** (NESM). But there is a serious difficulty. Recall that ESM rests on the so-called *fundamental postulate* of equilibrium statistical mechanics: all the accessible microstates of an isolated system in thermal equilibrium have equal à *priori* probabilities. Now, it must be said right away that NESM does not, as yet, have a fundamental guiding principle analogous to the fundamental postulate of ESM. This is what makes NESM much more difficult than ESM. We do know a great deal about how systems behave close to a state of thermal equilibrium: how, if slightly disturbed or shifted away from thermal equilibrium, they return to the state of equilibrium. But we do not have a complete understanding of all possible nonequilibrium states, especially those far away from thermal equilibrium. This is what makes the problems of NESM so challenging. Biological systems, for instance, typically operate under far-from-equilibrium conditions. A great deal remains to be done in this area.

In the light of these remarks, I believe that it is important for students to be introduced to at least the rudiments of nonequilibrium statistical mechanics, side by side with their exposure to the fundamentals of equilibrium statistical mechanics. This relatively short book is an attempt to do so. Brevity has been my guiding principle in this regard. I have not, therefore, made any attempt at a systematic exposition of NESM. In fact, I have chosen what is perhaps the simplest nonequilibrium problem to illustrate a few basic techniques. Indeed, it is so simple that it is not even a genuine nonequilibrium problem! Strictly speaking, it is concerned with the *approach* to (thermal) equilibrium in a specific instance. In simplified terms, the problem we address is as follows.

Consider a particle of mass m in a classical fluid that is in a state of thermal equilibrium at an absolute temperature T. The fluid as a whole is at rest. The mean velocity of the particle is zero. Let us consider any Cartesian component of its velocity, say the x-component, and denote this component by v. The mean squared value of v is obtained from the relation $\frac{1}{2}m\,(\mathrm{v}^2)_{\mathrm{eq}} = \frac{1}{2}\,k_B T$, where k_B is Boltzmann's constant. This is an expression of the equipartition theorem for the energy of the system. The probability distribution of v is given by the familiar **Maxwellian distribution of velocities**. The **probability density function** (PDF) of v in thermal equilibrium is a Gaussian function with a peak at v = 0. The width of the Gaussian is proportional to the square root of the absolute temperature. This much is well known from ESM.

Now suppose we measure the x-component of the velocity of the particle at some initial instant of time, say $t = 0$, and obtain the value v_0. That is, the PDF of its velocity is extremely sharply peaked about the value v_0 : in fact, it is just the Dirac delta function $\delta(\mathrm{v} - \mathrm{v}_0)$. As time elapses, it is evident that this velocity will change 'randomly', owing to collisions with the molecules of the fluid. After a sufficient amount of time has elapsed, the distribution of the velocity of the particle may be expected to tend to the equilibrium Maxwellian distribution. How, exactly, does the PDF of the velocity change from its initial δ-function form to its asymptotic Maxwellian form? What sort of statistical distribution does the velocity have at any arbitrary intermediate instant of time?

Simple as these questions appear to be, they lie outside the purview of equilibrium statistical mechanics. In fact, as posed above, they are quite difficult to answer in an elementary treatment. For technical reasons, it turns out to be easier to answer these questions in the case in which the mass m of the particle is much larger than the mass m_{mol} of a molecule of the fluid. This is the case we shall study in this book. In the pages that follow, we shall see how the answers to the questions raised above, and to other related questions, serve as a primer for the study of nonequilibrium statistical mechanics, fluctuations and **noise**. We shall use a *stochastic* approach based on the **Langevin equation**. The main reasons for doing so are as follows:

- The Langevin equation actually represents a certain model. This model serves as a basic paradigm for the application and analysis of random processes in the physical sciences.
- The assumptions and consequences of the model are, by and large, amenable to direct physical interpretation.
- An understanding of the Langevin equation and related topics enables one to analyze a variety of nonequilibrium phenomena in many other contexts.

This book is intended to be used as a supplement as well as a complement to a course on equilibrium statistical mechanics. A certain degree of familiarity with ESM is therefore assumed. So is a knowledge of the basic concepts of probability and statistics, including the rudiments of random processes. I did consider including a relatively expanded outline of these prerequisites, as this might have made the book more self-contained. But this would also have increased the length of the book very considerably. I therefore dropped the idea in order to keep the book focused on its primary objective, opting instead for brief summaries as well as more detailed accounts in appendixes on the relevant material. I have also devoted a chapter to a description of the properties of Markov processes, partly because of the great utility of this topic in all of nonequilibrium physics (including many aspects that are not discussed in this book).

Most of the topics dealt with in this book are included, at least at some basic level, in most university curricula in statistical mechanics. While ESM generally forms the bulk of the syllabus, this inclusion is an acknowledgment of the importance of nonequilibrium phenomena. These topics include (suspending the bold-face 'convention' for the moment!), in particular, the diffusion equation and its fundamental solution; Brownian motion; the Langevin equation; the Fokker-Planck equation and its solution; the Fokker-Planck equation in phase space; diffusion in a potential; diffusion of a charged particle in a magnetic field; fluctuationdissipation relations; the dynamic mobility; the generalized Langevin equation; memory kernels and frequency-dependent damping; and so on. In addition, in order to place the foregoing topics in their proper setting, several other topics are also discussed in this book: a brief account of Markov processes, both continuous and discrete (as already mentioned); the passage from a discrete random walk to the continuum limit of diffusion; diffusion in the presence of boundaries; first-passage time; the properties of the trajectory of a Brownian particle; the power spectra of random processes and pulse sequences, and the theorems associated with them (the Wiener-Khinchin theorem, Campbell's Theorem, Carson's theorem); analyticity properties and dispersion relations for generalized susceptibilities; a résumé of classical linear response theory; stable distributions; and more.

I have attempted to keep the discussion focused, concise, and guided by a certain logical flow of ideas[3]. For this reason, I have presented the material in the form of relatively short chapters for the most part, each devoted to a single theme or a small number of closely related aspects. In calculations, the details of the

3 In this sense, the whole book has been written in the spirit of an extended essay.

algebra have been omitted wherever these are straightforward, in order to save space (and to give the student an opportunity to fill in the steps concerned). However, all significant intermediate steps have been explained in some detail, so that the flow of the logic is not affected. The physical reasons for various mathematical assumptions have been explained carefully, wherever necessary. Detailed proofs of the mathematical theorems or results required have not been given, because doing so would have diverted attention away from the main theme. A set of appendixes covers several aspects and topics relevant to the material in the main text. And, at the risk of sounding pedantic, I have inserted a fairly large number of footnotes. The intention is to accommodate many additional remarks and sidelights without interrupting the text unduly. The reader could even skip the footnotes at first reading, and return to them subsequently[4]. The thrust of the book is on the concepts and principles involved. As a consequence, the discussion is somewhat theoretical and mathematical in nature. This does not imply that the subject itself is devoid of phenomenological content or experimental corroboration—far from it.

Several exercises have been included wherever relevant. These are not of the routine variety meant to test the application of formulas derived in the text. Here, too, I have tried to keep the focus on concepts and principles rather than numerical aspects. Moreover, I have taken the opportunity, in these exercises, to introduce and discuss many related additional topics of interest. In almost all cases, the exercises are of the "Show that..." variety. Therefore the results to be aimed at are also stated explicitly. While the exercises are not quite trivial, they are not very di cult, either. The diligent reader should be able to work out their solutions without too much diffculty.

As the material covered in this book is fairly standard, there is a substantial (indeed, vast) literature available on its various aspects. I have not attempted, therefore, to cite the relevant references or literature in each specific instance. Instead, I have presented a brief bibliography at the end of the book, as suggested additional reading. This bibliography may be consulted for detailed (and more technical) discussions of the matters touched upon in this book, and of course for many other topics as well.

The problems and challenges posed by nonequilibrium phenomena are as interesting as they are formidable. The subject is in its infancy. It is certain to grow in importance, scope and diversity of application. It is well worth getting acquainted with the elements of this fascinating field of inquiry.

---

4 On the other hand, it is not inconceivable that a more knowledgeable reader might find the footnotes interesting in their own right!

Chapter 1

Introduction

Fluctuations play a significant role in physical phenomena. In particular, there are important and deep connections between fluctuations at the microscopic level, and the irreversibility exhibited by phenomena at the macroscopic level.

1.1 Fluctuations

Fluctuations are an inevitable part of natural phenomena. Fluctuations lead to indeterminacy. This makes all of physical science inherently statistical in nature. There are two main sources of fluctuations and indeterminacy in physical systems that are rather obvious: (a) small but uncontrollable external perturbations, and (b) the inevitably finite precision of all measurements. However, over and above these, there are two deeper, more intrinsic, sources of fluctuations in nature: namely, (i) **quantum fluctuations** and (ii) **thermal fluctuations**. As a result, statistical techniques become unavoidable in the physical sciences[1].

At a basic level, such fluctuations are an inevitable consequence of a fundamental *granularity* in nature. If this granularity did not exist, there would be no fluctuations. If matter were not atomistic in nature, if the values of the charges and masses of elementary particles tapered down continuously to zero, if Planck's constant were zero, and so on, all natural phenomena would be describable in terms of partial differential equations like those of classical electromagnetism and hydrodynamics. Even then, **deterministic chaos** would be the rule rather than

[1]In this book, we shall be concerned almost exclusively with classical physics and thermal fluctuations.

the exception. Statistical techniques would again become mandatory.

In systems with a large number of degrees of freedom, in particular, one would expect randomness and irregularity to be endemic. What is astonishing, therefore, is not the occurrence of fluctuations, but rather the existence of well-defined, deterministic, macroscopic laws governing the average behavior of systems, *in spite of* the fluctuations. Ohm's Law in electricity, Hooke's Law in elasticity, Fick's Laws of diffusion and Fourier's Law for heat conduction, are just a few of the large number of physical examples of such macroscopic laws. What makes such simple laws possible in systems with complex internal dynamics?

There are deep reasons for this circumstance, and this fact is exploited in several branches of physics. In practical applications, one might regard fluctuations as noise that is an impediment to obtaining a clear signal. We would then attempt to eliminate or reduce the noise, so as to obtain a better signal-to-noise ratio. But it is also important to recognize that the fluctuations of a system really portray the effects of dynamics at a microscopic level. We can turn them to advantage as a cost-free probe into phenomena at a basic level. It turns out that there are far-reaching connections between fluctuations, on the one hand, and dissipation, instabilities, response to external stimuli, phase transitions, the irreversible behavior of macroscopic systems, etc., on the other (more on this, shortly). One may look upon such connections as *inner consistency conditions satisfied by systems possessing an extremely large number of degrees of freedom.* On the practical side, noise analysis techniques are important diagnostic tools for the study of complex systems such as electrical networks and even nuclear reactors—for instance, in the determination of their so-called '**transfer functions**' or '**response functions**'[2].

What do we mean by the phrase, 'response to an external stimulus'? Consider a system that is in thermal equilibrium. At a specified initial instant of time, we apply to it some 'force' that may be either constant in strength, or varying in time with some given frequency. This force could be manifestly 'external', such as an electric field switched on in the region occupied by the system; or it could be 'internal', such as a concentration gradient present within the system itself. When the force is turned on, the system responds by departing from its initial thermal equilibrium state. In certain cases it may eventually attain some **nonequilibrium steady state**. If the external force is sufficiently weak, the response will be linear (i. e., a linear functional of the applied force). Further, the

[2]It is this aspect of the subject of fluctuations and noise that is emphasized in the context of electrical and electronic engineering, control and automation, etc.

*non*equilibrium state of the system may actually be determinable in terms of the *equilibrium* properties of the system. This is the region of applicability of **Linear Response Theory** (LRT). We shall be concerned primarily with this regime.

What, exactly, is Linear Response Theory? LRT is essentially *equilibrium statistical mechanics combined with first-order time-dependent perturbation theory.* The fundamental importance of the linear response regime is as follows. If a system in thermal equilibrium is perturbed by a weak external force, LRT enables us to determine the departure from equilibrium to first order in the external force. Equivalently, when the external force is switched off, LRT tells us how the system relaxes back to thermal equilibrium. How does LRT achieve this? By making use of **correlation functions**. Autocorrelation and cross-correlation functions generalize the concept of the variance of a random variable, and serve as useful descriptors of the fluctuations in the variables. They will figure prominently in this book. There is a deep and general connection between (i) correlation functions in thermal equilibrium, in the absence of any time-dependent force or perturbation, and (ii) **response functions** in the time-domain or **generalized susceptibilities** in the frequency domain, in the presence of the perturbation. The latter quantify the response of a system to different applied stimuli. Examples of such quantities include frequency-dependent magnetic and electric susceptibilities, dielectric functions, elastic moduli and other mechanical response functions, mobilities, conductivities, and so on. LRT thus enables us to deduce the leading or first-order response of a system to an external stimulus in terms of its spontaneous fluctuations in the absence of such a stimulus.

However, we shall not develop the formalism of LRT itself in this book[3]—rather, as already stated in the prologue, we shall use a *stochastic* approach based on the **Langevin equation**. This is a *model* in which a random force or noise is introduced, with prescribed statistical properties. The reasons for adopting this approach have also been mentioned in the Prologue. It only remains to state that the results to be obtained using this approach will be completely consistent with LRT.

For completeness, we add here that, as the strength of the applied force is further increased, the response of the system may become nonlinear, and instabilities may occur: the possible mode of behavior of the system may bifurcate into several 'branches'. The most stable branch may be very far removed from

---

[3]Except for a brief account of LRT in the classical (as opposed to quantum mechanical) context, in Appendix I.

the original equilibrium state. This is what happens in a laser. As the optical pumping is increased in strength, the black-body radiation emitted in the thermal equilibrium state is perturbed into non-thermal radiation that continues to be weak and incoherent. As the laser threshold is crossed, however, a bifurcation of states occurs. The stable state beyond the threshold is a highly nonequilibrium state corresponding to the emission of monochromatic, coherent radiation of high intensity. In this book, we shall not go into the fascinating subject of NESM in the context of bifurcations and far-from-equilibrium situations. But we mention that extremely interesting **spatio-temporal phenomena** such as **nonlinear waves** and oscillations, **patterns** and **coherent structures** abound here, and are the subject of much current interest and research.

1.2 Irreversibility of macroscopic systems

Let us now turn to an aspect that is directly related to the theme of this book: the irreversible behavior of macroscopic systems that are governed, at the microscopic level, by dynamical laws that are reversible in time.

Consider, as a first example, a paramagnetic solid comprising $N\,(>>1)$ independent atomic or elementary magnetic moments μ. The time evolution of each moment is governed by a suitable 'microscopic' equation of motion—e. g., the rotational analog of Newton's equation for the motion of a classical magnetic moment vector, or perhaps the Schrödinger equation for the wave function of an atom from which the elementary magnetic moment arises, or the Heisenberg equation of motion for the magnetic moment operator. Suppose that, at $t = 0$, the system is in a state *well removed* from thermal equilibrium: for example, suppose all the magnetic moments are lined up in the same direction, so that the total magnetization is $M(0) = N\,\mu$. We shall discuss shortly how such an intuitively improbable state could have arisen. However this may be, we can ask: what is the subsequent behavior of the system? 'Experience' tells us that the net magnetic moment decays from its unusual initial value to its equilibrium value (zero, say) according to an equation of the form

$$M(t) = M(0)\,e^{-t/\tau}. \tag{1.1}$$

τ, called a **relaxation time**, is an important characterizer of the system. It may represent, for instance, the time scale on which magnetic energy is converted to the thermal or vibrational energy of the atoms constituting the paramagnet. What is pertinent here is that a solution of this type is *irreversible* in time: changing t to $-t$ completely alters the character of the solution and leads to an exponential *growth* of $M(t)$, which is totally contrary to our everyday experience. The differential equation obeyed by the macroscopic magnetic moment $M(t)$ is

$$\frac{dM}{dt} = -\frac{M(t)}{\tau}. \tag{1.2}$$

Equation (1.2) is a *first-order* differential equation in t, reflecting the inherently irreversible nature of the relaxation phenomenon.

A second example is provided by the phenomenon of **diffusion** of a molecular species in a fluid medium. If $C(\mathbf{r}\,,\,t)$ denotes the instantaneous concentration of the diffusing species at the point $\mathbf{r}$ at time t, the diffusion current is given by Fick's Second Law, namely,

$$\mathbf{j}(\mathbf{r}\,,\,t) = -D\,\nabla C(\mathbf{r}\,,\,t), \tag{1.3}$$

where D is the diffusion constant[4]. Combining this with the equation of continuity (or Fick's First Law)

$$\frac{\partial C}{\partial t} + \nabla\cdot\mathbf{j} = 0, \tag{1.4}$$

we obtain the **diffusion equation**

$$\frac{\partial C(\mathbf{r}\,,\,t)}{\partial t} = D\,\nabla^2 C(\mathbf{r}\,,\,t). \tag{1.5}$$

This is again of first order in t. Exactly analogous to this situation, we have the equation governing heat conduction in a medium, namely,

$$\frac{\partial T}{\partial t} = \kappa\,\nabla^2 T, \tag{1.6}$$

where κ stands for the thermal conductivity. This, too, is a first-order equation in t. It describes the irreversible phenomenon of heat conduction.

The following question then arises naturally: All these 'irreversible' equations must originate from more fundamental equations at the microscopic level, describing the motion of individual particles or atoms. But such equations, both in classical mechanics and in quantum mechanics, are themselves 'reversible', or *time-reversal invariant.* Examples are Newton's equation of motion for a classical particle,

$$m\frac{d^2\mathbf{r}}{dt^2} = \mathbf{F}(\mathbf{r})\,; \tag{1.7}$$

[4]As you know, the minus sign on the right-hand side of Eq. (1.3) indicates the fact that the diffusion current is directed so as to *lower* the concentration in regions of high concentration.

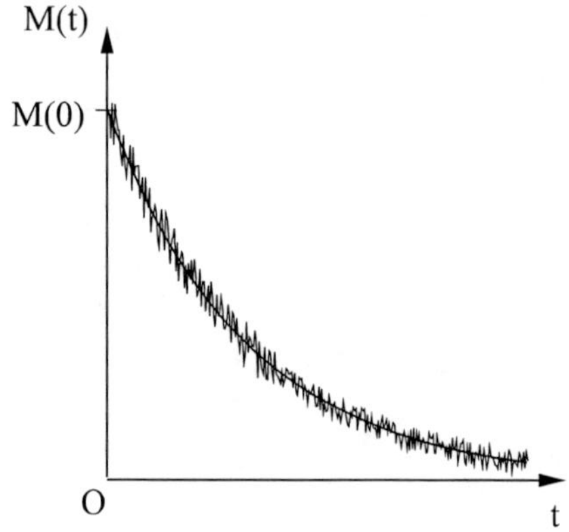

Figure 1.1: Schematic depiction of the decay (or 'relaxation') of the magnetization of a paramagnetic sample as a function of time. The initial magnetized state at $t = 0$ is achieved with the help of an applied field that is switched off at $t = 0$. The subsequent decay of the average or mean value of the magnetization follows the smooth curve shown. It can be modeled by a decaying exponential function of time. Riding on this curve are the thermal fluctuations that are always present, so that the instantaneous magnetization is actually a rapidly fluctuating quantity.

the wave equation for wave propagation in a medium,

$$\left(\frac{1}{c^2}\frac{\partial^2}{\partial t^2} - \nabla^2\right) u(\mathbf{r}\,,\, t) = 0\,; \tag{1.8}$$

the *pair* of Schrödinger equations for the wave function describing a quantum mechanical particle, and its complex conjugate,

$$\left(i\hbar\frac{\partial}{\partial t} - H\right)\psi(\mathbf{r}, t) = 0, \quad \left(i\hbar\frac{\partial\psi^*}{\partial t} + H\right)\psi^*(\mathbf{r}, t) = 0\,; \tag{1.9}$$

and so on. How, then, do the *macroscopic* equations, or rather, the equations satisfied by macroscopic quantities, become *irreversible*?

The answer lies, ultimately, in fluctuations. In the example of the paramagnet given above, it is the *average* magnetization $M(t)$ that decays smoothly to zero as t increases, as described by Eq. (1.1) or (1.2). But the *instantaneous* magnetization does not decrease to zero smoothly. It is in fact a very rapidly fluctuating function of t, being the sum of a very large number of individually and independently varying moments. These fluctuations persist even in thermal equilibrium. Figure

1.1 illustrates the situation (very) schematically. The smooth curve represents the average value $M(t)$ of the magnetization. The irregular, rapidly oscillating curve 'riding' on it represents the instantaneous magnetization of a given sample. The details of this curve are dependent on random fluctuations, and will change from sample to sample even in a set (or **ensemble**) of identical samples prepared so as to have the same initial value of the magnetization. The statistical properties of the 'fluctuations' or 'noise' riding over the mean value are, however, the same for all the samples in the ensemble.

We may regard the initial state with $M(0) = N\mu$ as having originated in one of two very different ways: either as an artificially *prepared* state created with the help of an aligning field that has been switched off; or as the result of a very improbable, but very large, *spontaneous* fluctuation that happened to line up all the moments at some instant of time. As far as its subsequent ($t > 0$) behavior is concerned, the system "does not care" which of these two possibilities led to the initial state. Taking the view that the initial state is a very *rare* spontaneous fluctuation, the system *appears to move irreversibly* (as t increases) toward more commonly occurring states, i. e., toward its most probable or equilibrium macrostate.

However, the system might, with non-vanishing probability, still suffer a fluctuation at some $t > 0$ that puts it back, momentarily, in its original 'rare' state. This would imply a *recurrence* of that state, and we can estimate the period of such a recurrence. For a classical dynamical system described by a set of generalized coordinates and momenta $(q_i\,,\,p_i)$, recurrence occurs when *all* the phase space variables return to some small, specified, neighborhood of their initial values[5]. It is therefore clear that the properties of recurrences— such as the **mean recurrence time**, also called the **Poincaré cycle**— will depend on the details of the way in which we partition phase space into neighborhoods or cells, i. e., on the **coarse-graining** of phase space. Under fairly general conditions, it can be shown that the mean time of recurrence to a given cell in phase space is inversely proportional to the **invariant measure** of the cell. This measure is the stationary *à priori* probability of finding the system in the given cell. Hence, the finer the partitioning of the phase space of a dynamical system, the smaller is the measure of any cell, and hence the larger is the mean time of recurrence to that cell. Further, as the number of degrees of freedom of the system increases to

[5]For strictly periodic motion, such as the trivial case of a single linear harmonic oscillator, the period of recurrence is simply the time period of the motion. But dynamical systems are typically *aperiodic* rather than periodic. This is why recurrence is defined as a return to a neighborhood in phase space, rather than to a precise starting point.

very large values, it becomes more and more improbable that *all* the coordinates and momenta would return simultaneously to the neighborhood of their initial values. As a consequence, the mean recurrence time becomes extremely large. For macroscopic systems, the Poincaré cycle usually turns out to be incomparably larger than the age of the universe! It increases roughly like e^N for N degrees of freedom, in general. For $N \sim 10^{23}$, this is almost inconceivably large—so large that, for all practical purposes, the system appears to behave irreversibly on all possible observational time scales[6]. There is another point that is relevant to the question of macroscopic irreversibility in spite of reversible dynamics at the microscopic level. This aspect will be discussed in Sec. 5.5.

Finally, let us also mention that, under certain special circumstances, spontaneous fluctuations can have another effect that is remarkable. They can alter the behavior of the system on a gross scale in quite a drastic manner. This is what happens in a **continuous phase transition**, at a **critical point**. Taking an example from magnetism once again, it is well known that the Weiss molecular field theory (an example of a **mean field theory**) represents a simple phenomenological model of ferromagnetism. In this theory, the magnetic susceptibility obeys the Curie-Weiss law

$$\chi = \frac{\text{const.}}{T - T_c} \tag{1.10}$$

just above the Curie temperature T_c. Further, the spontaneous (or zero-field) magnetization has the behavior

$$M_0(T) = \begin{cases} 0, & T > T_c \\ \text{const.}\,(T_c - T)^{\frac{1}{2}}, & T \lesssim T_c \end{cases} \tag{1.11}$$

in the vicinity of T_c. But mean field theory only deals with averages of quantities, and essentially neglects their fluctuations. When the latter are also taken into account appropriately, it is found that the susceptibility behaves as

$$\chi = \frac{\text{const.}}{(T - T_c)^{\gamma}} \quad (T \gtrsim T_c), \tag{1.12}$$

where the **critical exponent** γ (called the susceptibility exponent) is roughly equal to 1.3 (and not unity) for standard ferromagnetic materials. Further,

$$M_0(T) \sim (T_c - T)^{\beta} \quad (T \lesssim T_c), \tag{1.13}$$

[6]This is why, in order to see really large deviations from the most probable macrostate, it is quite impractical to wait for a recurrence to a rare state! Instead, one *prepares* the system artificially. For instance, in the case of a paramagnet, we would naturally use an external magnetic field to prepare the aligned state at $t = 0$.

where the critical exponent β (called the order-parameter exponent) is roughly equal to 0.3 (and not $\frac{1}{2}$). Fluctuations thus play a greatly enhanced role in the vicinity of such phase transitions—they alter even the *qualitative* behavior of the system as given by the values of the critical exponents. The correct incorporation of the effects of fluctuations in the critical region is a highly nontrivial problem in equilibrium statistical mechanics. It is solved by the **renormalization group** approach to critical phenomena. This is a method of handling situations in which fluctuations on all length and time scales get coupled to each other, and hence are of equal importance in determining the behavior of the system. Although it originated in quantum electrodynamics, and was first developed systematically in the context of equilibrium phase transitions, it is a general technique that has subsequently found application in many other areas such as percolation, chaotic dynamics and turbulence, to name just a few.

Chapter 2

The Langevin equation

As a prelude, some of the important properties of the Maxwellian distribution of velocities in thermal equilibrium are highlighted. The basic equation of motion of a particle in a fluid incorporating the effects of random molecular collisions, the Langevin equation, is introduced. We bring out the fundamental need to include dissipation in this equation.

2.1 Introduction

There is a very appealing approach to nonequilibrium statistical mechanics that is physically transparent, and is also of considerable use in a number of problems. We may term this the equation-of-motion method or, more accurately, the Langevin equation approach. One first writes an appropriate equation of motion for a subsystem, in which its interaction with the other degrees of freedom of the system is modeled in terms of a stochastic or random 'force' with suitable statistical properties. The task then is to extract the statistical properties of the subsystem of interest, starting from its equation of motion. The problem of the diffusive motion of a particle immersed in a fluid (the 'tagged' particle) offers the clearest illustration of the technique. This is the physical problem we shall pursue in this book.

It is worth understanding the genesis, or background, of the problem. Ideally, it would be very convenient if we could take the tagged particle to be a molecule of the fluid itself. After all, one of the important issues in nonequilibrium statistical mechanics concerns the time-dependent statistical properties of macroscopic systems such as fluids. However, the Langevin equation method in its original or simplest form turns out to be too drastic an approximation to be applicable, *as it stands*, to this case. At the molecular level, the behavior of fluids poses a complicated many-body problem. In principle, a rigorous approach to this problem leads to an infinite hierarchy of integro-differential equations for the statistical distribution functions involved, called the **BBGKY hierarchy**[1]. Under certain conditions, this infinite hierarchy can be replaced by a simpler system of equations, and subsequently further reduced to the **Boltzmann equation**. The latter is a convenient starting point in several physical situations. However, in spite of its apparently simple form, this approach is still quite complicated from a technical point of view. The Boltzmann equation itself can be replaced, in specific instances, by a more tractable equation, the **Fokker-Planck equation**. In particular, this approximation is valid (in the context of the fluid system) for the description of the motion of a particle of mass m in a fluid consisting of molecules of mass $m_{\mathrm{mol}} \ll m$. And, lastly, the Fokker-Planck equation is just another facet of the Langevin equation approach, as we shall see at length. Before getting down to the details, we reiterate:

- Over and above the motivation provided by the particular physical problem described above, the importance of the Langevin equation method lies in the adaptability and applicability of the stochastic approach to a wide class of nonequilibrium problems. This is its true significance.

It is desirable to have a good understanding of thermal equilibrium in order to appreciate nonequilibrium phenomena. We therefore begin with a brief recapitulation of the Maxwellian distribution of velocities in thermal equilibrium, and some of its important properties. Some other interesting properties of this distribution are considered in the exercises at the end of this chapter.

2.2 The Maxwellian distribution of velocities

Consider a particle of mass m immersed in a classical fluid in thermal equilibrium at an absolute temperature T. The distribution of velocities of the molecules of the fluid is given by the well-known Maxwellian distribution. This remains true for the tagged particle as well, with m_{mol} replaced by m in the distribution function. In all that follows, we concentrate (for simplicity) on any one Cartesian component of the velocity of the particle[2], and denote this component by v. The

[1] Named after Bogoliubov, Born, Green, Kirkwood and Yvon.

[2] The generalization of the formalism to include all three components of the velocity is straightforward. Whenever necessary, we shall also write down the counterparts of important formulas in which all the velocity components are taken into account. We shall use $\mathbf{v}$ to denote the velocity vector.

normalized equilibrium probability density function (PDF) of v, in a state of thermal equilibrium at a temperature T, is a Gaussian[3]. It is given by

$$p^{\rm eq}(v) = \left(\frac{m}{2\pi k_B T}\right)^{\frac{1}{2}} \exp\left(-\frac{mv^2}{2k_B T}\right). \tag{2.1}$$

The temperature-dependent factor in front of the exponential is a normalization constant. It ensures that

$$\int_{-\infty}^{\infty} dv\; p^{\rm eq}(v) = 1 \tag{2.2}$$

at any temperature. Figure 2.1 is a sketch of $p^{\rm eq}(v)$ as a function of v for two different temperatures T_1 and T_2, with $T_2 > T_1$. The total area under each curve is unity, in accord with Eq. (2.2). The mean value (or average value) of v and its mean squared value, in a state of thermal equilibrium, are given by

$$\langle v \rangle_{\rm eq} = \int_{-\infty}^{\infty} v\, p^{\rm eq}(v)\, dv = 0 \tag{2.3}$$

and

$$\left\langle v^2 \right\rangle_{\rm eq} = \int_{-\infty}^{\infty} v^2\, p^{\rm eq}(v)\, dv = \frac{k_B T}{m}, \tag{2.4}$$

respectively. Hence the **variance** of the velocity, defined as[4]

$$\left\langle \left(v - \langle v \rangle_{\rm eq}\right)^2 \right\rangle_{\rm eq} \equiv \left\langle v^2 \right\rangle_{\rm eq} - \langle v \rangle_{\rm eq}^2, \tag{2.5}$$

is again $k_B T/m$, and its **standard deviation** $(k_B T/m)^{1/2}$.

- The combination $(k_B T/m)^{1/2}$ represents the characteristic speed scale in the problem. It is often referred to as the **thermal velocity**.

The full-width-at-half-maximum (FWHM) of a Gaussian is directly proportional to its standard deviation. The FWHM of the Maxwellian distribution is given by $(8k_B T \ln 2/m)^{1/2} \simeq 2.355\,(k_B T/m)^{1/2}$. Equation (2.1) and its consequences follow directly from the **canonical Gibbs distribution** of equilibrium statistical mechanics.

---

[3]**Gaussian random variables** will occur very frequently in all that follows in this book. Some of the important properties of the Gaussian distribution are reviewed in Appendix D.

[4]Some basic features of the moments of a random variable and related quantities are listed in Appendix C.

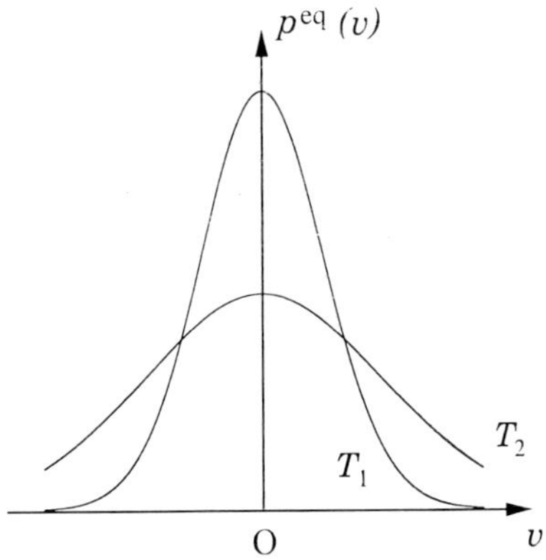

Figure 2.1: The normalized Maxwellian probability density function $p^{eq}(v)$ of a component of the velocity, Eq. (2.1), at two different temperatures T_1 and T_2 ($T_2 > T_1$). Each graph is a Gaussian centred at $v = 0$, with a width proportional to the square root of the absolute temperature. The total area under each curve is equal to unity.

Equation (2.1) gives the probability density function of (a Cartesian component of) the velocity. The corresponding **cumulative distribution function**[5] is given by

$$\mathcal{P}^{eq}(v) = \int_{-\infty}^{v} dv'\, p^{eq}(v'). \tag{2.6}$$

Thus, for each value of v, $\mathcal{P}^{eq}(v)$ is the area under the curve in Fig. 2.1 from $-\infty$ up to the value v of the abscissa. It is the total probability that the velocity of the tagged particle is less than or equal to the value v. $\mathcal{P}^{eq}(v)$ is a monotonically increasing function of v. It starts at the value $\mathcal{P}^{eq}(-\infty) = 0$, reaches the value $\frac{1}{2}$ at $v = 0$, and saturates to the value $\mathcal{P}^{eq}(\infty) = 1$. Using Eq. (2.1) for $p^{eq}(v)$ in Eq. (2.6), we find that the integral can be written in terms of an **error function**[6]. The result is

$$\mathcal{P}^{eq}(v) = \frac{1}{2}\left[1 + \text{erf}\left(v\sqrt{\frac{m}{2k_BT}}\right)\right]. \tag{2.7}$$

Figure 2.2 shows the cumulative distribution $\mathcal{P}^{eq}(v)$ as a function of v.

[5]In physics, probability *density* functions are often referred to as 'probability *distributions*' or just 'distributions', for short. This is somewhat loose terminology, but it should cause no confusion—although it is good to be careful with terminology.

[6]The error function is defined as $\text{erf}\,(x) = (2/\sqrt{\pi})\int_0^x du\, e^{-u^2}$. See Appendix A, Sec. A.2.

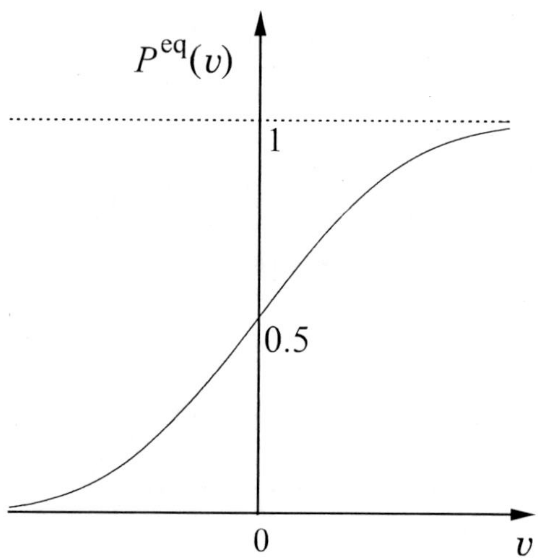

Figure 2.2: The cumulative distribution function $\mathcal{P}^{eq}(v)$ of a component of the velocity, at some fixed temperature. $\mathcal{P}^{eq}(v)$ is the probability that the velocity of the particle is less than or equal to v. The width of the 'kink' around the origin is again set by the characteristic speed scale in the problem, $(k_B T/m)^{1/2}$. As the probability density function $p^{eq}(v)$ is a Gaussian, the distribution function $\mathcal{P}^{eq}(v)$ can be expressed in terms of an error function (see Eq. (2.7)).

Equations (2.3) and (2.4) give the first two moments of the velocity in the state of thermal equilibrium. What can we say about the higher moments of v? The equilibrium PDF $p^{\mathrm{eq}}(v)$ is a Gaussian. Now, a Gaussian is a two-parameter distribution in which *all* the moments of the random variable are determined[7] in terms of the mean and the mean square (or, equivalently, the variance). The fact that $p^{\mathrm{eq}}(v)$ is a symmetric or even function of v implies that all the *odd* moments of v vanish identically. To determine the even moments, we may make use of the definite integral

$$\int_0^\infty dx\, x^r\, e^{-ax^2} = \Gamma\left(\tfrac{1}{2}(r+1)\right) \Big/ 2a^{(r+1)/2} \quad (a > 0\,,\, r > -1), \tag{2.8}$$

where $\Gamma(x)$ is the gamma function[8]. Using the elementary properties of the gamma function, we find the following result for the even moments of the velocity in thermal equilibrium:

$$\left\langle v^{2l} \right\rangle_{\mathrm{eq}} = \frac{(2l-1)!}{2^{l-1}\,(l-1)!} \left(\frac{k_B T}{m}\right)^l, \tag{2.9}$$

where $l = 1, 2, \ldots$. Hence

$$\left\langle v^{2l} \right\rangle_{\mathrm{eq}} = \frac{(2l-1)!}{2^{l-1}\,(l-1)!} \left\langle v^2 \right\rangle_{\mathrm{eq}}^l . \tag{2.10}$$

This relationship is characteristic of a Gaussian (with zero mean). *All* the even moments of the velocity are determined in terms of its second moment. In particular, note that

$$\frac{\left\langle v^4 \right\rangle_{\mathrm{eq}} - 3 \left\langle v^2 \right\rangle_{\mathrm{eq}}^2}{\left\langle v^2 \right\rangle_{\mathrm{eq}}^2} = 0. \tag{2.11}$$

For any general random variable, the counterpart of the combination on the left-hand side of Eq. (2.11) is called the **excess of kurtosis**. This quantity vanishes identically for a Gaussian distribution[9].

2.3 Equation of motion of the tagged particle

All the foregoing results are well known. They lie within the purview of equilibrium statistical mechanics. We would now like to go further. Suppose we measure the

[7] See Appendix D, Eqs. (D.1) and (D.2).

[8] See Appendix B, Eq. (B.9).

[9] See Appendix D, Sec. D.2.

x-component of the velocity of the tagged particle at some instant t_0, and find that its value is v_0. The tagged particle represents our subsystem. It is in thermal equilibrium with the molecules of the fluid (the heat bath in which the subsystem is immersed). The latter cause rapid but very small changes, or fluctuations, in the velocity of the tagged particle by means of collisions, and this is essentially a random process. Hence v is a random variable, and we may ask: what are its statistical properties? In particular, what is the probability distribution of the velocity v of the tagged particle at any later instant of time $t > t_0$? This is given by a *conditional* PDF which we denote by $p(v\,,\,t\,|\,v_0\,,\,t_0)$. As the system is in thermal equilibrium, the statistical properties of the velocity do not undergo any systematic change in time: in other words, the velocity is a **stationary random variable**. This implies that $p(v\,,\,t\,|\,v_0\,,\,t_0)$ is a function of the elapsed time, or the time *difference* $(t - t_0)$ alone, rather than a function of the two time arguments t and t_0 individually[10]. In effect, we may then set $t_0 = 0$, and simply write $p(v\,,t\,|\,v_0)$ for the conditional PDF. The initial condition on this PDF is of course a Dirac δ-function by definition,

$$p(v\,,0\,|\,v_0) = \delta(v - v_0), \tag{2.12}$$

since we started with the *given* initial velocity v_0. And, as $t \to \infty$, we might expect on physical or intuitive grounds that the average value of v would approach its equilibrium value of zero, *no matter* what v_0 was. In fact, we would expect the conditional PDF $p(v\,,t\,|\,v_0)$ itself to lose its dependence on the initial velocity v_0 gradually, as t becomes very large. In the limit $t \to \infty$, we would expect it to tend to the equilibrium PDF, i. e.,

$$\lim_{t\to\infty} p(v\,,t\,|\,v_0) = p^{\text{eq}}(v). \tag{2.13}$$

Are these guesses borne out? Can we give a physically realistic model for this approach to equilibrium? This is the problem we address now.

We begin with Newton's equation of motion[11] for the tagged particle,

$$m\dot{v}(t) = F(t), \tag{2.14}$$

where $F(t)$ is the *total* force on the particle. In order to see how equilibrium and nonequilibrium properties are inter-related, let us formally include the effect of an

[10] We shall say more about the stationarity of random variables subsequently.

[11] As already stated, we restrict ourselves to classical dynamics throughout this book. An extension of these considerations to quantum dynamics is possible to a certain extent, but quite nontrivial.

applied, external force as well. Then $F(t)$ is given by the sum

$$F(t) = F_{\text{int}}(t) + F_{\text{ext}}(t), \tag{2.15}$$

where $F_{\text{ext}}(t)$ is the resultant of all external, applied forces (if any), while $F_{\text{int}}(t)$ is the internal force arising from the bombardment of the molecules of the fluid (the 'heat bath' in which our subsystem is immersed). $F_{\text{int}}(t)$ is a *random* force: it is random because we focus on the subsystem (the tagged particle) alone, and we do not consider the details of the simultaneous evolution of the velocities of each of the individual particles of the fluid[12]. As a result, the velocity $v(t)$ of the tagged particle becomes a random variable. To proceed, we need more information about the random force F_{int}, or at least about its *statistical* properties.

It turns out that internal consistency requires that F_{int} be made up of two distinct parts, according to

$$F_{\text{int}}(t) = \eta(t) + F_{\text{sys}}(t). \tag{2.16}$$

Here $\eta(t)$ is a 'truly random' force, or noise, that is zero on the average, and is independent of the state of motion of the tagged particle. On the other hand, $F_{\text{sys}}(t)$ is a 'systematic' random force that depends on the state of motion of the tagged particle: it should, for instance, prevent very large velocity fluctuations from building up. This is to be expected on physical grounds. Roughly speaking, if the speed of the tagged particle is very large at some instant of time, more molecules would bombard it (per unit time) from the direction opposite to its direction of motion (i. e., from its 'front') than from its 'rear', and the overall

[12]This is an important footnote! It would, of course, be quite impractical, if not impossible, to keep track of the dynamics of $N \gtrsim 10^{23}$ particles. The price we pay for ignoring the details of the motion of the 'bath' degrees of freedom is the appearance of randomness in the dynamics of a single particle—even though we begin with Newtonian dynamics that is, in principle, quite deterministic and hence nonrandom.

But even more remarkable is the following fact. Even if the number of particles is quite small, and they are in interaction with each other, the system would, in general, display what is known as deterministic chaos. Incredibly enough, as low a value as $N = 3$ suffices to produce chaos! As a result of chaotic dynamics, computability and predictability are lost. The rate at which this happens is quantified by the **Liapunov exponents** of the dynamical system. It can be shown that the maximal Liapunov exponent increases like the number of degrees of freedom of the system. Hence there is absolutely no option but to take recourse to statistical methods (including, but not restricted to, computational techniques such as numerical simulation).

Regrettably, we cannot go into these aspects in this book, save for a few further remarks made in Sec. 5.5. But these and related matters concern subtle points that lie at the very foundations of statistical physics. They must be pondered upon seriously. The subject continues to be a topic of intensive current research in many directions. The student reader should find it encouraging to learn that many questions still await definitive answers.

effect would be a frictional or systematic force that slows it down. The simplest assumption we can make under the circumstances is based on our experience with macroscopic mechanics. We assume that $F_{\text{sys}}(t)$ is proportional to the instantaneous velocity of the tagged particle, and is directed *opposite* to it. Then

$$F_{\text{sys}}(t) = -m\,\gamma\,v(t), \tag{2.17}$$

where γ (a positive constant) is the **friction coefficient**. Its reciprocal has the physical dimensions of time. It must be kept in mind that Eq. (2.17) represents a *model*. Its validity must be checked by the validity of the consequences it predicts[13]. Using Eqs. (2.15)-(2.17) in Eq. (2.14), the equation of motion of the tagged particle becomes

$$m\dot{v}(t) = -m\,\gamma\,v(t) + \eta(t) + F_{\text{ext}}(t), \tag{2.18}$$

with the initial condition $v(0) = v_0$. This is the famous Langevin equation in its original form, once certain properties of $\eta(t)$ are specified. It is a linear **stochastic differential equation** (SDE) for the velocity of the tagged particle: v is the *driven* variable, while η is the *noise* that induces randomness in the velocity.

- The fundamental task is to extract the statistical properties of v, given those of η.

There is more than one way to do so. We shall use an approach that remains close to the physics of the problem. As a preliminary step, we write down the *formal* solution to Eq. (2.18). The linear nature of the differential equation makes this a straightforward task. The solution sought is

$$v(t) = v_0\,e^{-\gamma t} + \frac{1}{m}\int_0^t dt_1\,e^{-\gamma(t-t_1)}\,[\eta(t_1) + F_{\text{ext}}(t_1)]. \tag{2.19}$$

It will be recognised readily that the first term on the right-hand side is the 'complementary function', while the second term is the 'particular integral'. The former is chosen so as to satisfy the initial condition imposed on the solution. Equation (2.19) will be our starting point in the next chapter.

[13]A good question to ask is whether we could have started by ignoring F_{sys} altogether. This would have led us to an inconsistency (as we shall see in the Exercises at the end of Ch. 3). We could follow this wrong route and then back-track to the correct one. But it is less confusing to proceed along the correct lines from the start. We can always see *post facto* what dropping F_{sys} would have led to, by setting $\gamma = 0$ in the results to be deduced.

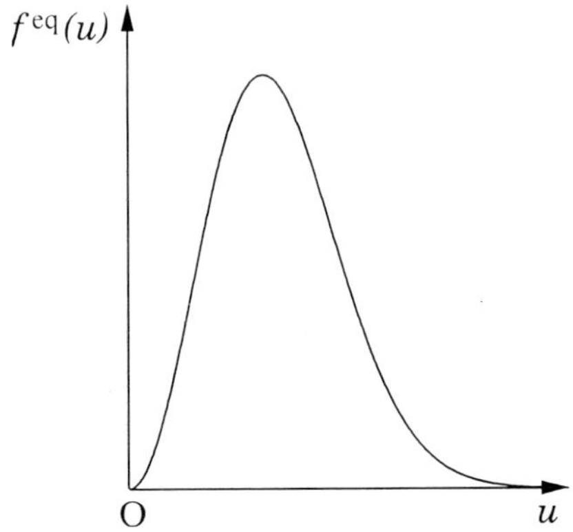

Figure 2.3: The probability density of the speed u of the tagged particle according to the Maxwellian distribution of velocities, Eq. (2.20). There is only one speed scale in the problem, and this is set by the combination $(k_BT/m)^{1/2}$. The PDF has a leading behavior proportional to u^2 at small values of u, i. e., for $u \ll (k_BT/m)^{1/2}$. It peaks at a speed $u_{\text{peak}} = (2k_BT/m)^{1/2}$, which is the most probable speed. The mean speed $\langle u \rangle_{\text{eq}} = (8k_BT/m\pi)^{1/2}$, while the root mean squared (or r.m.s.) speed $\langle u^2 \rangle_{\text{eq}}^{1/2} = (3k_BT/m)^{1/2}$. Hence $u_{\text{peak}} < \langle u \rangle_{\text{eq}} < \langle u^2 \rangle_{\text{eq}}^{1/2}$.

2.4 Exercises

2.4.1 Moments of the speed in thermal equilibrium

As already stated, Eq. (2.1) gives the PDF of each Cartesian component of the velocity of the tagged particle, in a state of thermal equilibrium. Let $u = |\mathbf{v}|$ denote the *speed* of the particle in three dimensions, and let $f^{\text{eq}}(u)$ be the corresponding normalized PDF of u. To find this PDF, we may write out the joint PDF in the three Cartesian components: this is a product of three PDFs, each of the form given by Eq. (2.1). Next, we change to spherical polar coordinates for the velocity and integrate over the angular coordinates. The result is

$$f^{\text{eq}}(u) = \left(\frac{m}{2\pi k_BT}\right)^{3/2} 4\pi u^2 \exp\left(-\frac{mu^2}{2k_BT}\right), \quad 0 \le u < \infty. \tag{2.20}$$

Figure 2.3 depicts $f^{\text{eq}}(u)$ as a function of u. Using the definite integral quoted in Eq. (2.8), it is easy to recover the well-known result

$$\langle u \rangle_{\text{eq}} = \int_0^\infty du\, u\, f^{\text{eq}}(u) = \left(\frac{8k_BT}{m\pi}\right)^{1/2} \tag{2.21}$$

for the mean speed of the particle.

(a) Verify that the even and odd moments of the speed are given by

$$\left\langle u^{2l}\right\rangle_{\text{eq}} = \frac{(2l+1)!}{2^l\, l!}\left(\frac{k_BT}{m}\right)^l \tag{2.22}$$

and

$$\left\langle u^{2l+1}\right\rangle_{\text{eq}} = \frac{2^{l+\frac{3}{2}}\,(l+1)!}{\sqrt{\pi}}\left(\frac{k_BT}{m}\right)^{l+\frac{1}{2}}, \tag{2.23}$$

respectively, where $l = 0,\ 1,\ 2,\ \ldots$.

(b) The mean value of the *reciprocal* of the speed may be found by setting $l = -1$ in the expression written down above[14] for $\langle u^{2l+1}\rangle_{\text{eq}}$. We get

$$\langle u^{-1}\rangle_{\text{eq}} = \left(\frac{2m}{\pi k_BT}\right)^{1/2}. \tag{2.24}$$

Therefore

$$\langle u^{-1}\rangle_{\text{eq}} > \langle u\rangle_{\text{eq}}^{-1}. \tag{2.25}$$

Similarly, the mean value of u^{-2} is found to be

$$\langle u^{-2}\rangle_{\text{eq}} = 4\pi\left(\frac{m}{2\pi k_BT}\right)^{3/2}\int_0^\infty du\,\exp\left(-\frac{mu^2}{2k_BT}\right) = \frac{m}{k_BT}. \tag{2.26}$$

On the other hand, the mean squared speed (recall that the motion is in three dimensions) is

$$\langle u^2\rangle_{\text{eq}} = \frac{3k_BT}{m}. \tag{2.27}$$

Once again, therefore,

$$\langle u^{-2}\rangle_{\text{eq}} > \langle u^2\rangle_{\text{eq}}^{-1}. \tag{2.28}$$

Establish the inequalities (2.25) and (2.28) *without* explicitly evaluating the averages involved in these. (Use an appropriate version of the Cauchy-Schwarz inequality in the space of a certain class of functions of u.)

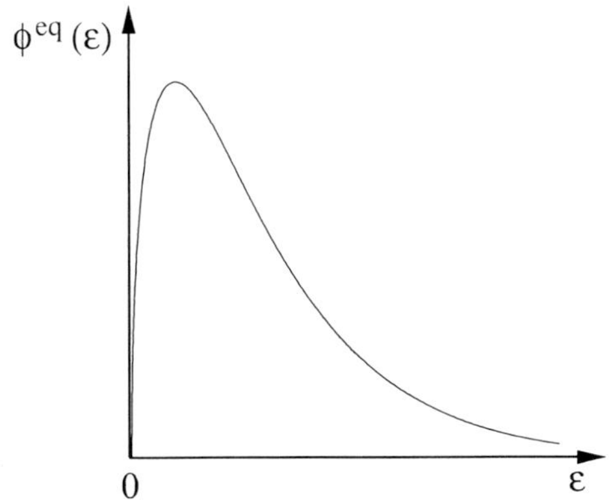

Figure 2.4: The probability density of the (kinetic) energy of the tagged particle, Eq. (2.29), according to the Maxwellian distribution of velocities. The PDF is essentially the product of the density of states, which is proportional to $\varepsilon^{1/2}$, and the Boltzmann factor, $\exp(-\varepsilon/k_B T)$. It has a leading behavior $\sim \varepsilon^{1/2}$ for $\varepsilon \ll k_B T$, and falls of exponentially rapidly for very large values of the energy ($\gg k_B T$).

2.4.2 Energy distribution of the tagged particle

From Eq. (2.20) for the PDF of the speed of the tagged particle, show that the probability density function $\phi^{\text{eq}}(\varepsilon)$ of its energy $\varepsilon = \frac{1}{2}mu^2$ is given by

$$\phi^{\text{eq}}(\varepsilon) = \frac{2}{\sqrt{\pi}} \left(\frac{1}{k_B T}\right)^{3/2} \varepsilon^{1/2} \exp\left(-\frac{\varepsilon}{k_B T}\right). \tag{2.29}$$

Figure 2.4 is a sketch of $\phi^{\text{eq}}(\varepsilon)$ as a function of ε. You should recognize the factor $\varepsilon^{1/2}$ as just the energy-dependence of the **density of states** of a free nonrelativistic particle moving in three-dimensional space.

2.4.3 Distribution of the relative velocity between two particles

Consider *two* different particles of mass m immersed in the fluid in thermal equilibrium. We assume that they move completely independent of each other, and that

[14]Recall that the integral in Eq. (2.8) is actually convergent for all $r > -1$. The integral involved in $\langle u^{-1}\rangle_{\text{eq}}$ corresponds to the case $r = 1$. Similarly, the integral involved in $\langle u^{-2}\rangle_{\text{eq}}$ corresponds to the case $r = 0$, and is also finite. However, $\langle u^{-l}\rangle_{\text{eq}}$ diverges (i. e., is infinite) for all $l \geq 3$.

they do not interact in any manner. Let v_1 and v_2 denote their respective velocities. Each of these random variables has a PDF given by Eq. (2.1). We may ask: what is the normalized probability density of the *relative* velocity $v_1 - v_2 \equiv v_{\text{rel}}$ of this pair of particles? Denoting this PDF by $F^{\text{eq}}(v_{\text{rel}})$, the formal expression for this function is

$$F^{\text{eq}}(v_{\text{rel}}) = \int_{-\infty}^{\infty} dv_1 \int_{-\infty}^{\infty} dv_2\, p^{\text{eq}}(v_1)\, p^{\text{eq}}(v_2)\, \delta\big(v_{\text{rel}} - (v_1 - v_2)\big)\,. \tag{2.30}$$

Study the expression on the right-hand side in Eq. (2.30), and convince yourself that it is indeed the correct 'formula' for the PDF required. Its validity depends on the fact that the two random variables v_1 and v_2 are statistically *independent* random variables. The Dirac δ-function in the integrand ensures that you integrate over all possible values of v_1 and v_2 *such that* their difference is equal to v_{rel}. The δ-function 'constraint' enables us to carry out one of the integrations directly[15]. The second integration also involves a Gaussian integral, but with a slightly modified form of integrand. In order to perform this integration, you will need the definite integral[16]

$$\int_{-\infty}^{\infty} dx\, \exp\left(-ax^2 + bx\right) = \sqrt{\frac{\pi}{a}}\, \exp\left(\frac{b^2}{4a}\right), \tag{2.31}$$

where $a > 0$ and b is an arbitrary number (not necessarily real). Show that

$$F^{\text{eq}}(v_{\text{rel}}) = \left(\frac{m}{4\pi k_B T}\right)^{\frac{1}{2}} \exp\left(-\frac{m v_{\text{rel}}^2}{4 k_B T}\right). \tag{2.32}$$

In other words, the PDF of the relative velocity is also a Gaussian, but with a variance that is twice that of the velocity of each individual particle.

2.4.4 Generalization to the case of many particles

A generalization of the result just derived is as follows. Consider n tagged particles of mass m in the fluid. Let us assume that the concentration of these particles is vanishingly small, and that they do not interact with one another, and move independently of each other. Let $V = (v_1 + v_2 + \cdots + v_n)/n$ be the velocity of

---

[15]Similar formulas for the PDFs of combinations of independent random variables will be used several times elsewhere in this book. In all cases, the δ-function enables you to 'do' one of the integrals at once. You may need to use the following properties of the δ-function: $\delta(x) = \delta(-x)$ and $\delta(ax) = (1/|a|)\,\delta(x)$, where x is the variable of integration.

[16]See Appendix A, Eq. (A.4).

their center of mass. The PDF of V is (using the same symbol $F^{\rm eq}$ as before, for convenience)

$$F^{\rm eq}(V) = \int_{-\infty}^{\infty} dv_1 \cdots \int_{-\infty}^{\infty} dv_n \; p^{\rm eq}(v_1) \cdots p^{\rm eq}(v_n) \; \delta\left(V - \frac{v_1 + \cdots + v_n}{n}\right). \tag{2.33}$$

Obviously, it is no longer very convenient to use the δ-function constraint to carry out one of the integrations, because the remaining ones get quite complicated as a result. What do we do? We know that "exponentiation converts addition to multiplication". The trick, therefore, is to convert the multiple integral to one that involves a *product* of individual integrals. This is done by using the following well-known representation of the Dirac δ function:

$$\delta(x) = \frac{1}{2\pi} \int_{-\infty}^{\infty} dk \, e^{ikx}. \tag{2.34}$$

Use this[17] to show that

$$F^{\rm eq}(V) = \left(\frac{mn}{2\pi k_B T}\right)^{\frac{1}{2}} \exp\left(-\frac{mnV^2}{2k_B T}\right). \tag{2.35}$$

What you have just established is a special case of the following important result: the PDF of a linear combination of independent Gaussian random variables is again a Gaussian[18]. This kind of 'addition theorem' is common to a whole family of probability distributions called **Lévy alpha-stable distributions** or **stable distributions** for short, of which the Gaussian is a limiting case[19]. In turn, this property leads to the celebrated **Central Limit Theorem** of statistics and its generalization to all stable distributions.

[17] You will also need to use the formula of Eq. (2.31), in this instance for an imaginary value of b.

[18] See Appendix D, Sec D.4.

[19] We shall encounter at least two other prominent members of the family of stable distributions in subsequent chapters. As this is an important and useful topic, Appendix K is devoted to a résumé of the properties of these distributions.

Chapter 3

The fluctuation-dissipation relation

We show that the requirement of internal consistency imposes a fundamental relationship between two quantities: the strength of the random force that drives the fluctuations in the velocity of a particle, on the one hand; and the coefficient representing the dissipation or friction present in the fluid, on the other.

3.1 Conditional and complete ensemble averages

What can we say about the random force $\eta(t)$ that appears in the Langevin equation (2.18)? It is reasonable to assume that its *average* value remains zero at all times. But what is the set over which this average is to be taken? As a result of our dividing up the system into a subsystem (the tagged particle) and a heat bath (the rest of the particles), there are two distinct averaging procedures involved:

- The first is a *conditional* average over all possible states of motion of the heat bath, i. e., of all the molecules of the fluid, but *not* the tagged particle. The latter is supposed to have a *given* initial velocity v_0. We shall use an overhead bar to denote averages over this *sub-ensemble*, namely, an ensemble of subsystems specified by the common initial condition $v(0) = v_0$ of the tagged particle. We shall call these conditional or partial averages.

- The second is a *complete* average, i. e., an average over all possible states of motion of all the particles, including the tagged particle. Such averages will be denoted by angular brackets, $\langle \cdots \rangle$. A complete average can be achieved by starting with the corresponding conditional average, and then performing a further averaging over all possible values of v_0, distributed according to some prescribed initial PDF $p_{\text{init}}(v_0)$. The latter depends on the preparation of the initial state.

- In writing down the Langevin equation (2.18) and its formal solution in Eq. (2.19), we have included a possible time-dependent external force, as we are also interested in the effects of such an applied force (or stimulus) upon our system. Without loss of generality, we take any such force to be applied from $t = 0$ onward, on a system that is initially in thermal equilibrium at some temperature T. It follows that we must identify $p_{\text{init}}(v_0)$ with the Maxwellian distribution [1] itself, i. e.,

$$p_{\text{init}}(v_0) \equiv p^{\text{eq}}(v_0) = \left(\frac{m}{2\pi k_B T}\right)^{\frac{1}{2}} \exp\left(-\frac{m v_0^2}{2k_B T}\right). \tag{3.1}$$

- We shall use the notation $\langle \cdots \rangle_{\text{eq}}$ for complete averages taken in a state of thermal equilibrium. Averages taken in a state perturbed *out* of thermal equilibrium by the application of a time-dependent external force will be denoted[2] simply by $\langle \cdots \rangle$

Returning to the mean value of the random force $\eta(t)$, the only assumption we need to make for the present is that its average value remains zero at all times, i. e.,

$$\overline{\eta(t)} = 0 \text{ for all } t. \tag{3.2}$$

From Eq. (2.19), it follows immediately that

$$\overline{v(t)} = v_0\, e^{-\gamma t} + \frac{1}{m}\int_0^t dt_1\, e^{-\gamma(t-t_1)}\, F_{\text{ext}}(t_1), \tag{3.3}$$

because F_{ext} is imposed from outside the system, and is unaffected by any averaging procedure. Equation (3.3) already yields valuable information, as we shall see very shortly. But first, an important point must be noted:

- From this point onward, up to Ch. 12, we shall consider the system *in the absence of any external force*. In Ch. 13, we shall extend the formalism to be developed to the case when a time-*independent* external force is present. However, the system continues to be in a state of thermal equilibrium. In

[1] Recall that we are dealing with a single Cartesian component of the velocity.

[2] The same notation will also be used in a more general context, for statistical averages of random processes or variables, whenever needed. No confusion should arise as a result, because the meaning of any average will be clear from the context, in all cases.

Ch. 15, we go on to the case of a time-*dependent* external force, in the context of dynamical response and the mobility. Such a force perturbs the system out of thermal equilibrium.

In particular, in the *absence* of any external force acting upon the system, Eq. (3.3) reduces to

$$\overline{v(t)} = v_0\, e^{-\gamma t}. \tag{3.4}$$

Therefore

$$\lim_{t\to\infty} \overline{v(t)} = 0 \quad \text{for any } v_0\,. \tag{3.5}$$

This suggests that the state of thermal equilibrium is indeed maintained with the help of the friction in the system: the dissipative mechanism that is always present in a system helps damp out the fluctuations and restore the system to equilibrium. If γ had been absent, we would have had the unphysical result $\overline{v(t)} = v_0$ for all t. This already shows that the systematic component $F_{\text{sys}} = -m\gamma v$ of F_{int} plays a crucial role. Further corroboration will come from the mean squared velocity, to be found shortly.

From Eq. (3.4), we may go on to find the complete average of v in the absence of F_{ext}. The complete average $\langle\cdots\rangle$ is nothing but the equilibrium average $\langle\cdots\rangle_{\text{eq}}$ in this situation. We have

$$\langle v(t)\rangle_{\text{eq}} \equiv \int_{-\infty}^{\infty} \overline{v(t)}\, p^{\text{eq}}(v_0)\, dv_0 = \int_{-\infty}^{\infty} v_0\, e^{-\gamma t}\, p^{\text{eq}}(v_0)\, dv_0 = 0 \tag{3.6}$$

for all $t \geq 0$. This follows quite simply from the fact that $p^{\text{eq}}(v_0)$ (a Gaussian centered at $v_0 = 0$) is an even function of v_0, while the factor v_0 is an odd function. Hence their product is an odd function, whose integral from $-\infty$ to ∞ vanishes.

3.2 Conditional mean squared velocity as a function of time

Let us now compute the mean squared velocity in the absence of an applied force. The solution for $v(t)$ given in Eq. (2.19) becomes, in the absence of F_{ext},

$$v(t) = v_0\, e^{-\gamma t} + \frac{1}{m}\int_0^t dt_1\, e^{-\gamma(t-t_1)}\, \eta(t_1). \tag{3.7}$$

Squaring this expression and taking the conditional average, we get

$$\overline{v^2(t)} = v_0^2\, e^{-2\gamma t} + \frac{1}{m^2}\int_0^t dt_1 \int_0^t dt_2\, e^{-\gamma(t-t_1)-\gamma(t-t_2)}\, \overline{\eta(t_1)\,\eta(t_2)}. \tag{3.8}$$

The cross-terms have vanished because $\overline{\eta(t)} = 0$. Now, the simplest assumption we can make is that $\eta(t)$ is a 'completely random' force, or a **white noise**—in technical terms, a **delta-correlated, stationary, Gaussian, Markov process**[3]. This means that the PDF of $\eta(t)$ is a Gaussian (with mean value zero, as we have found already). Further, $\eta(t)$ has no 'memory' at all: its **autocorrelation function** factorizes into a product of averages, and hence vanishes, according to

$$\overline{\eta(t_1)\,\eta(t_2)} = \big(\overline{\eta(t_1)}\,\big)\,\big(\overline{\eta(t_2)}\,\big) = 0 \quad \text{for all } t_1 \neq t_2\,. \tag{3.9}$$

Thus, the random forces at any two distinct instants of time t_1 and t_2 are assumed to be completely uncorrelated and independent of each other, no matter how close to each other t_1 and t_2 are[4]. Owing to stationarity, the average $\overline{\eta(t_1)\,\eta(t_2)}$ must be solely a function of the magnitude of time difference, $|t_1 - t_2|$. Hence the autocorrelation function of $\eta(t)$ must necessarily be of the form

$$\overline{\eta(t_1)\,\eta(t_2)} = \Gamma\,\delta(t_1 - t_2), \tag{3.10}$$

where Γ is a positive constant that represents the 'strength' (or normalization) of the noise $\eta(t)$. It has the physical dimensions of (force)^2 × (time) $= M^2L^4T^{-3}$. Using Eq. (3.10), the double integral in Eq. (3.8) can be evaluated easily. The result is

$$\overline{v^2(t)} = v_0^2\,e^{-2\gamma t} + \frac{\Gamma}{2m^2\gamma}\,\big(1 - e^{-2\gamma t}\big)\,. \tag{3.11}$$

To fix Γ itself, we need an additional physical input. Exactly as in the case of the mean value of $v(t)$ (recall Eqs. (3.4) and (3.5)), we impose the following requirement:

- As $t \to \infty$, the effect of the special choice of initial condition (namely, the specification of some particular value v_0 as the initial velocity) must be washed away, and the mean squared velocity $\overline{v^2(t)}$ must attain the value characteristic of thermal equilibrium at an absolute temperature T.

In other words, we require that the following two quantities be equal to each other identically: on the one hand, the limiting value as obtained from Eq. (3.11),

---

[3] We will explain, as we go along, the meaning of the various terms used here.

[4] We must recognize, though, that this is a (mathematical) idealization, and hence an approximation to the physical situation. No such force can actually exist. Any physical force must have a nonzero correlation time. However, the approximation is a very good one in the present context. Subsequently, in Sec. 4.1 and Sec. 11.2, we shall see why this is so.

The white noise approximation made here does lead to certain shortcomings of the Langevin equation. This aspect will be discussed in Ch. 17, where the Langevin equation will be appropriately generalized in order to overcome these drawbacks.

namely,

$$\lim_{t\to\infty} \overline{v^2(t)} = \frac{\Gamma}{2m^2\gamma}\,; \tag{3.12}$$

and, on the other, the mean squared velocity in thermal equilibrium, namely,

$$\langle v^2 \rangle_{\text{eq}} = \frac{k_B T}{m}. \tag{3.13}$$

This requirement of internal consistency at once constrains Γ to have the value

$$\Gamma = 2m\,\gamma\,k_B T. \tag{3.14}$$

This is an extremely important result.

- The relation $\Gamma = 2m\,\gamma\,k_B T$ fixes the amplitude of the *fluctuations* in the random force, as quantified by Γ, in terms of the temperature T and the *dissipation* or friction coefficient, γ. It is a classic example of the **fluctuation-dissipation theorem.**

In this instance, it may also be termed a **Nyquist relation**, similar to that for **thermal noise** or **Johnson noise** in a resistor[5]. Inserting Eq. (3.14) for Γ in Eq. (3.11), we get

$$\overline{v^2(t)} = \frac{k_B T}{m} + \left(v_0^2 - \frac{k_B T}{m}\right) e^{-2\gamma t}. \tag{3.15}$$

This result shows exactly how $\overline{v^2(t)}$ approaches the equilibrium value $(k_B T/m)$ of the mean squared velocity as $t \to \infty$. Figure 3.1 depicts the behavior of $\overline{v^2(t)}$ in the two cases $v_0^2 > k_B T/m$ and $v_0^2 < k_B T/m$, respectively. Once again, it is easily checked that a further averaging of $\overline{v^2(t)}$ over all values of v_0 gives

$$\int_{-\infty}^{\infty} \overline{v^2(t)}\, p^{\text{eq}}(v_0)\, dv_0 = \frac{k_B T}{m} = \langle v^2(t) \rangle_{\text{eq}} \quad \text{for all } t \geq 0. \tag{3.16}$$

Thus, averaging $\overline{v^2(t)}$ over the Maxwellian PDF $p^{\text{eq}}(v_0)$ leads to the recovery of the equilibrium value $\langle v^2 \rangle_{\text{eq}} = k_B T/m$, just as averaging $\overline{v(t)}$ over $p^{\text{eq}}(v_0)$ yielded the equilibrium value $\langle v \rangle_{\text{eq}} = 0$. It is evident that the fluctuation-dissipation relation (or 'FD relation', for short) between Γ and γ is crucial in the recovery of this result. In fact, the FD relation could also have been derived by the following alternative procedure:

[5]This will be discussed in Ch. 16, Sec. 16.1.3.

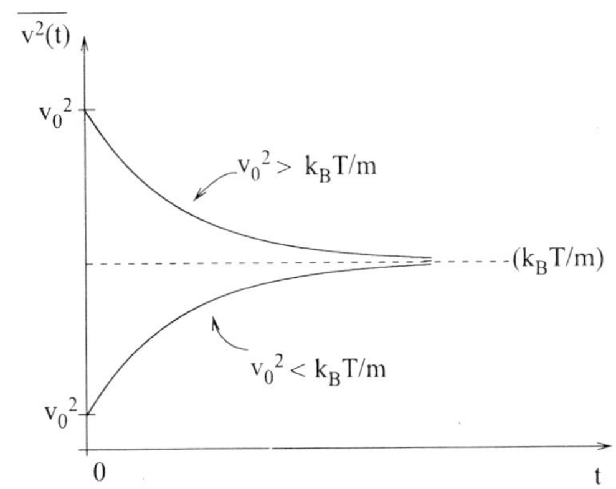

Figure 3.1: Approach of the conditional mean squared velocity to its equilibrium value, as given by Eq. (3.15). The upper and lower curves correspond to the cases in which the magnitude of the initial velocity of the tagged particle is, respectively, larger and smaller than $(k_BT/m)^{1/2}$, the root mean squared velocity in equilibrium.

- Take Eq. (3.11) for the conditional mean squared velocity. Average this expression over the initial velocity, i. e., find $\int_{-\infty}^{\infty} \overline{v^2(t)}\, p^{\text{eq}}(v_0)\, dv_0$. Impose the requirement that the result be identically equal to the mean squared velocity in equilibrium, namely, $\langle v^2 \rangle_{\text{eq}}$. It follows that Γ must be equal to $2m\,\gamma\, k_B\, T$.

3.3 Exercises

3.3.1 The contradiction that arises if dissipation is neglected

Suppose the internal force F_{int} due to molecular collisions did not have any systematic part F_{sys} at all. Then $F_{\text{int}}(t) = \eta(t)$. Work through the steps of this chapter once again in this case, to show that we then get

$$\overline{v(t)} = v_0\,, \quad \overline{v^2(t)} = v_0^2 + \frac{\Gamma t}{m^2}\,. \tag{3.17}$$

The first of these results is already quite unphysical. It implies that the initial velocity of the tagged particle persists for all time as its average velocity. Worse, the second result shows that the variance of the velocity *increases* with t. If we associate the temperature of the system with the variance of the velocity according to $\frac{1}{2}m\overline{v^2(t)} = \frac{1}{2}k_BT$, this implies that the temperature of the system *spontaneously* increases with time, and becomes unbounded as $t \to \infty$, even if the system is simply left alone in thermal equilibrium! The absurdity of these results brings out clearly the vital role played by the dissipation coefficient γ in maintaining the thermal equilibrium of the system.

3.3.2 Is the white noise assumption responsible for the contradiction?

As a diligent student, you might not be willing to agree so readily with the last statement. You might ask: Perhaps the assumption that the noise η is δ-correlated is to blame, rather than the absence of the dissipation coefficient γ? (We have already remarked that a noise with zero correlation time is, in principle, an unphysical assumption that cannot be correct at extremely short times.) Now suppose that, instead of the relation $\overline{\eta(t_1)\,\eta(t_2)} = \Gamma\,\delta(t_1 - t_2)$ of Eq. (3.10), we assumed that the noise had a finite correlation time, according to

$$\overline{\eta(t_1)\,\eta(t_2)} = K\,\exp\,(-\,|t_1 - t_2|/\tau)\,. \tag{3.18}$$

Here K is a positive constant, and τ represents the correlation time of the noise. (Roughly speaking, it is a measure of how long the noise η retains the 'memory' of its previous values.) It is easy to see that we would still find (in the absence of F_{sys}) the incorrect result $\overline{v(t)} = v_0$ for the mean value of the velocity. But what about the mean squared value $\overline{v^2(t)}$? Does its long-time behavior improve?

3.3.3 The generality of the role of dissipation

This is a slightly harder problem. Instead of the very specific decaying exponential form in Eq. (3.18), suppose you are given only that

(i) $\overline{\eta(t_1)\,\eta(t_2)}$ is a function of $|t_1 - t_2|$, the magnitude of the difference between the two time arguments; and further,

(ii) $\overline{\eta(t_1)\,\eta(t_2)}$ tends to zero as $|t_1 - t_2| \to \infty$.

Show that the inescapable conclusion based on properties (i) and (ii) is that $\overline{v^2(t)}$ increases linearly with t at long times, in the absence of γ. (The algebraic steps you need to show this are similar to those used in Ch. 15, Sec. 15.1.) Hence we may indeed assert that *the dissipation coefficient is an essential ingredient in maintaining thermal equilibrium in this system.*

Chapter 4

The velocity autocorrelation function

We calculate the autocorrelation function of the velocity of a tagged particle in a state of thermal equilibrium. This function quantifies the manner in which the memory in the velocity decays as a function of time. It is shown that the velocity autocorrelation function is a decaying exponential function of t. The correlation time of the velocity turns out to be precisely γ^{-1}, the reciprocal of the friction coefficient.

4.1 Velocity correlation time

In the preceding chapter, we have obtained useful information on the mean velocity and the scatter in the velocity (as measured by its variance, which follows from the mean squared velocity). The Langevin equation permits us to go even further. The autocorrelation function of the velocity is a generalization of its mean squared value (or its variance), and is the primary characterizer of the 'memory' that this random variable possesses. We now compute this function.

Consider the solution for the velocity given by Eq. (3.7), at any two instants of time t and t', where $t\,,\,t' > 0$. Taking the conditional average of the product $v(t)\,v(t')$, we have

$$\overline{v(t)\,v(t')} = v_0^2\,e^{-\gamma(t+t')} + \frac{1}{m^2}\int_0^t dt_1 \int_0^{t'} dt_2\, e^{-\gamma(t-t_1)-\gamma(t'-t_2)}\,\overline{\eta(t_1)\,\eta(t_2)}\,. \tag{4.1}$$

As before, the cross terms vanish because $\overline{\eta(t)} = 0$ at any instant of time. Putting

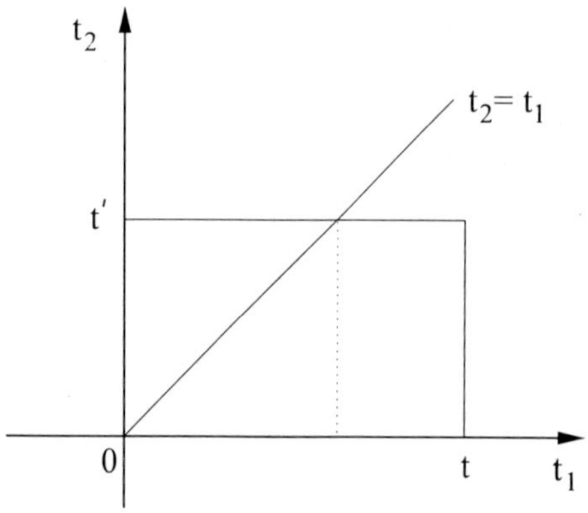

Figure 4.1: The region of integration over t_1 and t_2 in Eq. (4.2), in the case $t > t'$. The δ-function in the integrand restricts the integration to the 45° line $t_2 = t_1$. Hence the integration over t_1 effectively runs from 0 to t' rather than t. When $t < t'$, the roles of t_1 and t_2 are interchanged.

in the δ-function form given by Eq. (3.10) for $\overline{\eta(t_1)\,\eta(t_2)}$,

$$\overline{v(t)\,v(t')} = v_0^2 e^{-\gamma(t+t')} + \frac{\Gamma e^{-\gamma(t+t')}}{m^2}\int_0^t dt_1 \int_0^{t'} dt_2\, e^{\gamma(t_1+t_2)}\,\delta(t_1 - t_2). \qquad (4.2)$$

Suppose $t > t'$. Then the δ-function can be used to carry out the integration over t_2, but there is a nonvanishing contribution to the integral only as long as $0 \le t_1 \le t'$. (Figure 4.1 helps you see why this is so.) Therefore

$$\begin{aligned}\overline{v(t)\,v(t')} &= v_0^2\, e^{-\gamma(t+t')} + \frac{\Gamma}{m^2}\, e^{-\gamma(t+t')}\int_0^{t'} dt_1\, e^{2\gamma t_1} \\ &= v_0^2\, e^{-\gamma(t+t')} + \frac{\Gamma}{2m^2\gamma}\, e^{-\gamma t}\left(e^{\gamma t'} - e^{-\gamma t'}\right), \quad (t > t'). \qquad (4.3)\end{aligned}$$

To obtain the autocorrelation function of the velocity in thermal equilibrium, we can adopt either one of two procedures, as before:

(i) We may further average the conditional average $\overline{v(t)\,v(t')}$ over the Maxwellian PDF in v_0.

(ii) Or, alternatively, we may allow the transient contributions in Eq. (4.3) to die out by letting *both* t and t' tend to infinity, while keeping the difference $(t - t')$ finite.

Once again, the two alternative procedures yield exactly the same answer (as they ought to), if we use the fact that $\Gamma = 2m\gamma\, k_B T$. The result we obtain is

$$\big\langle v(t)\, v(t')\big\rangle_{\text{eq}} = \frac{k_B T}{m}\, e^{-\gamma(t-t')}, \quad t > t'. \tag{4.4}$$

If we had started with $t < t'$, instead, the same result would have been obtained, with t and t' simply interchanged. We therefore arrive at the important result

$$\big\langle v(t)\, v(t')\big\rangle_{\text{eq}} = \frac{k_B T}{m}\, e^{-\gamma|t-t'|} = \big\langle v^2\big\rangle_{\text{eq}}\, e^{-\gamma|t-t'|} \tag{4.5}$$

for the **equilibrium autocorrelation function** of the velocity[1]. The dependence of the equilibrium velocity autocorrelation on the difference of the two time arguments is another indication that the velocity is a stationary random process in the absence of a time-dependent external force. We may define the *normalized* velocity autocorrelation function as the dimensionless function

$$C(t) = \frac{\langle v(0)\, v(t)\rangle_{\text{eq}}}{\langle v^2\rangle_{\text{eq}}} = \frac{\langle v(0)\, v(t)\rangle_{\text{eq}}}{(k_B T/m)}\,. \tag{4.6}$$

In the present instance, $C(t)$ is just $\exp(-\gamma|t|)$. Figure 4.2 depicts this function for $t > 0$. The time-scale characterizing the exponential decay of the autocorrelation, γ^{-1}, is the **velocity correlation time**.

- The velocity correlation time, also called the **Smoluchowski time**, is the fundamental time scale in the problem under discussion. Roughly speaking, it represents the time scale over which the system returns to thermal equilibrium when disturbed out of that state by a small perturbation.

In other words, it is the time scale over which 'thermalization' occurs. As we have just seen, the autocorrelation function is just a decaying exponential function of time in the Langevin model that we are working with here. Hence the associated time constant is readily identified as the correlation time of the velocity. In more

[1]More precisely, the autocorrelation function of a random process $\xi(t)$ is the average value of the product $\delta\xi(t)\,\delta\xi(t')$, where $\delta\xi(t)$ is the *deviation* $\xi(t) - \langle\xi(t)\rangle$ of the random variable from its mean value at time t. Since $\langle v(t)\rangle_{\text{eq}} = 0$ in the present case, the left-hand side of Eq. (4.5) is indeed the autocorrelation function of the velocity (in the state of thermal equilibrium). Other definitions of the autocorrelation are sometimes used, that differ slightly from the one we have employed. In particular, one starts with a definition in terms of a time average, and then converts it to a statistical average using certain properties of the random process—see Ch. 5, Sec. 5.5. We shall not get into these technicalities here. We shall also use the term autocorrelation function for both $\langle\xi(t)\,\xi(t')\rangle$ as well as $\langle\delta\xi(t)\,\delta\xi(t')\rangle$, depending on the context. But the quantity referred to will always be written out explicitly, so that there is no confusion.

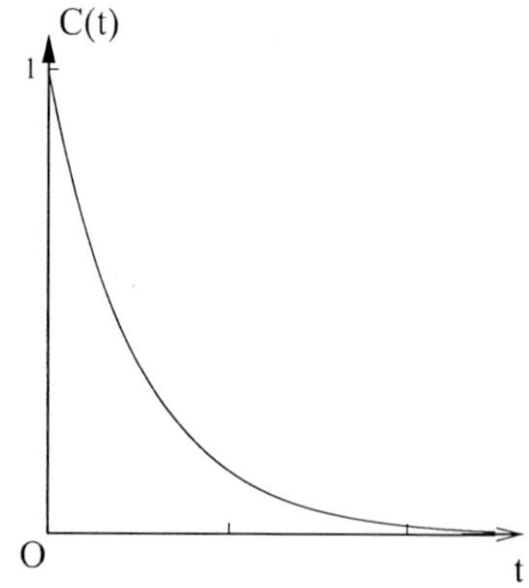

Figure 4.2: The normalized autocorrelation function of the velocity in thermal equilibrium, $\exp(-\gamma|t|)$, for $t > 0$. The time constant γ^{-1} represents the characteristic time scale over which thermal equilibration occurs in the system. $C(t)$ serves to define this characteristic time scale even in more complicated models: it is just the total area under the curve of $C(t)$ versus t for $t \geq 0$.

complicated situations, one goes beyond the original Langevin model to more involved equations of motion. The velocity autocorrelation function will then no longer be a single decaying exponential in time. An effective velocity correlation time can then be defined as

$$\tau_{\text{vel}} = \int_0^\infty C(t)\, dt. \tag{4.7}$$

We shall return to this point in Ch. 17, Sec. 17.5, in connection with the generalized Langevin equation.

The Smoluchowski time γ^{-1} is crucially relevant to the question of the validity of the Langevin equation for the motion of the tagged particle. Indeed, several different time scales are involved in the problem as a whole, and it is important to identify them clearly. But we can do so only after we develop the formalism some more. On the other hand, we do need to have some definite numbers at this stage, in order to get a feel for the physical situation under discussion. We therefore quote some typical orders of magnitude here, and defer a more detailed discussion to Ch. 11, Sec. 11.2 . Let us consider some typical values for the size and mass of the tagged particle: a radius a in the micron range, $1\,\mu$m to $10\,\mu$m, say, and a density that is essentially that of water under standard conditions, $\sim 10^3\,\text{kg/m}^3$. Hence the mass m lies in the range 10^{-15} kg to 10^{-12} kg, as compared to 10^{-26} kg for a molecule of water. In order to estimate γ, we equate the frictional force $m\gamma v$ on the particle to the viscous drag force as given by the Stokes' formula, $6\pi a\eta\, v$ (see Sec. 11.2), where η is the viscosity of water[2]. Putting in the room temperature value $\eta \sim 10^{-3}\,\mathrm{N\,s/m^2}$, we obtain a Smoluchowski time γ^{-1} that lies in the range 10^{-7} s to 10^{-5} s. The velocity correlation time of our micron-sized tagged particle is thus in the microsecond range. In marked contrast, the correlation time of the noise $\eta(t)$ can be estimated to be several orders of magnitude smaller than γ^{-1} (see Sec. 11.2). This is why the Langevin equation with a driving white noise is a satisfactory model over a wide temporal range.

4.2 Stationarity of a random process

We have used the phrase 'stationary random process' more than once in the foregoing, without actually defining this term formally. This is such an important concept that we must explain what it means, in some detail. Essentially,

- a stationary random process is one whose *statistical* properties do not change with time.

This means, to begin with, that its probability density function is independent of t. Hence the average value of the random variable is constant in time. Further, its mean squared value, and all its higher moments, are also constant in time. But it also means that *all* the multiple-time joint probability densities that describe the random process at any set of successive instants of time, are independent of the over-all origin in the time variable. All time arguments can therefore be shifted by the same amount without affecting any average values. Hence all joint probability densities, and all multiple-time averages, can only depend on relative time intervals.

- *In particular, the autocorrelation function of a stationary random process is a function only of the difference of the two time arguments involved, and not of the two arguments separately.*

The requirement that *all* joint probability densities be independent of an over-all shift in the time origin actually defines a **strictly stationary** random process. In practice, however, we have often to consider processes for which only the first few joint PDFs are known or deducible. In particular, we have to deal with random processes that are stationary at the level of the first moment (or the mean value), the second moment (or the mean squared value), and the two-point correlation

---

[2]This is the standard symbol for the viscosity. It should not be confused with the noise $\eta(t)$.

alone. This is the case in most physical applications. Such processes are said to be **wide-sense stationary**.

Equation (4.5) shows that the equilibrium autocorrelation of the velocity is, in fact, a *symmetric* or *even* function of its single time argument, since it is a function of the *modulus* of $(t - t')$. This property will be used more than once in the sequel. The evenness of the autocorrelation function is a general property of a stationary random process. The proof of this statement is quite simple. Taking the stationary process $v(t)$ itself an an example[3], we have

$$\langle v(t_0)\, v(t_0 + t)\rangle_{\text{eq}} \begin{cases} = \langle v(0)\, v(t)\rangle_{\text{eq}} \\ = \langle v(-t)\, v(0)\rangle_{\text{eq}} = \langle v(0)\, v(-t)\rangle_{\text{eq}} \,. \end{cases} \tag{4.8}$$

The first line above follows if we subtract t_0 from both time arguments on the left-hand side. The second line follows if we subtract $t_0 + t$ from both arguments, and then use the fact that we are dealing with classical (commuting) variables.

Returning to the problem under discussion, some noteworthy points that emerge from the calculation we have carried out in Sec. 4.1 are as follows:

- In the absence of an applied force F_{ext}, the internal random force $\eta(t)$ 'drives' the velocity $v(t)$.
- Although the driving process is δ-correlated (that is, its correlation time is zero), the driven process is *not* δ-correlated. It is exponentially correlated, with a nonzero correlation time γ^{-1}.
- The driving noise is a stationary random process. This property is carried over to the driven variable $v(t)$.

The velocity is, in fact, a strictly stationary process. We shall be able to see this more convincingly after Ch. 6. To the points listed above, we may add another, which will become clear in the sequel:

- The driving noise $\eta(t)$ is a Gaussian Markov process. Both these properties are also carried over to the velocity $v(t)$.

Before we turn our attention to this aspect of the problem, it is necessary to understand what is meant by a Markov process. We shall therefore digress to do so, in the next chapter[4].

---

[3] Note that we are considering a one-component process here. The generalization of this property to the case of a multi-component process will be discussed in Sec. 4.4.

[4] The general importance of Markov processes has already been referred to in the Prologue.

4.3 Velocity autocorrelation in three dimensions

As already stated in Ch. 2, our focus is generally on a single Cartesian component of the motion of the tagged particle, but we consider the full three-dimensional case whenever appropriate. It is useful, at this stage, to do so with regard to the autocorrelation of the velocity. Our motivation: we would like to extend the discussion to the physically important situation in which the tagged particle is electrically charged, and an applied magnetic field is also present.

4.3.1 The free-particle case

The extension of the foregoing to the full three-dimensional motion of the tagged particle is straightforward (in the absence of an applied force). The three-dimensional counterpart of the Langevin equation in the absence of an applied external force is

$$m\dot{\mathbf{v}}(t) = -m\gamma\mathbf{v}(t) + \boldsymbol{\eta}(t), \tag{4.9}$$

where each Cartesian component of the vector $\boldsymbol{\eta}(t)$ is an independent Gaussian white noise. The mean and autocorrelation of the noise are given by the following straightforward generalizations of Eqs. (3.2) and (3.10), respectively. These are

$$\overline{\eta_i(t)} = 0 \tag{4.10}$$

and

$$\overline{\eta_i(t_1)\,\eta_j(t_2)} = \Gamma\,\delta_{ij}\,\delta(t_1 - t_2), \tag{4.11}$$

where the Cartesian indices $i\,,j$ run over the values $1, 2$ and 3, and δ_{ij} is the Kronecker delta. Note that we have assumed that the noise strength Γ is the same for all the components of the random force. This is justified by the obvious isotropy[5] of the system under consideration. A straightforward repetition of the steps described in the foregoing then leads to the result

$$\big\langle v_i(t)\,v_j(t')\big\rangle_{\text{eq}} = \frac{k_BT}{m}\,\delta_{ij}\,e^{-\gamma|t-t'|}. \tag{4.12}$$

Hence[6]

$$\big\langle \mathbf{v}(t)\,\cdot\,\mathbf{v}(t')\big\rangle_{\text{eq}} = \big\langle v_i(t)\,v_i(t')\big\rangle_{\text{eq}} = \frac{3k_BT}{m}\,e^{-\gamma|t-t'|}. \tag{4.13}$$

Note also that

$$\big\langle \mathbf{v}(t)\,\times\,\mathbf{v}(t')\big\rangle_{\text{eq}} = 0, \tag{4.14}$$

[5] In other words, the equivalence of all directions in space.

[6] We use the usual summation convention for vector and tensor indices: a repeated index is to be summed over all its allowed values.

as we might expect: the different Cartesian components of the noise are independent of each other, and the corresponding components of the velocity remain uncorrelated at all times.

4.3.2 Charged particle in a constant magnetic field

We consider now the autocorrelation of the velocity of an electrically charged tagged particle in the presence of a constant, uniform magnetic field. The problem of the motion of a tagged particle in an applied electromagnetic field is of interest in several physical situations, especially in plasma physics. It serves as a case study for the analysis of dissipative effects in the presence of electromagnetic interactions. In this book, we restrict ourselves to the most elementary, but nevertheless very instructive, version of the problem: the motion of a single tagged particle in a constant, uniform magnetic field[7]. The velocity autocorrelation of the particle is studied here. Other aspects of the problem will be discussed in subsequent chapters: the conditional PDF of the velocity of the particle and the diffusive behavior of its displacement (Ch. 14), and the associated dynamic mobility (Ch. 15).

Let us first set down what we expect to find, on physical grounds. An applied magnetic field $\mathbf{B}$ exerts a Lorentz force $q(\mathbf{v} \times \mathbf{B})$ on a particle of charge q and mass m. If the particle moves in free space, it traces a helical path of constant pitch, in general. The axis of the helix is in the direction of $\mathbf{B}$. In the plane transverse to the field direction, the projection of the path is a circle. This circle is traversed with a uniform angular speed (or **cyclotron frequency**)

$$\omega_c = \frac{qB}{m}, \tag{4.15}$$

where $B = |\mathbf{B}|$. We are interested, however, in the behavior of the particle when it moves in a fluid in thermal equilibrium, under the combined influence of the Lorentz force, the dissipative frictional force $-m\gamma\mathbf{v}$, and the random noise due to molecular collisions. The first point to note is that the Lorentz force does no work upon the particle, as it is always directed perpendicular to the instantaneous velocity of the particle. The distribution of the energy of the tagged particle is therefore unaffected by the presence of the magnetic field. Thus we may expect the *equilibrium* or Maxwellian distribution of the velocity of the charged particle to remain unchanged[8]. However, the Lorentz force 'mixes up' the different components of the velocity of the particle. We may therefore expect the *approach* to

[7] The primary objective in the context of plasma physics is somewhat more involved. We are not concerned with this aspect here.

[8] This assertion requires a proof, which will be supplied in Ch. 14.

equilibrium, starting from a given initial velocity of the tagged particle, to differ from that in the field-free case. The behavior of the velocity autocorrelation function sheds light on precisely this aspect. We first compute this correlation function for $t > 0$, and then write down the result for all t based on a general argument[9].

In the presence of an applied magnetic field $\mathbf{B}$, the Langevin equation satisfied by the tagged particle at any time $t > 0$ is given by

$$m\dot{\mathbf{v}}(t) = -m\gamma\mathbf{v}(t) + q\,(\mathbf{v}\times\mathbf{B}) + \boldsymbol{\eta}(t). \tag{4.16}$$

Here $\boldsymbol{\eta}(t)$ comprises three independent Gaussian white noises as defined in the preceding section. It is at once evident that the motion of the particle *along* the direction of the magnetic field is not affected by the presence of the field[10]. To proceed, it is convenient to write the Langevin equation in component form. Let $\mathbf{n}$ be the unit vector in the direction of the magnetic field[11], i. e.,

$$\mathbf{B} = B\,\mathbf{n}. \tag{4.17}$$

Then, for each Cartesian index j,

$$\begin{aligned} m\dot{v}_j(t) &= -m\gamma v_j(t) + qB\,\epsilon_{jkl}\,v_k(t)\,n_l + \eta_j(t) \\ &= -m\gamma v_j(t) - qB\,\epsilon_{kjl}\,v_k(t)\,n_l + \eta_j(t), \end{aligned} \tag{4.18}$$

where ϵ_{jkl} is the usual totally antisymmetric Levi-Civita symbol in three dimensions[12]. In writing down the second of the equations above, we have used the fact

---

[9]The argument that enables us to extend the result to negative values of t involves **time reversal**. It is noteworthy in its own right. We shall therefore describe it in some detail, toward the end of this section.

[10]Take the dot product of each side of Eq. (4.16) with $\mathbf{B}$. The term involving the Lorentz force drops out because $\mathbf{B}\cdot(\mathbf{v}\times\mathbf{B}) \equiv 0$.

[11]A trivial point: we could simply say, "Let us choose the direction of the magnetic field to lie along the z-axis, without any loss of generality", and proceed. But I believe there is some merit in getting a bit of practice handling direction cosines (or unit vector components) and the index notation. (Moreover, every magnetic field does not have to be directed along the z-axis!) However, having derived general expressions for the reasons stated above, we too shall choose the direction of the magnetic field to be the z-direction when it is convenient to do so.

[12]Recall that ϵ_{jkl} is equal to $+1$ when jkl is an even permutation of the natural order 123, and -1 when jkl is an odd permutation of the natural order. (In three dimensions—and in three dimensions alone!— this happens to be the same as a 'cyclic permutation' of 123 and an 'anticyclic permutation', respectively.) It is equal to zero in all the twenty-one other cases, i. e., when any two of the indices take on the same value. ϵ_{jkl} is totally antisymmetric in its indices, i. e., it changes sign when any two of its indices are interchanged. In writing Eq. (4.18), we have also used the elementary fact that the components of a cross-product $\mathbf{a} = \mathbf{b}\times\mathbf{c}$ are written compactly as $a_j = \epsilon_{jkl}\,b_k\,c_l$ (with a summation implied over the repeated indices k and l).

that $\epsilon_{jkl} = -\epsilon_{kjl}$. Let us now define the constant (i. e., time-independent) matrix M, with elements given by

$$M_{kj} = \epsilon_{kjl}\, n_l\,. \tag{4.19}$$

Equation (4.18) then becomes

$$\dot{v}_j(t) = -\gamma v_j(t) - \omega_c\, v_k(t) M_{kj} + \frac{1}{m}\,\eta_j(t). \tag{4.20}$$

Our immediate purpose here is the determination of the equilibrium velocity autocorrelation function. We shall therefore assume some of the results that will be established in Ch. 14, and proceed by the shortest route to derive the expression we seek. The results we require are the following:

- The equilibrium distribution of the velocity remains the same Maxwellian distribution that it is in the field-free case.
- In thermal equilibrium, the velocity of the tagged particle is again a stationary random process.
- The fluctuation-dissipation relation $\Gamma = 2m\gamma k_B T$ also remains unchanged.

The fact that the velocity of the tagged particle is a stationary random process in equilibrium implies that the correlation function $\langle v_i(t')\, v_j(t)\rangle_{\rm eq}$ is a function of the time difference $(t-t')$ alone, rather than a function of the two time arguments separately, for every pair of indices i and j. The set of these correlation functions comprises the elements of the **velocity correlation tensor**. In order to keep the notation uniform, and consistent with that used earlier in the one-dimensional case (see Eq. (4.6)), we normalize these correlation functions by dividing each of them by the equilibrium mean squared value of any component of the velocity, namely, k_BT/m. Thus, we define the **velocity correlation matrix** $\mathsf{C}(t-t')$ as the (3×3) matrix with elements

$$C_{ij}(t-t') = \frac{\langle v_i(t')\, v_j(t)\rangle_{\rm eq}}{(k_BT/m)} = \frac{\langle v_i(0)\, v_j(t-t')\rangle_{\rm eq}}{(k_BT/m)}\,. \tag{4.21}$$

The differential equation satisfied by $\mathsf{C}(t)$ for all $t > 0$ is easily found. We have merely to pre-multiply both sides of the Langevin equation (4.20) by $v_i(0)$, and perform a complete average: namely, an average over the realizations of the noise, as well as an average over the initial velocity $v_i(0)$ with respect to the equilibrium Maxwellian distribution. This yields the equation

$$\frac{dC_{ij}(t)}{dt} = -\gamma\, C_{ij}(t) - \omega_c\, C_{ik}(t) M_{kj}\,, \quad (t > 0). \tag{4.22}$$

In writing down Eq. (4.22), we have used the fact that

$$\overline{v_i(0)\,\eta_j(t)} = v_i(0)\,\overline{\eta_j(t)} = 0 \tag{4.23}$$

for all $t > 0$. The justification for this step is important, and is as follows:

- The noise process drives the velocity process. According to the **principle of causality**, the cause cannot precede the effect. Hence $\boldsymbol{\eta}(t)$ is independent of $\mathbf{v}(0)$ for all $t > 0$. The average of the product therefore factorizes into the product of the averages, and the latter vanishes[13].

Let us return to Eq. (4.22). It is convenient to write it in matrix form: we have

$$\frac{d\mathsf{C}(t)}{dt} = -\mathsf{C}(t)\,(I\,\gamma + \mathsf{M}\,\omega_c), \quad (t > 0) \tag{4.24}$$

where I is the 3×3 unit matrix. Now, $(I\,\gamma + \mathsf{M}\,\omega_c)$ is a constant matrix (independent of t). The formal solution of Eq. (4.24) can therefore be written down at once: we get[14]

$$\mathsf{C}(t) = \mathsf{C}(0)\,e^{-(I\,\gamma + \mathsf{M}\,\omega_c)t} = e^{-\gamma t}\,\mathsf{C}(0)\,e^{-\mathsf{M}\,\omega_c\,t}, \tag{4.25}$$

for all $t > 0$. The initial correlation matrix has elements

$$C_{ij}(0) = \frac{\langle v_i(0) v_j(0)\rangle_{\rm eq}}{(k_B T/m)} = \delta_{ij}\,, \tag{4.26}$$

because the initial velocity is drawn from the Maxwellian distribution. Hence $\mathsf{C}(0) = I$, the (3×3) unit matrix. It remains to evaluate the final exponential factor in Eq. (4.25). We find[15]

$$e^{\pm\mathsf{M}\,\omega_c\,t} = I \pm \mathsf{M}\,\sin\,\omega_c t + \mathsf{M}^2\,(1 - \cos\,\omega_c t). \tag{4.27}$$

It follows from the definition of M, Eq. (4.19), that the matrix elements of M^2 are given by

$$\left(\mathsf{M}^2\right)_{kj} = \epsilon_{klr}\,n_r\,\epsilon_{ljs}\,n_s = n_k\,n_j - \delta_{kj}\,. \tag{4.28}$$

[13]We shall have more to say about causality and its consequences in the sequel, especially in Chs. 15, 16 and 17.

[14]Observe that the correlation matrix $\mathsf{C}(t)$ appears to the left of the matrix $(I\,\gamma + \mathsf{M}\,\omega_c)$ in Eq. (4.24). As a consequence, the initial correlation matrix $\mathsf{C}(0)$ appears to the *left* of the exponential in the solution (4.25). *À priori*, there is no reason to expect the matrices $\mathsf{C}(0)$ and M to commute with each other, although it will turn out that $\mathsf{C}(0)$ is just the unit matrix in the present instance.

[15]For the derivation of Eqs. (4.27) and (4.28), see the Exercises at the end of this chapter.

Substituting these expressions in Eq. (4.25), we arrive at the following solution for the elements of the velocity correlation matrix for $t \geq 0$:

$$\begin{aligned} C_{ij}(t) &= \frac{\langle v_i(0)v_j(t)\rangle_{\text{eq}}}{(k_BT/m)} \\ &= e^{-\gamma t}\,\big[n_i\,n_j + (\delta_{ij} - n_i\,n_j)\,\cos\,\omega_c t - \epsilon_{ijk}\,n_k\,\sin\,\omega_c t\big]. \end{aligned} \tag{4.29}$$

Note that, even though $\mathsf{C}(0)$ starts as a symmetric matrix (it is in fact just the unit matrix), $\mathsf{C}(t)$ has both a symmetric part (the terms in $C_{ij}(t)$ that are proportional to δ_{ij} and $n_i\,n_j$) and an antisymmetric part (the term in $C_{ij}(t)$ that involves $\epsilon_{ijk}n_k$). Equation (4.29) will be used in Sec. 15.4.2.

In order to gain some insight into what the general formula of Eq. (4.29) implies, let us choose a specific direction—the customary positive z-axis, say—for the magnetic field. Then $(n_1\,,\,n_2\,,\,n_3) = (0,0,1)$. We revert to the subscripts (x,y,z) for ease of understanding. Setting $i = j = 3$, Eq. (4.29) immediately yields

$$\langle v_z(0)v_z(t)\rangle_{\text{eq}} = \frac{k_BT}{m}\,e^{-\gamma t}, \tag{4.30}$$

precisely as expected: The motion in the longitudinal direction (i. e., along the direction of the magnetic field) remains unaffected by the field. On the other hand, in the transverse directions we find

$$\langle v_x(0)v_x(t)\rangle_{\text{eq}} = \langle v_y(0)v_y(t)\rangle_{\text{eq}} = \frac{k_BT}{m}\,e^{-\gamma t}\,\cos\,\omega_c t\,. \tag{4.31}$$

Further,

$$\langle v_x(0)v_y(t)\rangle_{\text{eq}} = \langle v_y(0)v_x(t)\rangle_{\text{eq}} = -\frac{k_BT}{m}\,e^{-\gamma t}\,\sin\,\omega_c t\,. \tag{4.32}$$

Equation (4.32) shows how the velocity components in the transverse directions get intertwined and correlated with each other as a result of the cyclotron motion induced by the magnetic field. We shall return in Sec. 16.5.4 to the implication of this feature for the **Hall mobility**. Finally, the longitudinal-transverse cross-correlations turn out to be

$$\langle v_x(0)v_z(t)\rangle_{\text{eq}} = \langle v_y(0)v_z(t)\rangle_{\text{eq}} = \langle v_z(0)v_x(t)\rangle_{\text{eq}} = \langle v_z(0)v_y(t)\rangle_{\text{eq}} = 0. \tag{4.33}$$

The velocity in the longitudinal direction thus remains uncorrelated with the velocity in any direction transverse to the magnetic field.

Next, let us deduce the expression for $C_{ij}(t)$ for *negative* values of t. We can do so without repeating the calculation. The argument required is an instructive one, and is as follows. In the case of the free particle, we found that the only modification required to extend the result to all values of t was the replacement of the damping factor $\exp(-\gamma t)$ by $\exp(-\gamma|t|)$. This factor represents the effect of dissipation, i. e., the aspect of the motion that is *irreversible*—that is, *not* invariant under time reversal. In the present instance, too, it is evident that the damping factor would undergo the same generalization. However, the factor in square brackets in Eq. (4.29) does not involve the dissipation coefficient γ at all. It represents the effect of the time-*reversible* part of the motion. Hence, to get the counterpart of this factor for negative values of t, we have to apply the time-reversal transformation to the expression that obtains for $t > 0$. That is, we must change t to $-t$ in every term. But it is very important to realize that this is not enough.

- *The magnetic field* **B** *is also reversed in direction under the time-reversal transformation*[16].

In other words, we must also let $n_k \to -n_k$ for every k. As a result, the term involving $n_k \sin \omega_c t$ in Eq. (4.29) *also* remains unchanged for negative values of t. The other terms in the square brackets remain unchanged in any case, because $\cos \omega_c t$ is an even function of t, and the quadratic combination $n_i n_j$ does not change sign, either. We thus arrive at the result

$$C_{ij}(t) = e^{-\gamma|t|}\,[n_i n_j + (\delta_{ij} - n_i n_j)\cos \omega_c t - \epsilon_{ijk} n_k \sin \omega_c t] \tag{4.34}$$

for *all* (real) values of t. We shall make use of this expression in Sec. 16.5.4.

[16]This is a standard result. An elementary way to see how it comes about is as follows. The basic premise is that the equation of motion of a charged particle moving in free space in an applied electromagnetic field remains invariant under time reversal. (At an even more fundamental level, this supposition rests on the assumption—ultimately based on experimental observation—that electromagnetic interactions are time-reversal invariant.) The equation of motion is $m\,d\mathbf{v}/dt = q\,[\mathbf{E} + (\mathbf{v} \times \mathbf{B})]$. As both $\mathbf{v}$ and t change sign under time reversal, the left-hand side of the equation remains unchanged under the transformation. Hence the right-hand side must behave in the same way. Since $\mathbf{v}$ changes sign under time reversal, $\mathbf{B}$ must also do so. In passing, note that the same argument shows that the electric field $\mathbf{E}$ does *not* change sign under time reversal.

A heuristic, 'physical' argument is the following: suppose the magnetic field is produced by a current flowing in a circuit. Under time reversal, all velocities are reversed, so that the current flows in the opposite sense. This reverses the direction of the magnetic field. As *all* magnetic fields (no matter how they arise) must transform in the same way under any given transformation, *any* magnetic field is reversed in direction under time reversal.

4.4 Time reversal property of the correlation matrix

Finally, let us deduce the behavior, under time reversal, of the correlation matrix $\mathsf{C}(t)$ for a general stationary multi-dimensional (or vector) random process like $\mathbf{v}(t)$. We have seen in Eqs. (4.8) that the correlation in the one-dimensional case, $\langle v(0)\, v(t)\rangle_{\text{eq}}$, is a symmetric (or even) function of t. This arises from the stationarity property of the process $v(t)$ in equilibrium. The generalization of this result to the case of a multi-component process like $\mathbf{v}(t)$ is easily found. Repeating the argument following Eqs. (4.8), we now have

$$\langle v_i(t_0)\, v_j(t_0+t)\rangle_{\text{eq}} \begin{cases} = \langle v_i(0)\, v_j(t)\rangle_{\text{eq}} \\ = \langle v_i(-t)\, v_j(0)\rangle_{\text{eq}} = \langle v_j(0)\, v_i(-t)\rangle_{\text{eq}}\,. \end{cases} \tag{4.35}$$

Hence

$$C_{ij}(t) = C_{ji}(-t), \quad \text{or, in matrix form,} \quad \mathsf{C}(t) = \mathsf{C}^{\mathrm{T}}(-t). \tag{4.36}$$

It is easy to see that this property is satisfied both in the free-particle case, Eq. (4.12), and in the presence of a magnetic field, Eq. (4.34). We will have occasion to use this symmetry in Ch. 16, Sec. 16.5.2. It follows from Eq. (4.36) that

$$\frac{C_{ij}(t) \pm C_{ji}(t)}{2} = \frac{C_{ij}(t) \pm C_{ij}(-t)}{2}\,. \tag{4.37}$$

Hence the *symmetric* part of the tensor $C_{ij}(t)$ is an *even* function of t, while the *antisymmetric* part of this tensor is an *odd* function of t. Observe that the expression in Eq. (4.34) for the velocity correlation tensor in the presence of a magnetic field satisfies the foregoing symmetry properties.

4.5 Exercises

4.5.1 Higher-order correlations of the velocity

Consider the tagged particle in the absence of any external force field. We revert to the case of a single Cartesian component of the velocity, for convenience. We have calculated the autocorrelation function (or 'two-point function' or 'two-point correlation') of the velocity. We may proceed along the same lines to calculate all its higher-order correlations as well. These quantities are the equilibrium averages

$$\langle v(t_1)\, v(t_2)\, \ldots\, v(t_n)\rangle_{\text{eq}}\,,$$

for $n \geq 3$. The inputs required for this purpose are the corresponding higher-order correlations of the noise $\eta(t)$. We have stated that $\eta(t)$ is a stationary, Gaussian,

Markov process with mean value zero[17]. It follows from these properties that all the odd-order correlation functions of $\eta(t)$ vanish. Further, all even-order correlation functions factorize into linear combinations of two-point correlations. In particular, the three-point correlation vanishes, i. e.,

$$\overline{\eta(t_1)\,\eta(t_2)\,\eta(t_3)} = 0. \tag{4.38}$$

The four-point correlation of the noise factors into a linear combination of two-point correlations[18] according to

$$\begin{aligned}\overline{\eta(t_1)\,\eta(t_2)\,\eta(t_3)\,\eta(t_4)} &= \big(\overline{\eta(t_1)\,\eta(t_2)}\big)\,\big(\overline{\eta(t_3)\,\eta(t_4)}\big)\\ &+ \big(\overline{\eta(t_1)\,\eta(t_3)}\big)\,\big(\overline{\eta(t_2)\,\eta(t_4)}\big)\\ &+ \big(\overline{\eta(t_1)\,\eta(t_4)}\big)\,\big(\overline{\eta(t_2)\,\eta(t_3)}\big).\end{aligned} \tag{4.39}$$

But the noise is δ-correlated, as in Eq. (3.10). Hence

$$\begin{aligned}\overline{\eta(t_1)\,\eta(t_2)\,\eta(t_3)\,\eta(t_4)} &= \Gamma^2\,\big[\,\delta(t_1-t_2)\,\delta(t_3-t_4)+\delta(t_1-t_3)\,\delta(t_2-t_4)\\ &\quad +\delta(t_1-t_4)\,\delta(t_2-t_3)\,\big].\end{aligned} \tag{4.40}$$

(a) Use Eq. (4.38) to show that

$$\langle v(t_1)\,v(t_2)\,v(t_3)\rangle_{\text{eq}} = 0. \tag{4.41}$$

(b) Use Eq. (4.40) to show that

$$\begin{aligned}\langle v(t_1)\,v(t_2)\,v(t_3)\,v(t_4)\rangle_{\text{eq}} &= \left(\frac{k_BT}{m}\right)^2\Big\{e^{-\gamma|t_1-t_2|-\gamma|t_3-t_4|}+\\ &+\; e^{-\gamma|t_1-t_3|-\gamma|t_2-t_4|}+e^{-\gamma|t_1-t_4|-\gamma|t_2-t_3|}\Big\}.\end{aligned} \tag{4.42}$$

To establish these results, first calculate the corresponding conditional averages. Next, integrate the expressions so obtained over the initial velocity v_0, with a weight factor given by the equilibrium density $p^{\text{eq}}(v_0)$. Alternatively, simply let all the time arguments t_i tend to infinity, keeping all the differences between them finite.

Note that Eq. (4.42) implies the factorization

$$\begin{aligned}\langle v(t_1)\,v(t_2)\,v(t_3)\,v(t_4)\rangle_{\text{eq}} &= \langle v(t_1)\,v(t_2)\rangle_{\text{eq}}\,\langle v(t_3)\,v(t_4)\rangle_{\text{eq}}\\ &+ \langle v(t_1)\,v(t_3)\rangle_{\text{eq}}\,\langle v(t_2)\,v(t_4)\rangle_{\text{eq}}\\ &+ \langle v(t_1)\,v(t_4)\rangle_{\text{eq}}\,\langle v(t_2)\,v(t_3)\rangle_{\text{eq}}\,.\end{aligned} \tag{4.43}$$

[17] Markov processes will be discussed in the next chapter, as already mentioned.

[18] Observe that the right-hand side of Eq. (4.39) is a symmetric, equally-weighted linear combination of products of all the possible two-point correlations.

Again, this result is a consequence of the fact that the velocity is a Gaussian random process. In general, the equilibrium n-point correlation of the velocity vanishes when n is odd, and reduces to a symmetric linear combination of products of all possible two-point correlations when n is even.

4.5.2 Calculation of a matrix exponential

(a) Establish the identity in Eq. (4.27), namely,

$$e^{\pm \mathsf{M}\,\omega_c t} = I \pm \mathsf{M}\,\sin \omega_c t + \mathsf{M}^2\,(1 - \cos \omega_c t). \tag{4.44}$$

Here is a quick way to do so. Note that the eigenvalues of M are 0, i and $-i$. Recall the Cayley-Hamilton Theorem, which states that

- every matrix satisfies its own characteristic equation.

Hence $\mathsf{M}(\mathsf{M} - iI)(\mathsf{M} + iI) = 0$, which gives $\mathsf{M}^3 = -\mathsf{M}$. Use this fact to compute the exponential in closed form.

(b) Given that the matrix elements of M are given by $M_{kj} = \epsilon_{kjl}\, n_l$, show that those of M^2 are given by $(\mathsf{M}^2)_{kj} = n_k\, n_j - \delta_{kj}$. You need to use the well-known identity $\epsilon_{lmn}\,\epsilon_{lrs} = \delta_{mr}\,\delta_{ns} - \delta_{ms}\,\delta_{nr}$.

4.5.3 Correlations of the scalar and vector products

(a) From the expression in Eq. (4.34) for $\langle v_i(0) v_j(t)\rangle_{\text{eq}}$, show that

$$\langle \mathbf{v}(0) \cdot \mathbf{v}(t)\rangle_{\text{eq}} = \frac{k_B T}{m}\, e^{-\gamma|t|}\,(1 + 2\cos \omega_c t). \tag{4.45}$$

(b) Similarly, show that

$$\langle \mathbf{v}(0) \times \mathbf{v}(t)\rangle_{\text{eq}} = -\frac{2k_B T\,\mathbf{n}}{m}\, e^{-\gamma|t|}\,\sin \omega_c t\,, \tag{4.46}$$

where $\mathbf{n}$ is the unit vector in the direction of the magnetic field. Note that these expressions reduce correctly to those of Eqs. (4.13) and (4.14) in the absence of a magnetic field, on setting $\omega_c = 0$.

Chapter 5

Markov processes

We summarize some of the salient properties of Markov processes, including the master equation satisfied by the conditional PDF of such a process. The Kramers-Moyal expansion of the master equation is given, and its reduction to the Fokker-Planck equation under specific conditions is discussed. The master equation for a discrete Markov process and its formal solution are also considered. We discuss the corresponding stationary distribution, and write down its explicit form in the case when the detailed balance condition is satisfied. General remarks (that are relevant to understanding statistical mechanics) are made on the concepts of ergodicity, mixing and chaos in dynamical systems. Some special Markov jump processes that are important in applications are described in the exercises at the end of the chapter.

5.1 Continuous Markov processes

Let $\xi(t)$ be a variable that varies randomly with the time t, the latter being taken to be a continuous variable. $\xi(t)$ is called a **stochastic process**. The set of values that ξ can take is called the **sample space** (or **state space**). Unless otherwise specified, we shall take this to be a continuous set of (real) values. The statistical properties of the random process are specified completely by an infinite set of multiple-time joint probability densities

$$p_1(\xi, t),\ p_2(\xi_2\,, t_2\,;\, \xi_1\,, t_1),\ p_3(\xi_3\,, t_3\,;\, \xi_2\,, t_2\,;\, \xi_1\,, t_1),\ \ldots \quad \textit{ad inf}.$$

Here $t_1 < t_2 < t_3 < \ldots$. The n-time joint PDF p_n has the following meaning: when it is multiplied by the product of infinitesimals $d\xi_1\, d\xi_2\, \ldots\, d\xi_n$, the quantity

$$p_n(\xi_n\,, t_n\,;\, \xi_{n-1}\,, t_{n-1}\,;\, \ldots\,;\, \xi_1\,, t_1)\, d\xi_1\, d\xi_2\, \ldots\, d\xi_n$$

is the probability that the random variable ξ has a value in the range $(\xi_1\,,\,\xi_1+d\xi_1)$ at time t_1, a value in the range $(\xi_2\,,\,\xi_2+d\xi_2)$ at time $t_2\,,\cdots$, and a value in the range $(\xi_n\,,\,\xi_n+d\xi_n)$ at time t_n. This n-time joint PDF is expressible, by definition, as the product of an n-time *conditional* PDF and an $(n-1)$-time joint PDF, according to

$$\begin{aligned} p_n(\xi_n\,,\,t_n\,;\,\xi_{n-1}\,,\,t_{n-1}\,;\;\ldots\;;\,\xi_1\,,\,t_1) = \;& p_n(\xi_n\,,\,t_n\,|\,\xi_{n-1}\,,\,t_{n-1}\,;\ldots;\,\xi_1\,,\,t_1) \\ \times\;& p_{n-1}(\xi_{n-1}\,,\,t_{n-1}\,;\,\xi_{n-2}\,,\,t_{n-2}\,;\,\ldots\,;\,\xi_1\,,\,t_1). \end{aligned} \tag{5.1}$$

The first factor on the right-hand side is a conditional PDF: it is the probability density at the value ξ_n of the random variable ξ, *given* that the events to the right of the vertical bar have occurred at the successively earlier instants $t_{n-1}\,,\,t_{n-2}\,,\,\ldots\,,\,t_2$ and t_1, respectively.

A Markov process is one with a 'memory' that is restricted, at any instant of time, to the immediately preceding time argument *alone*. That is,

$$p_n(\xi_n\,,\,t_n\,|\,\xi_{n-1}\,,\,t_{n-1}\,;\,\ldots\,;\,\xi_1\,,\,t_1) = p_2(\xi_n\,,\,t_n\,|\,\xi_{n-1}\,,\,t_{n-1}) \tag{5.2}$$

for all $n \geq 2$.

- The single-time-step memory characterizing a Markov process is equivalent to saying that the *future* state of the process is only dependent on its *present* state, and not on the history of *how* the process reached the present state.

This implies at once that all the joint PDFs of a Markov process are expressible as products of just two independent PDFs: a *single-time* PDF $p_1(\xi\,,\,t)$, and a *two-time* conditional PDF $p_2(\xi\,,\,t\,|\,\xi'\,,\,t')$ (where $t' < t$), according to

$$\begin{aligned} p_n(\xi_n\,,\,t_n\,;\,\xi_{n-1}\,,\,t_{n-1}\,;\ldots;\,\xi_1\,,\,t_1) \;=\;& p_2(\xi_n\,,\,t_n\,|\,\xi_{n-1}\,,\,t_{n-1}) \\ \times\;& p_2(\xi_{n-1}\,,\,t_{n-1}\,|\,\xi_{n-2}\,,\,t_{n-2})\cdots \\ \times\;& p_2(\xi_2\,,\,t_2\,|\,\xi_1\,,\,t_1)\,p_1(\xi_1\,,\,t_1). \end{aligned} \tag{5.3}$$

If, further, the Markov process is a *stationary* process, then

$$p_1(\xi\,,\,t) = p_1(\xi) \quad \text{(independent of } t\text{)}, \tag{5.4}$$

and

$$p_2(\xi\,,\,t\,|\,\xi'\,,\,t') = p_2(\xi\,,\,t-t'\,|\,\xi') \quad \text{(a function of the difference } t-t'\text{)}. \tag{5.5}$$

We drop the subscripts 1 and 2 in p_1 and p_2, and write the two different functions as simply $p(\xi)$ and $p(\xi\,,\,t-t'\,|\,\xi')$, for notational simplicity[1]. Therefore, for a stationary Markov process,

$$p_n(\xi_n\,,\,t_n\,;\,\xi_{n-1}\,,\,t_{n-1}\,;\,\ldots\,;\,\xi_1\,,\,t_1) = \left[\prod_{j=1}^{n-1} p(\xi_{j+1}\,,\,t_{j+1}-t_j\,|\,\xi_j)\right] p(\xi_1), \quad (5.6)$$

for every $n \geq 2$. From this point onward, we shall discuss the case of stationary processes, unless otherwise specified. This is entirely for convenience, because it is possible to simplify the notation somewhat in the case of stationary processes[2].

Next, we need inputs for the stationary PDF $p(\xi)$ and the conditional PDF $p(\xi\,,\,t\,|\,\xi')$. These are, *à priori*, independent quantities. Now, the dynamics underlying the random process ξ generally enjoys a sufficient degree of **mixing**[3]. This feature ensures that the property already suggested in Eq. (2.13) holds good: that is,

$$\lim_{t\to\infty} p(\xi\,,\,t\,|\,\xi') = p(\xi). \quad (5.7)$$

In other words, the memory of the initial value ξ' is 'lost' as $t \to \infty$, and the conditional PDF simply tends to the stationary PDF $p(\xi)$. As a consequence,

- the single conditional density function $p(\xi\,,\,t\,|\,\xi')$ completely determines *all* the statistical properties of such a stationary Markov process.

The next step is to find this conditional density. We may bear in mind the following general (and perhaps obvious) fact: In all physical applications of probability and random processes, we can only write equations for, or model, or obtain unambiguously, *conditional* probabilities or probability densities—rather than their absolute counterparts[4].

---

[1] No confusion should arise as a result. The first density is time-independent, while the second is not. We shall refer to the two functions as the stationary density and the conditional density, respectively.

[2] Moreover, one of the main random processes with which we shall be concerned, the velocity process described by the Langevin equation, also turns out to be a stationary Markov process. But it must be understood that much of the formalism can be applied or extended to nonstationary random processes as well, with obvious changes of notation. Nonstationary processes also occur often in physical applications. A notable example is the *displacement* of the tagged particle, which will turn out to be a nonstationary Markov process under certain conditions.

[3] Further remarks on the concept of mixing will be made in Sec. 5.5.

[4] Try to find a counter-example!

5.2 Master equations for the conditional density

5.2.1 The Chapman-Kolmogorov equation

A starting point for the derivation of an equation satisfied by the conditional PDF $p(\xi\,,\,t\,|\,\xi\,')$ is the **Bachelier-Smoluchowski-Chapman-Kolmogorov equation** or chain equation satisfied by Markov processes:

$$p(\xi\,,\,t\,|\,\xi_0) = \int d\xi\,'\,p(\xi\,,\,t-t'\,|\,\xi\,')\,p(\xi\,'\,,\,t'\,|\,\xi_0) \quad (0 < t' < t). \tag{5.8}$$

In physical terms, this equation says the following:

- The probability of 'propagating' from an initial value ξ_0 to a final value ξ is the product of the probabilities of propagating from ξ_0 to any intermediate value $\xi\,'$, and subsequently from $\xi\,'$ to ξ, summed over all possible values of the intermediate state[5].

The problem with Eq. (5.8) is that it is a *nonlinear* integral equation for the conditional PDF. It can be reduced to a linear equation by defining a **transition probability** density per unit time. Let $w(\xi\,|\,\xi\,')\,d\xi$ be the probability per unit time of a transition from a given value $\xi\,'$ to the range of values $(\xi\,,\,\xi+d\xi)$ of the random variable. Equation (5.8) can then be reduced to the integro-differential equation

$$\frac{\partial}{\partial t}p(\xi\,,\,t\,|\,\xi_0) = \int d\xi\,'\Big\{\,p(\xi\,',\,t\,|\,\xi_0)\,w(\xi\,|\,\xi\,') - p(\xi\,,\,t\,|\,\xi_0)\,w(\xi\,'\,|\,\xi)\,\Big\}. \tag{5.9}$$

The (obvious) initial condition on the PDF is of course $p(\xi\,,\,0\,|\,\xi_0) = \delta(\xi-\xi_0)$. Equation (5.9) is usually referred to as the **master equation**. The discrete version of such an equation, applicable to a Markov process in which the sample space of the random process is a discrete set of values (or 'states'), is perhaps more familiar: it occurs as a **rate equation** in numerous applications. We shall consider this case in Sec. 5.3. It is clear from the form of the master equation that the first term on the right is a 'gain' term, while the second is a 'loss' term.

5.2.2 The Kramers-Moyal expansion

Even though the integro-differential equation (5.9) is a linear equation, solving it is far from an easy task, in general. One possible approach is to convert it to a

[5] Contrary to a common impression, the chain equation (often called the **Chapman-Kolmogorov equation**) is *not* unique to Markov processes. It is in fact an identity that is satisfied by a more general family of stochastic processes than just Markov processes.

differential equation. This is done as follows. Let us write $w(\xi\,|\,\xi\,')$ as a function of the starting state $\xi\,'$ and the size of the transition or jump, $\xi-\xi\,'\equiv\Delta\xi$, so that

$$w(\xi\,|\,\xi\,')\equiv W(\xi\,',\,\Delta\xi),$$

say. Equation (5.9) becomes

$$\begin{aligned}\frac{\partial}{\partial t}p(\xi\,,\,t\,|\,\xi_0) &= \int d(\Delta\xi)\,\Big\{p(\xi-\Delta\xi\,,\,t\,|\,\xi_0)\,W(\xi-\Delta\xi\,,\,\Delta\xi) \\ &\quad -p(\xi\,,\,t\,|\,\xi_0)\,W(\xi\,,\Delta\xi)\Big\}.\end{aligned} \tag{5.10}$$

We now assume that all the moments of the transition probability exist, i. e., that the quantities

$$A_n(\xi)=\int_{-\infty}^{\infty}d(\Delta\xi)\,(\Delta\xi)^n\,W(\xi\,,\,\Delta\xi) \tag{5.11}$$

are finite for all positive integral values of n. Then, a formal expansion of the right-hand side of Eq. (5.10) can be made in terms of these moments. The result is the partial differential equation

$$\frac{\partial}{\partial t}p(\xi\,,\,t\,|\,\xi_0)=\sum_{n=1}^{\infty}\frac{1}{n!}\left(-\frac{\partial}{\partial\xi}\right)^n\big[A_n(\xi)\,p(\xi\,,\,t\,|\,\xi_0)\big]. \tag{5.12}$$

This is the **Kramers-Moyal expansion** of the master equation. As Eq. (5.12) is a partial differential equation of infinite order, it is no more tractable than the original master equation, as it stands[6]. A degree of simplification occurs in those cases in which the transition probability density $w(\xi\,|\,\xi\,')$ is only a function of the *separation* or difference between the two values ξ and $\xi\,'$: that is, $W(\xi\,',\,\Delta\xi)=W(\Delta\xi)$, a function only of the jump $\Delta\,\xi$. Then the moments A_n become constants that are independent of ξ, and the Kramers-Moyal expansion becomes

$$\frac{\partial}{\partial t}p(\xi\,,\,t\,|\,\xi_0)=\sum_{n=1}^{\infty}\frac{A_n}{n!}\left(-\frac{\partial}{\partial\xi}\right)^n p(\xi\,,\,t\,|\,\xi_0). \tag{5.13}$$

5.2.3 The forward Kolmogorov equation

Going back to Eq. (5.12), the first two terms on the right-hand side represent, respectively, the 'drift' and 'diffusion' contributions to the process. In many situations, it turns out to be a satisfactory approximation to neglect the higher

[6]It also presumes the existence of the derivatives involved on the right-hand side of Eq. (5.12). On the other hand, the master equation (5.9) itself is also applicable to Markovian *jump* processes as well. See the exercises at the end of this chapter.

order terms if the third and higher moments of the transition probability density are sufficiently small. There is also another general result, known as **Pawula's Theorem**, which asserts that

- if *any* of the even moments A_{2l} where $l \geq 1$ vanishes, then *all* the A_n for $n \geq 3$ vanish, and the Kramers-Moyal expansion terminates after the second term[7].

As a consequence of Pawula's Theorem, there are only three possibilities associated with the Kramers-Moyal expansion, in a strict sense:

(i) $A_2 = 0$, in which case the expansion terminates with the first-order derivative on the right-hand side. This is a trivial case in which there is only a drift, and the dynamics is essentially deterministic.

(ii) One of the even moments A_{2l} (where $l \geq 2$) vanishes, and the expansion terminates with the second-order derivative on the right-hand side.

(iii) The expansion does not terminate at any finite order, and remains an infinite series, in principle[8].

The second possibility above is the most significant one. An important class of processes called **diffusion processes** (also termed **diffusions** in the mathematical literature), for which the expansion terminates at the second derivative term, will be discussed in Ch. 6. In all such cases, the master equation reduces to the second order partial differential equation

$$\frac{\partial}{\partial t}\, p(\xi\,,\, t\,|\,\xi_0) = -\frac{\partial}{\partial \xi}\left[A_1(\xi)\, p(\xi\,,\, t\,|\,\xi_0)\right] + \frac{1}{2}\frac{\partial^2}{\partial \xi^2}\left[A_2(\xi)\, p(\xi\,,\, t\,|\,\xi_0)\right]. \tag{5.14}$$

This is called the **forward Kolmogorov equation** or the **Fokker-Planck equation** (FPE). It is the most commonly used form of the master equation for the conditional PDF in physical applications of continuous Markov processes.

---

[7]This is an interesting, even surprising, result—but it is not magic. It is derived by an application of the Cauchy-Schwarz inequality to the moments A_n.

[8]Although in specific instances it may be a good approximation, from the numerical point of view, to truncate the expansion at some finite order.

5.2.4 The backward Kolmogorov equation

It is interesting to note that the differential operator on the right-hand side of the forward Kolmogorov equation is not self-adjoint, in general[9]. This feature has important physical consequences. The *adjoint* of the FPE (5.14) also plays a role in diffusion processes. The derivatives of the conditional PDF $p(\xi\,,\,t\,|\,\xi_0\,,\,t_0)$ with respect to the *initial* value ξ_0 and the initial time t_0 appear in the **backward Kolmogorov equation**. This equation reads, in the general (nonstationary) case[10],

$$\frac{\partial p}{\partial t_0} = -A_1(\xi_0\,,\,t_0)\,\frac{\partial p}{\partial \xi_0} - \frac{1}{2}A_2(\xi_0\,,\,t_0)\,\frac{\partial^2 p}{\partial \xi_0^2}\,, \qquad (5.15)$$

where p stands for $p(\xi\,,\,t\,|\,\xi_0\,,\,t_0)$. Again, in the case of a stationary process, the conditional PDF is a function of $(t - t_0)$ alone, rather than t and t_0 separately. Hence we can use the fact that $\partial/\partial t_0 = -\partial/\partial t$, and the backward equation reduces to

$$\frac{\partial}{\partial t}\,p(\xi\,,\,t\,|\,\xi_0) = A_1(\xi_0)\,\frac{\partial}{\partial \xi_0}\,p(\xi\,,\,t\,|\,\xi_0) + \frac{1}{2}A_2(\xi_0)\,\frac{\partial^2}{\partial \xi_0^2}\,p(\xi\,,\,t\,|\,\xi_0)\,. \qquad (5.16)$$

The backward Kolmogorov equation is useful, for instance, in connection with the **first-passage time distribution** and its moments, for a general process described by a Fokker-Planck equation[11].

5.3 Discrete Markov processes

5.3.1 The master equation and its solution

In the modeling of physical phenomena, there is another important class of Markov processes that is of considerable utility,. The sample spaces of these processes are discrete: the random variable can only assume values ξ_j, where $j = 1, 2, \ldots, N$,

[9] In this respect, the Fokker-Planck equation differs significantly from another important partial differential equation with an analogous mathematical structure (first order in time, second order in the spatial variable)—namely, the Schrödinger equation for the position-space wave function of a particle moving in a potential.

[10] The forward Kolmogorov equation is also applicable when the process concerned is nonstationary, although we have written it in Eq. (5.14) for a stationary PDF $p(\xi\,,\,t\,|\,\xi_0)$. In the more general case the PDF is of course $p(\xi\,,\,t\,|\,\xi_0\,,\,t_0)$. In the case of the backward equation we write down the nonstationary counterpart first, in order to bring out the dependence on t_0.

[11] In Ch. 10, we shall discuss the first-passage time problem associated with the diffusion equation, but not in terms of the backward Kolmogorov equation. In any case, the forward and backward equations are essentially the same in structure in the case of diffusion in the absence of an external force. The corresponding Fokker-Planck operator is just the second derivative operator $\partial^2/\partial\xi^2$, which is self-adjoint (for the boundary conditions usually considered).

say[12]. The positive integer j thus labels these values (or 'states'). If the *time* variable is also discrete-valued, then we have a (finite or infinite) **Markov chain**. There is a vast literature on this subject. However, in the applications to nonequilibrium phenomena that are generally encountered, t is a continuous variable, more often than not. We shall therefore restrict ourselves here to this case. We shall also borrow some terminology that belongs, properly speaking, to the case of Markov chains. The results discussed below are appropriate extensions of the corresponding ones for Markov chains. Let $P(j, t\,|\,i)$ denote the normalized probability that the system is in the state j at time t, where $1 \leq j \leq N$, given that it started in state i at $t = 0$. Note that we now deal with probabilities rather than probability densities, because the sample space is discrete rather than continuous. Further, let $w(j\,|\,k)$ be the transition probability per unit time to jump from the state k to the state j. It is convenient to use the notation

$$w(j\,|\,k) \equiv w_{jk} \tag{5.17}$$

for this transition rate[13]. Then the master equation satisfied by the set of probabilities $\{P(j, t\,|\,i)\}$, $j = 1, \ldots, N$, is[14]

$$\frac{d}{dt}P(j,t\,|\,i) = \sum_{k=1}^{N}{}'\big\{w_{jk}\,P(k,t\,|\,i) - w_{kj}\,P(j,t\,|\,i)\big\}, \quad 1 \leq j \leq N, \tag{5.18}$$

where the prime on the summation symbol is meant to indicate that the value $k = j$ is omitted in the sum over k. The initial condition on the probabilities is of course $P(j, 0\,|\,i) = \delta_{ij}$. Let $\mathsf{P}(t)$ denote the column vector with elements $P(j, t\,|\,i)$. Although the initial state i is not explicit in this notation, we must remember that the master equation is a time evolution equation for *conditional* probabilities (or *conditional* probability densities). The master equation (5.18) can then be written as the matrix equation

$$\frac{d\mathsf{P}}{dt} = \mathsf{W}\,\mathsf{P}, \tag{5.19}$$

where W is called the **relaxation matrix** in the physics literature. From the mathematical point of view, W is actually the *infinitesimal generator* of the

[12] The formalism that follows also applies when the number of possible states, N, is infinite, provided certain convergence conditions are satisfied. We do not consider these questions here.

[13] Cautionary remark: The notation $w(j\,|\,k) \equiv w_{kj}$ is also used frequently in the literature. A simple but useful exercise: translate all the formulas and results given here to that case, in order to help you read the literature on the subject more easily!

[14] i, j, k are merely state labels here, and not tensor indices. We therefore indicate sums over these labels explicitly. No summation over repeated indices is implied in this context.

continuous-time discrete Markov process. This $(N \times N)$ matrix has off-diagonal elements

$$\mathsf{W}_{jk} = w_{jk} \quad (j \neq k) \tag{5.20}$$

and diagonal elements

$$\mathsf{W}_{jj} = -\sum_{k=1}^{N}{}' w_{kj} \,. \tag{5.21}$$

It is important to remember that W is *not* a symmetric matrix, in general. The formal solution to Eq. (5.19) is

$$\mathsf{P}(t) = e^{\mathsf{W}t}\, \mathsf{P}(0), \tag{5.22}$$

where $\mathsf{P}(0)$ is a column vector whose i^{th} row is unity, and all other elements are zero. Here the exponential of the matrix $\mathsf{W}t$ is *defined* as an infinite sum of matrices according to[15]

$$e^{\mathsf{W}t} = \sum_{n=0}^{\infty} \frac{(\mathsf{W}t)^n}{n!} \,. \tag{5.23}$$

The time-dependence of the probability distribution is essentially determined, in principle, by the eigenvalue spectrum of the matrix W (see below). However, finding the exponential of the matrix $\mathsf{W}t$ explicitly is not always a simple task. An equivalent, but more convenient, way of dealing with the problem is to take the Laplace transform of both sides of Eq. (5.19) (or Eq. (5.22)). Denoting the Laplace transform of $\mathsf{P}(t)$ by $\widetilde{\mathsf{P}}(s)$, we get[16]

$$\widetilde{\mathsf{P}}(s) = \big(sI - \mathsf{W}\big)^{-1}\, \mathsf{P}(0), \tag{5.24}$$

where I is the $(N \times N)$ unit matrix. The matrix $(sI - \mathsf{W})^{-1}$, which is the *resolvent* of the transition matrix W, essentially determines the Laplace transform $\widetilde{\mathsf{P}}(s)$ of the probability vector $\mathsf{P}(t)$. Thus, the problem of exponentiating a matrix is replaced by the simpler one of finding the inverse of a related matrix.

The known properties of the relaxation matrix W make it possible to say something about its eigenvalue spectrum. We restrict ourselves to the most general

---

[15]The definition (5.23) involves an infinite sum, and one must therefore worry about its convergence. Recall that the infinite sum defining e^z (where z is any complex number) converges absolutely for all finite z, i. e., for all $|z| < \infty$. Similarly, the sum representing the exponential of an $(N \times N)$ matrix converges for all finite values of the *norm* of the matrix. Hence the convergence of the series is certainly guaranteed for all such matrices with finite elements.

[16]Recall that the Laplace transform of $df(t)/dt$ is given by $s\widetilde{f}(s) - f(0)$, where $\widetilde{f}(s)$ is the Laplace transform of $f(t)$.

statements, without going into the details of the necessary and sufficient conditions for their validity, exceptional cases, and so on. The sum of the elements of each column of W is equal to zero. It follows at once that $\det \mathsf{W} = 0$, so that 0 is an eigenvalue of this matrix. (The physical implication of this fact will become clear shortly.) The elements of W are real. Hence its eigenvalues occur in complex conjugate pairs. Moreover—and most importantly—, we can assert that the real parts of all the eigenvalues are negative[17], except for the eigenvalue 0. Hence the time evolution of the probabilities, which is governed by $e^{\mathsf{W}t}$, is given by decaying exponentials in t, multiplied by possible factors involving sinusoidal functions of t. This justifies the term 'relaxation matrix' for W.

5.3.2 The stationary distribution

What, then, do the probabilities $P(j,t\,|\,i)$ 'relax' to as $t \to \infty$? A very important question connected with the master equation is the existence and uniqueness of a time-independent or **stationary distribution** P^{st} (not to be confused with a stationary random *process*). This distribution is given by the solution of

$$\frac{d\mathsf{P}^{\text{st}}}{dt} = \mathsf{W}\,\mathsf{P}^{\text{st}} = 0. \tag{5.25}$$

Therefore the stationary distribution is nothing but the set of elements of the *right* eigenvector of the matrix W corresponding to the eigenvalue zero[18]. Going back to the explicit form of the master equation, Eq. (5.18), it is clear that the stationary distribution is given by the solution of the set of N homogeneous simultaneous equations

$$\left(\sum_{k=1}^{N}{}' w_{kj}\right) P_j^{\text{st}} - \sum_{k=1}^{N}{}' w_{jk}\, P_k^{\text{st}} = 0, \quad (j = 1, 2, \ldots, N) \tag{5.26}$$

[17] A quick way of seeing this is by applying **Gershgorin's Circle Theorem** to the matrix W. In the complex plane, mark the position of the diagonal element w_{11}. It lies on the negative real axis. With this point as the center, draw a circle of radius equal to the sum $\sum_j{}' w_{1j}$ of all the off-diagonal elements in the first column. It is obvious that this circle passes through the origin, and lies in the left half-plane, with the imaginary axis as a tangent. Repeat this procedure for each of the diagonal elements $w_{22}, \ldots, w_{NN}$. All the eigenvalues of W are guaranteed to lie in the union of these Gershgorin discs (including their boundaries). Hence all the eigenvalues (except the eigenvalue 0) must have negative real parts.

[18] As each column of W adds up to zero, it is evident that the uniform row vector $(1 \;\; 1 \;\; \cdots \;\; 1)$ is the (unnormalized) *left* eigenvector of W with eigenvalue zero. Since W is not symmetric in general, the right and left eigenvectors are not adjoints of each other. It is also necessary to show that (i) the right eigenvector is unique, and (ii) its elements are positive numbers (after normalization they add up to unity). We do not go into the conditions under which these requirements are met.

taken together with the normalization condition $\sum_{j=1}^{N} P_j^{\text{st}} = 1$. The stationary distribution P^{st} is thus expressible in terms of the set of transition rates.

5.3.3 Detailed balance

An important special case of the foregoing corresponds to the situation in which the **detailed balance** condition applies, namely, when *each* term in the summand in Eq. (5.18) vanishes by itself for the stationary distribution: that is, when we have

$$w_{jk}\, P_k^{\text{st}} = w_{kj}\, P_j^{\text{st}} \tag{5.27}$$

for *every pair* of distinct values of j and k. The detailed balance condition has its origin in time reversal invariance[19]. When detailed balance applies, the corresponding stationary distribution may be termed the equilibrium distribution, with probabilities denoted by P_k^{eq}. These probabilities can be found easily in terms of the transition rates. We have $P_k^{\text{eq}} = (w_{kj}/w_{jk})\, P_j^{\text{eq}}$. Using the normalization condition $\sum_k P_k^{\text{eq}} = 1$, we obtain the compact formula

$$P_j^{\text{eq}} = \left[1 + \sum_{k=1}^{N}{}' \left(\frac{w_{kj}}{w_{jk}}\right)\right]^{-1}. \tag{5.28}$$

Observe that when W is a symmetric matrix, so that $w_{kj} = w_{jk}$, the normalized equilibrium distribution reduces to the uniform distribution, $P_j^{\text{eq}} = 1/N$ for every j.

An application of the detailed balance condition in the case of a *continuous* Markov process will be encountered in Sec. 14.1.2, in connection with the motion of a charged tagged particle in an applied magnetic field.

5.4 Autocorrelation function

In the chapters that follow, correlation functions will play a prominent role. It is therefore useful to have, for ready reference, a 'formula' for the autocorrelation function of a random process in terms of its one-time and two-time PDFs[20]. This formula is nothing but an explicit expression of the definition of the autocorrelation function. It is valid for both Markov and non-Markov processes.

---

[19]We shall not enter here into a discussion of how this comes about.

[20]In Ch. 4, by contrast, we computed the autocorrelation of the velocity of the tagged particle directly from the Langevin equation.

We have, by definition,

$$\begin{aligned}\langle \xi(t_1)\,\xi(t_2)\rangle &= \int d\xi_1 \int d\xi_2\; \xi_1\,\xi_2\; p_2(\xi_2\,, t_2\,;\,\xi_1\,,\,t_1)\\ &= \int d\xi_1 \int d\xi_2\; \xi_1\,\xi_2\; p_2(\xi_2\,, t_2\,|\,\xi_1\,,\,t_1)\,p_1(\xi_1\,,\,t_1).\end{aligned} \tag{5.29}$$

For a stationary random process this becomes, on writing $t_2 - t_1$ as simply t,

$$\langle \xi(0)\,\xi(t)\rangle = \int d\xi_1 \int d\xi_2\; \xi_1\,\xi_2\; p(\xi_2\,, t\,|\,\xi_1)\,p(\xi_1), \tag{5.30}$$

in terms of the stationary and conditional PDFs. As we have mentioned in Ch. 4, when the mean value of the random process is nonzero, the autocorrelation function is defined in terms of the deviation from the mean value, namely, $\delta\xi = \xi - \langle\xi\rangle$. It is easy to show that the autocorrelation function then simplifies to

$$\langle \delta\xi(0)\,\delta\xi(t)\rangle = \int d\xi_1 \int d\xi_2\; \xi_1\,\xi_2\; p(\xi_2\,, t\,|\,\xi_1)\,p(\xi_1) - \langle\xi\rangle^2\,, \tag{5.31}$$

where $\langle\xi\rangle = \int d\xi\,\xi\,p(\xi)$. The integrations in the formulas above are replaced by summations in the case of a discrete sample space.

5.5 Remarks on ergodicity, mixing, and chaos

We conclude this chapter with some remarks on a noteworthy aspect of the stationary random processes considered in this chapter and in the rest of this book. These remarks may appear to be technical in nature, but they are important for a better understanding of statistical mechanics[21].

Throughout, we assume that the stationary processes that we deal with are **ergodic**: that is, given sufficient time, the random variable concerned can start from *any* of its initial values and attain *all* its possible values at some time or the other. In other words, the sample space of the random variable is explored completely, and does not decouple eventually into disjoint sets[22]. As a consequence of this property, the **time average** from $t = 0$ up to $t = T$ of any function of the random variable can be replaced, in the limit $T \to \infty$, by an **ensemble average**—namely, a statistical average over a certain probability distribution. For an ergodic

[21]This is why I have elaborated upon them to some extent, and placed them in a separate section. You may find it useful to return to this section subsequently. Some of the references included in the bibliography provide in-depth discussions of the matters touched upon here.

[22]In the case of Markov chains, this means that we consider only **irreducible Markov chains**.

stationary random process, this means that the moments of the random variable are given by[23]

$$\lim_{T\to\infty}\frac{1}{T}\int_0^T dt\,\xi^n(t) = \langle \xi^n\rangle = \int d\xi\,\xi^n\,p(\xi). \tag{5.32}$$

Similarly, the two-point correlation is given by

$$\lim_{T\to\infty}\frac{1}{T}\int_0^T dt_0\;\xi(t_0)\,\xi(t_0+t) = \langle \xi(0)\,\xi(t)\rangle\,, \tag{5.33}$$

where the right-hand side is precisely the statistical average given in Eq. (5.30).

The stochasticity in the random variables with which we are primarily concerned in this book arises from thermal fluctuations. These, in turn, originate in the underlying dynamics of all the constituents of a system at the microscopic or particle level. Recall that our attention is restricted to classical dynamics. In the general context of **classical dynamical systems**, ergodicity has the following specific meaning. Consider an infinitesimal 'volume' element in the phase space of the system, whose points represent a set of (neighboring) initial conditions, i. e., the initial values of all the independent dynamical variables of the system. Ergodicity implies that this volume element 'visits' all regions of the accessible phase space, given sufficient time. More precisely, points belonging to this initial set come arbitrarily close, as time progresses, to all points in the accessible phase space. By 'accessible' we mean that the conditions extant—either imposed externally on the system, or prevailing by virtue of the initial conditions on the dynamical variables—may be such that only a part of the phase space is available for such excursions. For instance: suppose there is a constant of the motion, i. e., a function of the dynamical variables that remains constant in value during the motion. Then the accessible phase space is obviously restricted to *that* subset of the full phase space, on which the function maintains its given initial value. The numerical value of any constant of the motion is determined, of course, by the initial conditions.

These considerations are quite basic to the understanding of statistical mechanics—both equilibrium and nonequilibrium. At this point, it is useful to digress briefly to recall the situation in equilibrium statistical mechanics. As we know, the fundamental postulate of the latter subject is:

- *All the accessible microstates of an isolated system in thermal equilibrium are equally probable.*

[23]The integrals on the left-hand sides of Eqs. (5.32) and (5.33) are replaced by sums in the case of discrete-time processes.

Underlying this postulate is the understanding that a generic or typical Hamiltonian system with a large number of degrees of freedom is ergodic on each of its **energy surfaces**. By an energy surface we mean a region of the phase space in which the Hamiltonian $H(q,p)$ of the system has a given, constant, numerical value—namely, the specified total energy E of the isolated system. The phase space probability density describing the ensemble—the **microcanonical ensemble** in this instance—is just $\delta\big(H(q,p)-E\big)$. This implies a uniform probability density everywhere on the energy surface. Roughly speaking, any initial volume element on such a subset of phase space visits all parts of it equally frequently, given sufficient time. A little more accurately, the volume element spends, on the average, equal fractions of the total time in equal-sized portions of the energy surface. Incidentally, it is tacitly understood that the foregoing statements apply to a system that is in a given state of motion as a whole—i. e., for given values of the other constants of the motion, such as the total linear momentum $\mathbf{P}$ and the total angular momentum $\mathbf{L}$ of the system[24]. Thus the energy surface one speaks of is really the common intersection of the level surfaces of all the constants of the motion, including the Hamiltonian. The ten **Galilean invariants** ($H, \mathbf{P}, \mathbf{L}$ and $\mathbf{R}-\mathbf{P}t/M$) are, in fact, the only (smooth, 'analytic') constants of the motion that a *generic* N-particle Hamiltonian system possesses. The number of constants of the motion that are in involution with each other[25] (for instance, H and the three components of $\mathbf{P}$) is far smaller than the number ($3N$) required for integrability. Hence such a Hamiltonian is (highly!) non-integrable. The dynamics, although restricted (for almost all sets of initial conditions) to some energy surface, is chaotic, in the sense explained below.

Let us return to our general discussion. During the process of evolution, an initial volume element in phase space may get badly distorted—in particular, points that started out close to each other may find themselves quite far apart as time progresses. The degree to which the scrambling up or *mixing* of an initial volume element in phase space occurs is precisely quantifiable. (But we shall not go into these details here.) It is this property of mixing that we referred to in writing down Eq. (5.7). Ergodicity *per se* does not require this sort of scrambling

[24]Typically, we are concerned with a system comprising N ($\gg 1$) particles in three-dimensional space. We assume further that the potential energy term in the Hamiltonian is unchanged under a shift of the origin of coordinates, or a rotation of the coordinate axes. (The kinetic energy term is obviously unchanged by these transformations.) This invariance leads, by **Noether's Theorem**, to the conservation of $\mathbf{P}$ and $\mathbf{L}$. In addition, the components of $\mathbf{R}-\mathbf{P}t/M$, where $\mathbf{R}$ is the position of the center of mass and M is the total mass of the system, are also (explicitly t-dependent) constants of the motion.

[25]That is, all their **Poisson brackets** with each other vanish identically.

up. In the context of dynamical systems, ergodic behavior alone (i. e., without mixing) means that any initial volume element merely traverses all parts of the accessible phase space with little or no distortion, unlike what happens when we have mixing as well.

- Mixing implies ergodicity, but the converse is not true (ergodicity does not *imply* mixing).

Evidently, mixing may occur in many different ways. The separation of initially neighboring points in phase space may increase like a power of t. Or, going a step further, it may increase *exponentially* with t. In that case, the system displays **exponential sensitivity to initial conditions**. In general, this sort of exponential instability leads to dynamical chaos[26]. As we have already mentioned in Sec. 2.3, a thermodynamic system such as a fluid is certainly highly chaotic when viewed as a dynamical system comprising a very large number of interacting degrees of freedom. There is a hierarchy of increasing randomness in dynamical systems, in which each successive property implies the preceding ones, but the converse is not necessarily true. The hierarchy goes something like this:

- Ergodicity $\Longleftarrow$ mixing $\Longleftarrow$ exponential sensitivity on the average in phase space $\Longleftarrow$ exponential sensitivity everywhere in phase space (global chaos).

An interesting question now arises, from the point of view of the subject matter of this book. In the case of ordinary diffusion (to be studied in greater detail in subsequent chapters), we shall see that the root mean squared displacement of the tagged particle increases with time like $t^{1/2}$. An initial small separation between two diffusing particles can also be shown to increase like $t^{1/2}$, on the average. Does this mere power-law separation (as opposed to an exponential separation) mean that the underlying dynamics is mixing, but not chaotic? Not at all. We must remember that diffusion, regarded as a stochastic process, does not fall under the purview of deterministic dynamics. The microscopic dynamics in a fluid in thermal equilibrium, that leads to the diffusion of a tagged particle, *is* chaotic, as we have already stated. It is helpful, at this stage, to read once again the comments made in a lengthy footnote in Sec. 2.3. In particular, recall that the maximal Liapunov exponent of the dynamical system represented by the fluid is proportional to its number of degrees of freedom (typically, $\gtrsim 10^{23}$), which is very large indeed[27].

[26] As you know, chaotic dynamics is a major field of study in its own right. A convenient 'definition' of chaos may be in order, merely to have a semblance of completeness. The technical conditions for chaos are: (i) a bounded phase space, (ii) a dense set of unstable periodic orbits in this phase space, and (iii) exponential sensitivity to initial conditions. The last feature is characterized by the existence of one or more *positive* Liapunov exponents.

[27] However, in the interests of technical accuracy and completeness, we mention that it *is*

5.6 Exercises

5.6.1 The dichotomous Markov process

The **dichotomous Markov process** (or DMP) is a very basic stationary random process that occurs in various forms in numerous applications. The DMP is a Markovian jump process in which the sample space comprises just two values, ξ_1 and ξ_2. The random variable flips back and forth between these values (or states) at random instants of time. Let the mean rate of a transition from ξ_1 to ξ_2 be λ_1, and let that from ξ_2 to ξ_1 be λ_2. Successive transitions are supposed to be completely uncorrelated with each other.

Suppose the system is in the state ξ_1 at any instant of time. The probability that it will make a transition to the state ξ_2 in the next infinitesimal time interval δt is just $\lambda_1\,\delta t$. Now, δt is taken to be so small that the probability of two or more transitions occurring within this time interval is negligible. Hence the probability that it will *not* make a transition in the interval δt is $(1-\lambda_1\,\delta t)$. From this fact, it is easy to calculate the probability that, if the system has just arrived at the value ξ_1 at some instant of time t_0, it remains in that state at time $t\,(\geq t_0)$ without having made any transitions in between. All we have to do is to divide the interval $(t-t_0)$ into n sub-intervals, each of duration δt. Since the process is Markovian, there is no history-dependence. The no-transition or zero-transition probability sought is therefore $(1-\lambda_1\,\delta t)^n$, in the limit $n\to\infty$, $\delta t\to 0$ such that the product $n\,\delta t = t-t_0$. But this limit is just $\exp\left[-\lambda_1\,(t-t_0)\right]$. The mean time between transitions in state ξ_1, or the **mean residence time** in that state between two successive transitions, is clearly $\tau_1=\lambda_1^{-1}$.

Exactly the same statements apply to the state ξ_2, with λ_1 replaced by λ_2, and the roles of the states ξ_1 and ξ_2 interchanged. Figure 5.1 shows a typical realization of a DMP.

(a) Show that the mean and variance of ξ are given by

$$\langle\xi\rangle = \frac{\tau_1\,\xi_1+\tau_2\,\xi_2}{\tau_1+\tau_2} = \frac{\xi_1\,\lambda_2+\xi_2\,\lambda_1}{\lambda_1+\lambda_2} \tag{5.34}$$

---

possible to extend concepts such as mixing to the case of stochastic dynamics. But this is a nontrivial and fairly technical topic, and is outside the scope of this book. It turns out that quantities such as *average* Liapunov exponents can indeed be defined rigorously for stochastic dynamics. When the diffusion is stable in a certain technical sense (roughly speaking, when there is no exponentially growing drift), the average Liapunov exponent vanishes (as in the case of ordinary diffusion) or becomes negative (as in the case of the Ornstein-Uhlenbeck process, to be defined in Ch. 6.)

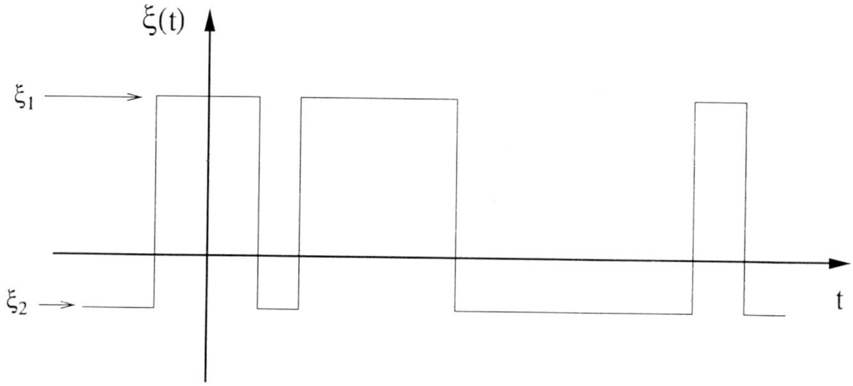

Figure 5.1: A realization of a dichotomous Markov process with values ξ_1 and ξ_2. Successive transitions between the two states are uncorrelated with each other. The mean durations of the two states are τ_1 and τ_2, respectively. The mean transition rates from ξ_1 to ξ_2, and vice versa, are the reciprocals of these mean durations. The DMP is a stationary, discrete, Markov process with an autocorrelation function that is an exponentially decaying function of time. The correlation time τ_{corr} of the process is the harmonic mean of τ_1 and τ_2.

and

$$\Big\langle \big(\xi - \langle \xi \rangle\big)^2 \Big\rangle = \frac{\tau_1\,\tau_2\,(\xi_1-\xi_2)^2}{(\tau_1+\tau_2)^2} = \frac{\lambda_1\,\lambda_2\,(\xi_1-\xi_2)^2}{(\lambda_1+\lambda_2)^2}. \tag{5.35}$$

(b) The master equation for a DMP is very easily written down. Denoting the respective probabilities of states ξ_1 and ξ_2 by P_1 and P_2, we have

$$\frac{d\mathsf{P}}{dt} = \mathsf{W}\,\mathsf{P} \quad \text{where} \quad \mathsf{P}(t) = \begin{pmatrix} P_1(t) \\ P_2(t) \end{pmatrix} \quad \text{and} \quad \mathsf{W} = \begin{pmatrix} -\lambda_1 & \lambda_2 \\ \lambda_1 & -\lambda_2 \end{pmatrix}. \tag{5.36}$$

The formal solution to the master equation is $\mathsf{P}(t) = e^{\mathsf{W}t}\,\mathsf{P}(0)$. The exponential of the matrix $\mathsf{W}t$ is calculated quite easily in this case. A simple way is to note that $\mathsf{W}^2 = -(\lambda_1+\lambda_2)\,\mathsf{W}$, so that the summation in Eq. (5.23) can be carried out very easily. Show that

$$e^{\mathsf{W}t} = \frac{1}{2\lambda}\begin{pmatrix} \lambda_2 + \lambda_1\,e^{-2\lambda t} & \lambda_2\,(1-e^{-2\lambda t}) \\ \lambda_1\,(1-e^{-2\lambda t}) & \lambda_1 + \lambda_2\,e^{-2\lambda t} \end{pmatrix}, \tag{5.37}$$

where $\lambda = \frac{1}{2}(\lambda_1+\lambda_2)$ is the mean transition rate for the DMP. Hence show that the four normalized conditional probability densities characterizing the DMP are given by

$$
\begin{aligned}
P(\xi_1\,,t\,|\,\xi_1) &= (\lambda_2+\lambda_1\,e^{-2\lambda t})/(\lambda_1+\lambda_2)\,,\\
P(\xi_2\,,t\,|\,\xi_1) &= \lambda_1\,(1-e^{-2\lambda t})/(\lambda_1+\lambda_2)\,,\\
P(\xi_1\,,t\,|\,\xi_2) &= \lambda_2\,(1-e^{-2\lambda t})/(\lambda_1+\lambda_2)\,,\\
P(\xi_2\,,t\,|\,\xi_2) &= (\lambda_1+\lambda_2\,e^{-2\lambda t})/(\lambda_1+\lambda_2)\,.
\end{aligned}
\tag{5.38}
$$

Observe that

$$
\lim_{t\to\infty} e^{\mathsf{W}t} = \frac{1}{2\lambda}\begin{pmatrix}\lambda_2 & \lambda_2\\ \lambda_1 & \lambda_1\end{pmatrix}.
\tag{5.39}
$$

Hence verify that the stationary probability distribution of the DMP is simply

$$
\mathsf{P}^{\text{eq}} = \begin{pmatrix}P_1^{\text{eq}}\\ P_2^{\text{eq}}\end{pmatrix} = \begin{pmatrix}\lambda_2/(\lambda_1+\lambda_2)\\ \lambda_1/(\lambda_1+\lambda_2)\end{pmatrix},
\tag{5.40}
$$

whatever the initial distribution $\mathsf{P}(0)$ may be. We may also write the stationary distribution in the form

$$
\begin{aligned}
P(\xi) &= \frac{\lambda_2}{\lambda_1+\lambda_2}\,\delta(\xi-\xi_1)+\frac{\lambda_1}{\lambda_1+\lambda_2}\,\delta(\xi-\xi_2)\\
&= \frac{\tau_1}{\tau_1+\tau_2}\,\delta(\xi-\xi_1)+\frac{\tau_2}{\tau_1+\tau_2}\,\delta(\xi-\xi_2).
\end{aligned}
\tag{5.41}
$$

Clearly, the final expression in Eq. (5.41) is precisely what we may expect on physical grounds for the stationary distribution, in terms of the fractions of the total time spent (on the average) in the two states ξ_1 and ξ_2 over a very long interval of time.

(c) Show that a DMP is exponentially correlated, i. e.,

$$
\langle\delta\xi(0)\,\delta\xi(t)\rangle = \frac{\lambda_1\,\lambda_2\,(\xi_1-\xi_2)^2}{(\lambda_1+\lambda_2)^2}\,e^{-2\lambda t}\,,
\tag{5.42}
$$

where $\delta\xi(t)\equiv\xi(t)-\langle\xi\rangle$. Note that the correlation time is

$$
\tau_{\text{corr}} = \frac{1}{2\lambda} = \frac{1}{\lambda_1+\lambda_2} = \frac{\tau_1\,\tau_2}{\tau_1+\tau_2}\,,
\tag{5.43}
$$

rather than λ^{-1}, as one might guess at first sight.

The symmetric DMP : If $\xi_1=-\xi_2=c$, say, and $\lambda_1=\lambda_2=\lambda$, we have a symmetric DMP, and all the foregoing expressions simplify considerably. The instants at which the system makes a transition from one of the states to the other constitute a **Poisson pulse process** with a mean pulse rate[28] λ: that is, the sequence of instants is completely uncorrelated, and the probability that exactly n transitions occur in a given time interval t is Poisson-distributed, being given by $(\lambda t)^n \, e^{-\lambda t}/n!$. We get, in this case,

$$\langle \xi \rangle = 0, \quad \langle \xi^2 \rangle = c^2, \quad \langle \xi(0) \, \xi(t) \rangle = c^2 \, e^{-2\lambda t}. \tag{5.44}$$

Further,

$$P(\pm c \,, t \,|\, \pm c) = e^{-\lambda t} \cosh \lambda t \,, \quad P(\pm c \,, t \,|\, \mp c) = e^{-\lambda t} \sinh \lambda t \,. \tag{5.45}$$

5.6.2 The Kubo-Anderson process

Going back to the general case of a Markov process and the master equation (5.9), suppose the transition probability density per unit time $w(\xi \,|\, \xi\,')$ is a function of the final state ξ alone, independent of the starting state $\xi\,'$: that is, $w(\xi \,|\, \xi\,') = \lambda \, u(\xi)$, say, where λ is a constant with the physical dimensions of $(\text{time})^{-1}$ (recall that w involves a time rate of transitions).

(a) Show that the normalized solution for $p(\xi \,, t \,|\, \xi_0)$ reduces in this case to

$$p(\xi \,, t \,|\, \xi_0) = \delta(\xi - \xi_0) \, e^{-\lambda t} + u(\xi) \left(1 - e^{-\lambda t}\right). \tag{5.46}$$

The physical interpretation of the function $u(\xi)$ is now obvious: passing to the $t \to \infty$ limit in the solution above, we find $\lim_{t\to\infty} p(\xi \,, t \,|\, \xi_0) = u(\xi)$. It follows that $u(\xi) = p(\xi)$, where $p(\xi)$ is simply the normalized PDF of the stationary Markov process ξ. Therefore the conditional PDF of the process is given by

$$p(\xi \,, t \,|\, \xi_0) = \delta(\xi - \xi_0) \, e^{-\lambda t} + p(\xi) \left(1 - e^{-\lambda t}\right). \tag{5.47}$$

The process may also be described as follows: successive transitions occur at instants of time that constitute a Poisson process, the mean time between successive jumps being given by λ^{-1}. At each jump, the new state is chosen at random, from the distribution represented by the stationary PDF $p(\xi)$.

Such a Markov process is called a **Kubo-Anderson** process. It finds numerous applications in statistical physics—in particular, in the modeling of relaxation phenomena. Figure 5.2 shows a realization of the Kubo-Anderson process in the

---

[28] This rate is called the **intensity** of the Poisson process in the statistics literature.

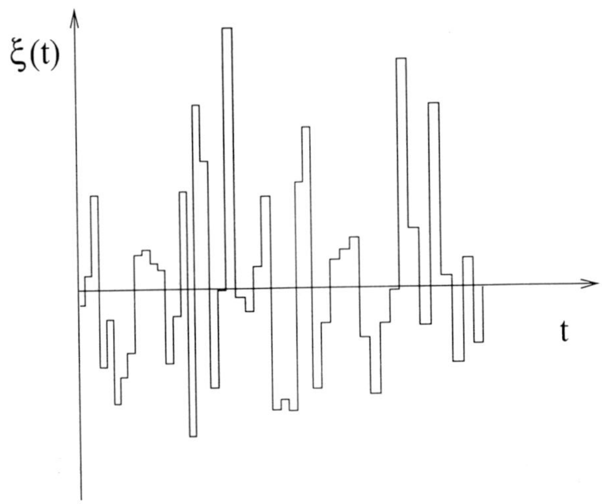

Figure 5.2: A realization of a Kubo-Anderson process. This is also a stationary Markovian jump process with an exponentially decaying autocorrelation function. The successive jumps or transitions comprising the process are uncorrelated with each other. Transitions occur at a constant mean rate λ. The probability that n transitions occur in a time interval t is given by the Poisson distribution, $(\lambda t)^n \exp(-\lambda t)/n!$ The mean number of transitions in any interval t is therefore equal to λt. The state or value to which the process jumps at any transition is independent of the current value of the random variable, and is selected at random, weighted according to a given stationary PDF $p(\xi)$.

case when $p(\xi)$ is a normalized Gaussian with zero mean.

(b) Show that the Kubo-Anderson process is exponentially correlated, i. e., show that the autocorrelation function is given by

$$\langle \delta\xi(0)\,\delta\xi(t)\rangle = \big\langle (\delta\xi)^2\big\rangle\, e^{-\lambda t}, \quad (t \geq 0) \tag{5.48}$$

where $\big\langle (\delta\xi)^2\big\rangle = \langle \xi^2\rangle - \langle\xi\rangle^2$ and $\langle \xi^n\rangle = \int d\xi\, \xi^n\, p(\xi)$.

5.6.3 The kangaroo process

Generalizing the Kubo-Anderson process, it is conceivable that the transition rate λ is actually dependent on the starting state ξ'. In this case we must write $w(\xi\,|\,\xi') = \lambda(\xi')\,u(\xi)$. In other words, $w(\xi\,|\,\xi')$ is assumed to be a product of functions of its two arguments. The master equation (5.9) becomes

$$\frac{\partial}{\partial t}p(\xi\,,t\,|\,\xi_0) = u(\xi)\int d\xi'\,\lambda(\xi')\,p(\xi',t\,|\,\xi_0) - \lambda(\xi)\,p(\xi\,,t\,|\,\xi_0)\int d\xi'\,u(\xi'). \tag{5.49}$$

To determine $u(\xi)$ self-consistently, take the limit $t\to\infty$, so that the conditional PDF $p(\xi\,,t\,|\,\xi_0)$ tends to the stationary PDF $p(\xi)$. We find

$$u(\xi) = \frac{\lambda(\xi)\,p(\xi)}{\langle\lambda\rangle} \tag{5.50}$$

apart from a multiplicative constant, where

$$\langle\lambda\rangle = \int d\xi\,\lambda(\xi)\,p(\xi) \tag{5.51}$$

represents the mean transition rate. Equation (5.49) reduces to

$$\frac{\partial}{\partial t}p(\xi\,,t\,|\,\xi_0) = \frac{\lambda(\xi)\,p(\xi)}{\langle\lambda\rangle}\int d\xi'\,\lambda(\xi')\,p(\xi',t\,|\,\xi_0) - \lambda(\xi)\,p(\xi\,,t\,|\,\xi_0). \tag{5.52}$$

The jump process ξ described by this master equation is called a **kangaroo process** [29].

To solve Eq. (5.52), take the Laplace transform of both sides. Let $\widetilde{p}(\xi\,,s)$ denote the Laplace transform of $p(\xi\,,t\,|\,\xi_0)$. This transforms Eq. (5.52) to the following inhomogeneous integral equation for $\widetilde{p}(\xi\,,s)$:

$$\widetilde{p}(\xi\,,s) = \frac{\delta(\xi-\xi_0)}{s+\lambda(\xi)} + \frac{\lambda(\xi)\,p(\xi)}{\langle\lambda\rangle\,\big(s+\lambda(\xi)\big)}\int d\xi'\,\lambda(\xi')\,\widetilde{p}(\xi',s). \tag{5.53}$$

[29] As we have mentioned earlier, the Kubo-Anderson process is a special case of the kangaroo process. In turn, the DMP is a special case of the Kubo-Anderson process.

The integral equation (5.53) is easy to solve, as the kernel of the integral operator is a separable kernel of rank one: that is, it is the product of a single function of ξ and a single function of $\xi\,'$, rather than a finite sum of such products.

(a) Show that the solution is

$$\widetilde{p}(\xi\,,\,s) = \frac{\delta(\xi-\xi_0)}{s+\lambda(\xi)} + \frac{p(\xi)\,\lambda(\xi)\,\lambda(\xi_0)}{\langle\lambda\rangle\;\big(s+\lambda(\xi)\big)\,\big(s+\lambda(\xi_0)\big)\,\phi(s)}\,, \tag{5.54}$$

where

$$\phi(s) = 1 - \frac{1}{\langle\lambda\rangle}\int d\xi\,\frac{\lambda^2(\xi)\,p(\xi)}{s+\lambda(\xi)} \equiv 1 - \frac{1}{\langle\lambda\rangle}\left\langle\frac{\lambda^2}{s+\lambda}\right\rangle. \tag{5.55}$$

(b) Verify that the expression in Eq. (5.47) for the conditional PDF of a Kubo-Anderson process is recovered correctly from Eq. (5.54) when $\lambda(\xi)$ is a constant, λ.

(c) A formal expression for the (normalized) autocorrelation function of a kangaroo process may now be obtained. We have, by definition,

$$C(t) = \frac{\langle\xi(0)\,\xi(t)\rangle}{\langle\xi^2\rangle} = \int d\xi_0\int d\xi\;\xi_0\,\xi\;p(\xi\,,\,t\,|\,\xi_0)\,p(\xi_0)\Big/\int d\xi\,\xi^2\,p(\xi)\,. \tag{5.56}$$

The Laplace transform of $C(t)$ is therefore

$$\widetilde{C}(s) = \frac{1}{\langle\xi^2\rangle}\int d\xi_0\int d\xi\;\xi_0\,\xi\;\widetilde{p}(\xi\,,\,s)\,p(\xi_0). \tag{5.57}$$

Suppose ξ ranges from $-\infty$ to ∞, and further suppose that both $p(\xi)$ and $\lambda(\xi)$ are even functions of their arguments. Show that the autocorrelation of a kangaroo process is then given by the very suggestive form

$$C(t) = \frac{1}{\langle\xi^2\rangle}\int_{-\infty}^{\infty} d\xi\,e^{-\lambda(\xi)\,t}\,\xi^2\,p(\xi). \tag{5.58}$$

Hence $C(t)$ is not just a single damped exponential in time. Rather, it is a continuous superposition of such exponentials, and therefore can take on diverse functional forms. **Non-exponential correlation functions** arise in numerous physical applications. We have already emphasized the importance of correlation functions as indicators of the nature of the random processes going on in a system. In particular, the correlation function may have an asymptotic (large-t) behavior that falls off more slowly than a decaying exponential: for instance, the decay may be like a *stretched exponential* $\exp(-t^{\alpha})$, where $0<\alpha<1$; or like a *power law* $t^{-\alpha}$, where $\alpha > 0$. Kangaroo processes provide a convenient and versatile way to model a wide variety of situations in which such non-exponential correlation functions occur, by suitable choices of the state-dependent transition rate $\lambda(\xi)$.

(d) Let $-\infty < \xi < \infty$, and suppose $\lambda(\xi) \propto |\xi|^{\beta}$ where β is a positive number. That is, the transition probability density per unit time $w(\xi \,|\, \xi\,') \propto |\xi\,'|^{\beta}\, p(\xi)$. This means that the larger the value (in magnitude) of the starting state $\xi\,'$, the greater is the rate of transitions or jumps out of that state. This is plausible on physical grounds: if the system momentarily finds itself in a state that deviates considerably from some range of moderate values of the variable, it is likely to make a transition out of that state more rapidly than otherwise. Show that the corresponding correlation function exhibits a power law decay, $C(t) \sim t^{-3/\beta}$.

(e) More generally, the asymptotic (or large-t) behavior of $C(t)$ can be extracted from Eq. (5.58) by a Gaussian approximation to the integral involved. For definiteness, let $-\infty < \xi < \infty$, and let $\lambda(\xi)$ and $p(\xi)$ be even functions of ξ. Further, let $\lambda(\xi)$ be a monotonically increasing function of ξ in $(0\,,\,\infty)$. Show that, if $\lambda(\xi)$ is such that the equation $\xi\,\lambda'(\xi)\, t = 2$ (where the prime denotes the derivative) has a unique nontrivial root $\overline{\xi}(t)$, and further $\overline{\xi}^{\,2}\,\lambda''(\overline{\xi})\, t + 2 > 0$, then the asymptotic behavior of $C(t)$ is given by

$$C(t) \sim \frac{\big(\overline{\xi}(t)\big)^3 \exp\big[-t\lambda\big(\overline{\xi}(t)\big)\big]}{\Big[\big(\overline{\xi}(t)\big)^2\,\lambda''\big(\overline{\xi}(t)\big)\, t + 2\Big]^{1/2}}\,. \tag{5.59}$$

It is evident that $C(t)$ can display many different kinds of asymptotic behavior, depending on the details of the variation of the transition rate $\lambda(\xi)$ as a function of ξ—these details determine the t-dependences of $\overline{\xi}(t)$ and $\lambda''\big(\overline{\xi}(t)\big)$ in Eq. (5.59).

Chapter 6

The Fokker-Planck equation and the Ornstein-Uhlenbeck distribution

We write down the stochastic differential equation (SDE) defining a general diffusion process, and the corresponding Fokker-Planck equation (FPE) for the conditional PDF of the process. A basic SDE$\leftrightarrow$FPE correspondence is introduced. This is applied to the case of the Langevin equation for the velocity process, and the Ornstein-Uhlenbeck distribution for the velocity is obtained. Another way of viewing the fluctuation-dissipation relation is brought out.

6.1 Diffusion processes

Let us resume our discussion of the random process represented by the velocity in the Langevin equation. We have found the mean squared value of $v(t)$, as well as its equilibrium autocorrelation function, a decaying exponential in time. But we have yet to find the conditional PDF $p(v\,,\,t\,|\,v_0)$ itself. We now consider this question.

In the theory of Markov processes, there is a celebrated result known as **Doob's Theorem**, which essentially states that:

- There is only one (nontrivial) stationary, Gaussian, Markov process. This is the **Ornstein-Uhlenbeck process**, which has an exponentially decaying autocorrelation function.

We shall define an Ornstein-Uhlenbeck (or OU) process shortly, and identify the associated PDF. We have already stated (without proof) that $v(t)$ is indeed a stationary, Gaussian, Markov process in the absence of an external force. And, as already pointed out, Eq. (4.5) shows explicitly that the velocity autocorrelation function is a decaying exponential. The theorem, if applicable, would therefore fix the form of $p(v\,,\,t\,|\,v_0)$. Even more helpful in this regard is the fact that Doob's Theorem also has a 'converse' of sorts:

- The only stationary Gaussian process with an exponential autocorrelation function is a certain Markov process, namely, the Ornstein-Uhlenbeck process.

This last result enables us, based on just the assumption that the velocity is a stationary, Gaussian random process, to identify the velocity not only as a Markov process, but also as an OU process[1].

We must now establish this conclusion more rigorously. For this purpose, we appeal to a basic result in the theory of stochastic differential equations. We shall have occasion to use this result more than once in the chapters that follow[2]. The stochastic differential equation satisfied by $v(t)$, namely, the Langevin equation

$$\dot{v}(t) = -\gamma\, v(t) + \frac{1}{m}\,\eta(t), \tag{6.1}$$

---

[1]As an aside, we remark that the statements of Doob's Theorem and its converse should be understood carefully, as it is easy to misinterpret them. For instance, there are plenty of *discontinuous* or *jump* processes that are stationary, Markov processes with exponentially decaying autocorrelation functions. The dichotomous Markov process and the Kubo-Anderson process considered in the exercises in Ch. 5 are examples of such cases. These are certainly not OU processes. In fact, there also exist *continuous*, stationary, Markov processes with decaying-exponential autocorrelation functions that are *not* Gaussian processes, and hence are not OU processes. The full set of continuous, stationary, Markov processes with an autocorrelation function that is a single decaying exponential in time consists of all those processes for which the stationary PDF $p(\xi) \equiv \lim_{t\to\infty} p(\xi\,,\,t\,|\,\xi\,')$ satisfies the following conditions (E. Wong, Proc. AMS Symp. Appl. Math. **16**, 264 (1963)):

(i) The differential equation $dp(x)/dx = [A(x)/B(x)]\,p(x)$, where $A(x) = a_0 + a_1 x$ and $B(x) = b_0 + b_1 x + b_2 x^2$, respectively. Such functions $p(x)$ are said to be of the **Pearson type**.

(ii) $B(x)\,p(x)$ vanishes at the end-points in x.

(iii) $p(x)$ has finite first and second moments.

The Gaussian corresponds to the case $a_0 = b_1 = b_2 = 0$ and $a_1 < 0$, $b_0 > 0$. There are four other nontrivial possibilities.

[2]In keeping with our general approach, we do not prove the result concerned. Rather, it is merely applied to the problem at hand.

implies that $v(t)$ is an OU process by definition, as we shall see below. Such a process is a special case of a more general family of continuous Markov processes known as diffusion processes, that we have already mentioned in Ch. 5, Sec. 5.2. We shall define a diffusion process shortly. It is convenient to factor out the strength Γ of the noise $\eta(t)$ in Eq. (6.1) above, and to write this equation as

$$\dot{v}(t) = -\gamma\, v(t) + \frac{\sqrt{\Gamma}}{m}\, \zeta(t). \tag{6.2}$$

Here $\zeta(t)$ is a stationary, Gaussian, δ-correlated Markov process with zero mean and a δ-function autocorrelation of unit strength:

$$\langle \zeta(t) \rangle = 0 \quad \text{and} \quad \langle \zeta(t)\, \zeta(t') \rangle = \delta(t - t'). \tag{6.3}$$

Before we proceed, a matter of notation must be clarified. As it involves an important physical aspect of the problem, we explain the point in some detail.

The attentive reader may have noticed that Eqs. (3.2) and (3.10) involved overhead bars, i. e., partial or *conditional* averages, whereas Eqs. (6.3) involve *complete* averages. The point is that further averaging of Eqs. (3.2) and (3.10) over v_0 does not change the right-hand sides of these equations. Hence we have

$$\overline{\eta(t)} = \langle \eta(t) \rangle_{\text{eq}} = 0\,, \tag{6.4}$$

and similarly

$$\overline{\eta(t_1)\, \eta(t_2)} = \langle \eta(t_1)\, \eta(t_2) \rangle_{\text{eq}} = \Gamma\, \delta(t_1 - t_2)\,. \tag{6.5}$$

From a physical point of view, this is understood as follows. The random force $\eta(t)$ arises from the heat bath, i. e., the particles of the fluid in which the tagged particle (for which we have written down the Langevin equation) is immersed. The heat bath has an enormously larger number of degrees of freedom than the tagged particle (the subsystem of interest). The assumption is that the heat bath is not affected significantly by the state of the subsystem[3]. On the other hand, the converse is definitely *not* true: the subsystem is indeed affected *strongly* by the heat bath.

To complete the explanation: in Eqs. (6.3), we have used $\langle \cdots \rangle$ instead of $\langle \cdots \rangle_{\text{eq}}$ in expressing the mean and autocorrelation of the noise $\zeta(t)$. This is because we have factored out the system-dependence of the noise by writing $\eta(t) \equiv \sqrt{\Gamma}\, \zeta(t)$,

[3]The incorporation of the systematic part of the random force $(-m\gamma v)$ in the Langevin equation is itself a way, in some sense, of taking into account the 'back reaction' of the motion of the tagged particle on the medium.

so that $\zeta(t)$ is a scale-free noise defined entirely by the properties listed. It has nothing to do with the presence or absence of any external force. We therefore replace $\langle\cdots\rangle_{\text{eq}}$ with $\langle\cdots\rangle$ as far as $\zeta(t)$ is concerned, not only for notational simplicity but also to indicate its generality.

We are ready, now, to define a diffusion process in the mathematical sense:

- A diffusion process is a Markovian random process $\xi(t)$ that satisfies a stochastic differential equation (SDE) of the form

$$\dot{\xi} = f(\xi, t) + g(\xi, t)\,\zeta(t), \tag{6.6}$$

  where f and g are given functions of their arguments, and $\zeta(t)$ is a Gaussian white noise as specified above.

Further, we note that:

- For the output or driven variable $\xi(t)$ to be a *stationary* random process as well, it is necessary (but *not* sufficient!) that f and g have no explicit t-dependence.

The case $g =$ constant corresponds to **additive noise**, whereas any other form of g corresponds to **multiplicative noise**.

6.2 The SDE↔FPE correspondence

Given Eq. (6.6), it can be shown from the theory of stochastic differential equations that the master equation satisfied by the conditional density $p(\xi\,,\,t\,|\,\xi_0\,,\,t_0)$ is just the Fokker-Planck equation, Eq. (5.14) (generalized to the nonstationary case). Moreover, the coefficients A_1 and A_2 in that equation are determined in terms of the functions f and g occurring in Eq. (6.6), in the following manner:

- The FPE corresponding to Eq. (6.6) is[4]

$$\frac{\partial}{\partial t}p(\xi\,,t\,|\,\xi_0\,,\,t_0) = -\frac{\partial}{\partial \xi}\left[f(\xi,t)\,p(\xi\,,t\,|\,\xi_0\,,\,t_0)\right] + \frac{1}{2}\,\frac{\partial^2}{\partial \xi^2}\left[g^2(\xi,t)\,p(\xi\,,t\,|\,\xi_0\,,\,t_0)\right]. \tag{6.7}$$

We shall refer to the two-way implication between Eqs. (6.6) and (6.7) as 'the **SDE↔FPE correspondence**'. We shall use it several times in the sequel.

[4]We have used the so-called **Itô interpretation** (as opposed to the **Stratonovich interpretation**) of the stochastic differential equation (6.6) in writing down the general FPE of Eq. (6.7). This distinction is not relevant to the problem at hand—see the next footnote.

- If, further, (i) f is a linear function of ξ, and (ii) g is a constant (i. e., it is independent of ξ), then ξ is an Ornstein-Uhlenbeck process.

Compare Eqs. (6.2) and (6.6): It is now evident that $v(t)$ is indeed an OU process, because the function $f(v)$ is the linear function $-\gamma v$ in this case, while the function $g(v)$ is just the constant $\sqrt{\Gamma}/m$, where Γ is the constant multiplying the Dirac δ-function in the autocorrelation of the white noise $\eta(t)$. We note that the noise is additive in the present instance[5]. In the special case of the Langevin equation (6.2), the FPE therefore reduces to

$$\frac{\partial p}{\partial t} = \gamma \frac{\partial}{\partial v}(vp) + \frac{\Gamma}{2m^2}\frac{\partial^2 p}{\partial v^2}\,. \tag{6.8}$$

Using the FD relation between Γ and γ, the Fokker-Planck equation for the conditional PDF of v becomes

$$\frac{\partial p}{\partial t} = \gamma \frac{\partial}{\partial v}(vp) + \frac{\gamma k_B T}{m}\frac{\partial^2 p}{\partial v^2}. \tag{6.9}$$

The initial condition on p is given, as we know, by Eq. (2.12), i. e., $p(v\,,0\,|\,v_0) = \delta(v - v_0)$. The solution to this equation with **natural boundary conditions**, namely, $p \to 0$ as $|v| \to \infty$, is called the **Ornstein-Uhlenbeck distribution**[6]. The task now is to find this distribution explicitly. This is done in the next section.

Equation (6.9) is the *original* Fokker-Planck equation, being perhaps the first example of an equation of this class (apart from the diffusion equation itself, of course). It is also called **Rayleigh's equation**. The same equation can be arrived at in another way[7]. Instead of working with an equation of motion such as the Langevin equation, the underlying random process can be studied by means of the Boltzmann equation for the conditional PDF $p(v\,,\,t\,|\,v_0)$. If we assume that the velocity is only altered slightly in any single collision (the so-called weak-collision approximation), the Boltzmann equation can be simplified, to obtain the Fokker-Planck equation (6.9).

---

[5]The distinction between the Itô and Stratonovich interpretations of the stochastic differential equation (6.6) disappears in the case of an additive noise, in the sense that both of them lead to the same FPE. As this is the case of interest to us at present, we do not go any further into this interesting aspect of stochastic differential equations here. Some more comments will be made in this regard in Ch. 9, Sec. 9.9, in connection with the so-called **Itô calculus**.

[6]As we have already cautioned in the case of the Maxwellian distribution, probability densities are often called probability distributions in the physics literature.

[7]Recall the comments made in Ch. 2, Sec. 2.1.

Finally, note that the FPE (6.8) can be written in the form of an **equation of continuity**, namely,

$$\frac{\partial p}{\partial t} + \frac{\partial J}{\partial v} = 0, \tag{6.10}$$

where the probability current density is given by

$$J(v,t) = -\left[\gamma\, vp(v,t) + \frac{\Gamma}{2m^2}\,\frac{\partial p(v,t)}{\partial v}\right]. \tag{6.11}$$

The advantage of writing the FPE in this form will be clear when we discuss stationary and equilibrium distributions in Sec. 6.4 below, and again in Ch. 14.

6.3 The Ornstein-Uhlenbeck distribution

There is more than one way to solve Eq. (6.9) rigorously. But there is a much simpler way to *write down* the solution we seek. The formal solution to the Langevin equation (6.2) is

$$v(t) = v_0\, e^{-\gamma t} + \frac{\sqrt{\Gamma}}{m}\int_0^t dt'\, e^{-\gamma(t-t')}\, \zeta(t'). \tag{6.12}$$

The right-hand side of this equation represents a *linear transformation* of the Gaussian noise ζ. We have, by now, verified in more than one instance that a linear combination of independent Gaussian random variables is also a Gaussian random variable[8]. Now, $\zeta(t_1)$ and $\zeta(t_2)$ at any two different instants of time t_1 and t_2 are uncorrelated and independent random variables. Thus, the expression on the right-hand side in Eq. (6.12), which involves an integral or sum over ζ at different instants of time, is just such a linear combination. Hence,

- the PDF $p(v\,,\,t\,|\,v_0)$ must remain Gaussian in form at all times.

It then follows that the PDF is completely determined by the mean and variance of the velocity. In the present instance, these two quantities are certain known functions of t. We have already found the (conditional) mean velocity and mean squared velocity in Eqs. (3.4) and (3.15), respectively. The conditional variance of the velocity is therefore

$$\overline{v^2(t)} - \left(\overline{v(t)}\right)^2 = \frac{k_B T}{m}\left(1 - e^{-2\gamma t}\right). \tag{6.13}$$

Note how *the conditional variance is actually independent of the initial value* v_0. We now refer to Eq. (D.1) of Appendix D, which expresses the PDF of a Gaussian

[8] See also Appendix D, Sec. D.4.

random variable in terms of its mean and variance. The expressions in Eq. (3.4) for the mean, and Eq. (6.13) for the variance, then imply that the Gaussian PDF we seek is

$$p(v\,,\,t\,|\,v_0) = \left[\frac{m}{2\pi k_B\,T(1-e^{-2\gamma t})}\right]^{\frac{1}{2}} \exp\left[\frac{-m(v-v_0e^{-\gamma t})^2}{2k_BT(1-e^{-2\gamma t})}\right]. \tag{6.14}$$

This is the normalized PDF corresponding to the OU distribution.

- The OU distribution of Eq. (6.14) is a Gaussian whose peak, located at $\overline{v(t)} = v_0\,e^{-\gamma t}$, starts initially at v_0 and drifts to zero as $t \to \infty$.
- Simultaneously, the width of the Gaussian broadens, as the standard deviation of the velocity increases monotonically from an initial value zero to a limiting equilibrium value $(k_B\,T/m)^{1/2}$.
- In the limit $t \to \infty$, the OU distribution becomes the equilibrium Maxwellian distribution
$$p^{\rm eq}(v) = \left(\frac{m}{2\pi k_BT}\right)^{\frac{1}{2}} \exp\left(-\frac{mv^2}{2k_BT}\right). \tag{6.15}$$

Figure 6.1 shows how the Ornstein-Uhlenbeck distribution starts out as a δ-function at some value v_0, and evolves asymptotically to the Maxwellian distribution.

6.4 The fluctuation-dissipation relation again

We note, at this stage, that we could have deduced the FD relation $\Gamma = 2m\,\gamma\,k_BT$ in yet another way. Starting with the FPE of Eq. (6.8) (which follows, as we have said, from the Langevin equation (6.2)), let us pass to the limit $t \to \infty$. In this limit, we expect $p(v\,,t\,|\,v_0)$ to lose all memory of the initial velocity v_0, and to tend to some PDF in v that we denote by $p_\infty(v)$, say. As this PDF is independent of time, Eq. (6.8) reduces to an ordinary differential equation, namely,

$$\frac{d}{dv}\left\{\frac{\Gamma}{2m^2}\,\frac{dp_\infty}{dv} + \gamma\,v\,p_\infty\right\} = 0. \tag{6.16}$$

Referring to Eqs. (6.10) and (6.11), we see that the expression in curly brackets in the above is just the negative of the *stationary* probability current $J^{\rm st}(v)$. Equation (6.16) says that the divergence (in velocity space) of this stationary current must vanish[9]. Hence $J^{\rm st}$ must be a constant, i. e., it must be independent of v. But if

[9]Recall that we are dealing with the one-dimensional case.

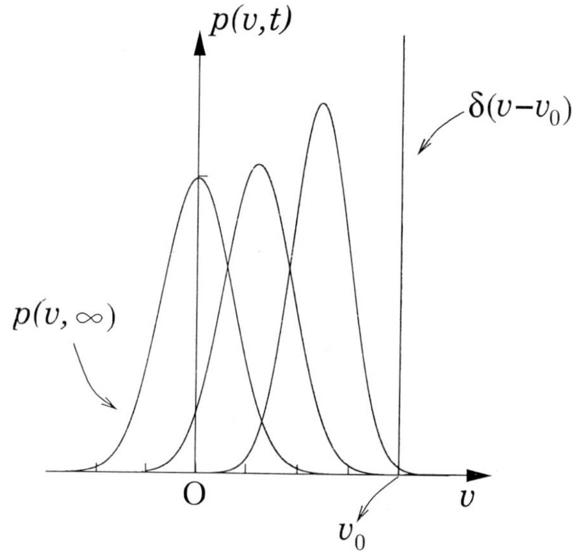

Figure 6.1: Time evolution (from right to left) of the normalized PDF (Eq. (6.14)) corresponding to the Ornstein-Uhlenbeck velocity distribution from an initial δ-function, $\delta(v - v_0)$, to the limiting Maxwellian form $p^{\text{eq}}(v)$ as $t \to \infty$. The PDF is a Gaussian at all times, with a variance proportional to $(1 - e^{-2\gamma t})$. The OU process is the only continuous, stationary, Gaussian, Markov process, and it has an exponentially decaying autocorrelation function.

p_∞ is to be a normalizable density with finite moments, then it must vanish faster than any negative power of v as $|v| \to \infty$. Hence its derivatives with respect to v must also vanish as $|v| \to \infty$. Therefore $J^{\rm st}$ must in fact be zero[10]. Thus $p_\infty(v)$ satisfies the first-order ordinary differential equation

$$\frac{\Gamma}{2m^2}\frac{dp_\infty}{dv} + \gamma\, v\, p_\infty = 0. \tag{6.17}$$

This equation is solved easily, to yield the normalized PDF

$$p_\infty(v) = \left(\frac{m^2\gamma}{\pi\Gamma}\right)^{\frac{1}{2}} \exp\left(-\frac{m^2\,\gamma\, v^2}{\Gamma}\right). \tag{6.18}$$

Now impose the requirement specified in Eq. (2.13): namely, that

$$p_\infty(v) = \lim_{t\to\infty} p(v\,,t\,|\,v_0)$$

be identical to the *equilibrium* density $p^{\rm eq}(v)$. A comparison of the Gaussian in Eq. (6.18) with the Maxwellian PDF of Eq. (6.15) shows at once that this condition is met if and only if $\Gamma = 2m\,\gamma\,k_BT$.

6.5 Exercises

6.5.1 Verification

Verify that the solution presented in Eq. (6.14) satisfies the Fokker-Planck equation (6.9).

6.5.2 Green function for the Fokker-Planck operator

Equation (6.14) represents the solution to the Fokker-Planck equation (6.9) with the δ-function initial condition (or *sharp* initial condition)

$$p(v\,,t)\big|_{t=0} = \delta(v - v_0).$$

Therefore it does more than merely solve the partial differential equation: it is essentially the **Green function** corresponding to the differential operator involved in the Fokker-Planck equation, namely,

$$\mathcal{D}_{\rm FP} = \frac{\partial}{\partial t} - \gamma\frac{\partial}{\partial v}v - \frac{\gamma k_BT}{m}\frac{\partial^2}{\partial v^2}\,. \tag{6.19}$$

[10]We will have occasion, in Ch. 14, Sec. 14.1.2, to recall this statement when discussing the detailed balance condition in a more general case.

That is, Eq. (6.14) is the solution (for all $t > 0$) to the equation

$$\mathcal{D}_{\text{FP}}\, G(v\,, t\,;\, v_0\,, 0) = \delta(v - v_0)\,\delta(t) \tag{6.20}$$

with natural boundary conditions (i. e., it is the Green function that vanishes as $v \to \pm\infty$). We can now write down the solution to the equation

$$\mathcal{D}_{\text{FP}}\, p(v, t) = 0 \tag{6.21}$$

with the *general* initial condition $p(v, 0) = p_{\text{init}}(v)$, where $p_{\text{init}}(v)$ is any given, arbitrary, normalized initial distribution of velocities. The solution satisfying the natural boundary conditions $p(v, t) \to 0$ as $v \to \pm\infty$ is simply

$$p(v\,,\,t) = \int_{-\infty}^{\infty} dv_0 \left[\frac{m}{2\pi k_B\, T(1 - e^{-2\gamma t})}\right]^{\frac{1}{2}} \exp\left\{\frac{-m(v - v_0 e^{-\gamma t})^2}{2k_B T(1 - e^{-2\gamma t})}\right\} p_{\text{init}}(v_0), \tag{6.22}$$

for all $t > 0$. By letting $t \to \infty$ in the solution above, we see at once that, independent of the precise form of $p_{\text{init}}(v_0)$, the time-dependent distribution $p(v\,,\,t)$ tends to the equilibrium Maxwellian distribution in this limit. In this sense, we may conclude that

- *the state of thermal equilibrium is an asymptotically stable state.*

The **Gibbs measure** of this state is $p^{\text{eq}}(v)\, dv$. The Maxwellian $p^{\text{eq}}(v)$ is the **global attractor** in the space of probability density functions. The **basin of attraction** of this attractor comprises all normalizable probability densities with finite variances.

(a) Now suppose we choose the initial distribution to be the equilibrium Maxwellian distribution itself. That is, put in $p^{\text{eq}}(v_0)$ for $p_{\text{init}}(v_0)$ in Eq. (6.22), and carry out the integration[11] over v_0 to find $p(v\,,\,t)$ in this case. Does the final result surprise you?

(b) We assert that:

(i) The result for $p(v\,,\,t)$ in part (a) above could have been written down without an actual calculation.

[11] The basic integral you need for this purpose is just the one quoted in Eq. (2.31) (see also Appendix A, Eq. (A.4)), namely,

$$\int_{-\infty}^{\infty} dx\, \exp\left(-ax^2 + bx\right) = \sqrt{(\pi/a)}\, \exp\left(b^2/4a\right) \quad \text{where } a > 0.$$

(ii) No other form of $p_{\text{init}}(v_0)$ can yield this result.

If you understand why these statements are valid, then you have gained a somewhat deeper understanding of the state of thermal equilibrium.

6.5.3 Joint distribution of the velocity

The two-time joint probability density of the velocity is, by definition,

$$p_2(v_2\,,\,t_2\,;\,v_1\,,\,t_1\,) \equiv p_2(v_2\,,\,t_2\,|\,v_1\,,\,t_1\,)\,p_1(v_1\,,\,t_1),$$

where $t_2 > t_1$.

(a) Use the stationarity property of the velocity in thermal equilibrium, to show that the two-time joint probability density of the velocity is given by

$$\begin{aligned} p_2(v_2\,,\,t_2\,;\,v_1\,,\,t_1\,) &= \left[\frac{m}{2\pi k_B T\,(1-e^{-2\gamma(t_2-t_1)})^{1/2}}\right] \\ &\times \exp\left\{\frac{-m\big(v_1^2 - 2v_1\,v_2\,e^{-\gamma(t_2-t_1)} + v_2^2\big)}{2k_B T\,(1-e^{-2\gamma(t_2-t_1)})}\right\}, \end{aligned} \tag{6.23}$$

where $t_2 > t_1$. This expression is a generalized Gaussian[12] in the variables v_1 and v_2 : the exponent is a non-positive quadratic function of these two variables. Note how this joint PDF factorizes into $p^{\text{eq}}(v_1)\,p^{\text{eq}}(v_2)$ when the time difference $(t_2 - t_1)$ is sufficiently large, i. e., when $\gamma(t_2 - t_1) \gg 1$, as expected.

(b) Using the Markov property of the velocity, write down the n-time joint probability density $p_n(v_n\,,\,t_n\,;\,v_{n-1}\,,\,t_{n-1}\,;\ldots\,;\,v_1\,,\,t_1)$ in the form of a generalized Gaussian in the variables $v_1\,,\,v_2\,,\,\ldots\,,\,v_n$.

[12]See Appendix D, Sec. D.6.

Chapter 7

The diffusion equation

We revert to the phenomenological diffusion equation for the probability density of the position of a tagged particle, and deduce expressions for the mean and higher moments of the displacement. The fundamental Gaussian solution of the one-dimensional diffusion equation is obtained and discussed. The 'diffusion regime' in which the diffusion equation is valid is identified. We also consider the three-dimensional diffusion equation and its fundamental solution. Other interesting properties associated with the diffusion equation and its solution are discussed in the exercises at the end of the chapter.

7.1 Introduction

In this chapter and in the next three chapters, we step back from the Langevin model and revert to an even more phenomenological description of the random motion of a tagged particle in a fluid—namely, the diffusion equation. Subsequently, in Ch. 11, we shall make contact with the Langevin model and place the diffusion equation in its proper context.

The random motion of the tagged particle leads to its diffusion in the fluid. We have already referred to the phenomenological diffusion equation (Eq. (1.5)) that follows from Fick's Laws for the concentration of a diffusing species in a fluid medium. What is the origin of this phenomenological equation for the local concentration of a diffusing molecular species? It lies in the corresponding diffusion equation for the positional probability density $\mathsf{p}(\mathbf{r}, t)$ of an individual particle such as our tagged particle[1]. The latter is, in fact, the equation originally

[1]We shall use the symbol p for the normalized positional PDF in order to distinguish it from

derived by Einstein in his analysis of the physical problem of Brownian motion—the irregular, random motion of micron-sized particles (such as colloidal particles) suspended in a liquid, under the incessant and random buffeting of the molecules of the liquid[2]. Restricting ourselves to motion in the x-direction for simplicity, the diffusion equation is given by

$$\frac{\partial \mathsf{p}(x,t)}{\partial t} = D\,\frac{\partial^2 \mathsf{p}(x,t)}{\partial x^2}\,. \tag{7.1}$$

D is the diffusion coefficient of the tagged particle. The normalization condition on the PDF $\mathsf{p}(x,t)$ is given by

$$\int_{-\infty}^{\infty} dx\ \mathsf{p}(x\,,\,t) = 1 \quad \text{for all } t \geq 0. \tag{7.2}$$

We must emphasize again that, at this level, Eq. (7.1) is purely phenomenological in nature. However, because we have a microscopic model for the motion of a tagged particle (the Langevin equation and its consequences), we may expect to connect up the diffusion equation to that model. Specifically, we would like to investigate

- whether the model of random motion based on the Langevin equation enables us to identify the conditions under which the diffusion equation (7.1) is valid; and
- whether the parameter D, which is also a transport coefficient, is determinable from first principles based on dynamics at the microscopic level.

These expectations will indeed be borne out, as we shall see in Ch. 11. For the moment, however, let us accept Eq. (7.1) as it stands, and work out its consequences. Further implications of the diffusion equation will be taken up in Ch. 9. In particular, we will be concerned there with the stochastic process represented by the position $x(t)$.

---

p, which we have used for the PDF of the velocity.

[2]The early history of the phenomenon of Brownian motion and its explanation is well known. I assume that the reader is familiar with at least the broad features of this fascinating story, and so it will not be repeated here. It suffices to say that its explanation and experimental verification in the early years of the 20th century clinched the physical reality of the atomic nature of matter. It also demonstrated another important fact: namely, that we can experimentally observe, and determine the values of, physical quantities that are astronomically large or extremely small, relative to the magnitudes we encounter in our day-to-day macroscopic world, or the 'world of middle dimensions'—Avogadro's number ($\sim 10^{23}$) and the size of small molecules ($\sim 10^{-10}$ m), respectively. Incidentally, the original observations of Brown pertain to the small particles contained *inside* pollen grains, and not to pollen grains themselves—the latter are too big in size and mass to be Brownian particles under the conditions involved.

7.2 The mean squared displacement

Consider diffusion on the infinite line, $-\infty < x < \infty$. Then, without loss of generality, we may choose the starting point of the diffusing tagged particle to be the origin on the x-axis[3]. The mean value of the displacement of the particle at time t is, by definition,

$$\overline{x(t)} = \int_{-\infty}^{\infty} dx\, x\, \mathsf{p}(x\,,\, t\,|\,0). \tag{7.3}$$

In keeping with our notational convention, an overhead line has been used to denote the average because it is a *conditional* one, taken with respect to the conditional PDF $\mathsf{p}(x\,,\,t\,|\,x_0)$ with x_0 set equal to zero, as stated above[4]. To find this mean value, we multiply both sides of the diffusion equation (7.1) by x, and integrate over x from $-\infty$ to $+\infty$. The left-hand side is simply $(d/dt)\,\overline{x(t)}$. On the right-hand side, we integrate by parts, and use the fact that p is a normalized PDF. The latter requirement implies that $\mathsf{p}(x,t)$ must necessarily tend to zero as $x \to \pm\infty$. We also assume that $x\,\partial\mathsf{p}(x,t)/\partial x \to 0$ as $x \to \pm\infty$. This assumption will be justified subsequently, when we write down the explicit solution for the PDF $\mathsf{p}(x,t)$. We thus obtain the following ordinary differential equation for the mean displacement at time t:

$$\frac{d}{dt}\,\overline{x(t)} = 0. \tag{7.4}$$

The solution to Eq. (7.4) is trivial: $\overline{x(t)} = 0$, on using the fact that this quantity is zero at $t = 0$, by definition.

Next, we repeat this procedure for the mean *squared* displacement at any time $t > 0$. Multiply both sides of Eq. (7.1) by x^2 and integrate over x from $-\infty$ to $+\infty$. Integrating the right-hand side by parts twice in succession, we find the differential equation satisfied by the mean squared displacement:

$$\frac{d}{dt}\,\overline{x^2(t)} = 2D. \tag{7.5}$$

To arrive at this equation, we need to assume further that $x^2\,\partial\mathsf{p}(x,t)/\partial x \to 0$ as $x \to \pm\infty$. Once again, this will be justified shortly, immediately after Eq. (7.11). Using the initial value $\overline{x^2(0)} = 0$, we obtain

$$\overline{x^2(t)} = 2D\,t. \tag{7.6}$$

[3]This restriction is easily removed. If the starting point is $x(0) = x_0$, we have merely to replace $x(t)$ by $X(t) \equiv x(t) - x_0$ in all the results in the sequel.

[4]For notational simplicity, we shall often simply write $\mathsf{p}(x,t)$ for this conditional PDF, except in those instances when we wish to call attention to the initial condition specifically.

This is a fundamental result. An experimental determination of the mean squared displacement of a tagged particle in a time interval t thus enables us to determine the diffusion coefficient.

The third and higher moments of $x(t)$ may also be found by the procedure used above. But it is easier to calculate these moments directly from the PDF $\mathsf{p}(x, t)$, as we shall see shortly.

7.3 The fundamental Gaussian solution

Next, we turn to the solution of the diffusion equation (7.1) itself. This partial differential equation is of first order in the time t and second order in the spatial variable x. It is a **parabolic differential equation**, in the classification of second-order partial differential equations. According to the general theory of these equations, specifying an initial condition and two boundary conditions leads to a unique solution of such an equation. As we have taken the particle to be at $x = 0$ initially, the corresponding initial condition on $\mathsf{p}(x, t)$ is $\mathsf{p}(x, 0) = \delta(x)$. Moreover, we also require that $\mathsf{p}(x, t)$ be a normalized probability density (Eq. (7.2)). A necessary condition for this is $\mathsf{p}(x, t) = 0$ at $x = \pm\infty$ for all $t \geq 0$ (natural boundary conditions). The task is to find the solution that satisfies the foregoing conditions. There is more than one way to do so. A convenient method is as follows.

The Fourier transform of $\mathsf{p}(x, t)$ with respect to x, and its Laplace transform with respect to t, is defined as

$$\widetilde{\mathsf{p}}(k\,,\,s) = \int_{-\infty}^{\infty} dx\, e^{-ikx} \int_0^{\infty} dt\, e^{-st}\; \mathsf{p}(x\,,\,t)\,. \tag{7.7}$$

Now take the Fourier transform with respect to x, and the Laplace transform with respect to t, of both sides of the diffusion equation (7.1). Recall that the Laplace transform of the time derivative of a function of t involves the value of that function at $t = 0$.[5] In the present instance, this is simply $\mathsf{p}(x, 0) = \delta(x)$. Moreover, the Fourier transform of $\delta(x)$ is just unity. We thus obtain a simple algebraic equation for $\widetilde{\mathsf{p}}(k\,,\,s)$, which yields

$$\widetilde{\mathsf{p}}(k\,,\,s) = \frac{1}{s + Dk^2}\,. \tag{7.8}$$

[5] The formula, already quoted in Ch. 5, Sec. 5.3.1, is $\mathbb{L}[\dot{f}(t)] = \mathbb{L}[f(t)] - f(0)$, where $\mathbb{L}$ denotes the Laplace transform.

The formal inversion of the transform in Eq. (7.7) is given by

$$\mathsf{p}(x\,,\,t) = \frac{1}{2\pi}\int_{-\infty}^{\infty} dk\, e^{ikx}\, \frac{1}{2\pi i}\int_{c-i\infty}^{c+i\infty} ds\, e^{st}\, \widetilde{\mathsf{p}}(k\,,\,s), \tag{7.9}$$

where c lies to the right of the pole of $\widetilde{\mathsf{p}}(k\,,\,s)$ at $s = -Dk^2$, i. e., $c > 0$. It is obviously simpler to invert the Laplace transform first. All we need is the fact that the inverse Laplace transform of $(s+a)^{-1}$ (where a is a positive number) is e^{-at}. Let us continue to use a tilde to denote the Fourier transform of the PDF $\mathsf{p}(x\,,\,t)$ with respect to x. We have

$$\widetilde{\mathsf{p}}(k\,,\,t) = \exp\,(-Dk^2 t). \tag{7.10}$$

This quantity is called the **characteristic function**[6] of the random variable x. In order to invert the Fourier transform, we again need the integral given in Eq. (2.31) (or Eq. (A.4), Appendix A). Finally, we obtain the fundamental solution sought. It is the well-known normalized Gaussian[7]

$$\mathsf{p}(x,t) = \frac{1}{\sqrt{4\pi Dt}}\,\exp\left(-\,\frac{x^2}{4Dt}\right), \quad t > 0. \tag{7.11}$$

Owing to the exponential factor, it is now obvious that p vanishes faster than any negative power of x as $|x| \to \infty$, as do all its derivatives with respect to x. The assumptions made in deriving the values of the mean and mean squared displacement, namely, that both $x\,\partial\mathsf{p}(x,t)/\partial x$ and $x^2\,\partial\mathsf{p}(x,t)/\partial x \to 0$ as $x \to \pm\infty$, are thus justified.

Figure 7.1 shows the evolution of the Gaussian PDF $\mathsf{p}(x,t)$ as t increases. Note that this positional PDF does not tend to any nontrivial limiting or equilibrium PDF as $t \to \infty$. Rather, we find that, for all x, $\mathsf{p}(x,t) \to 0$ as $t \to \infty$.[8] This is in marked contrast to the case of the velocity as specified by the Langevin equation: the OU distribution in v tends to the equilibrium Maxwellian distribution, as we have seen earlier.

---

[6] See Appendix C, Eq. (C.8).

[7] Note that the characteristic function in Eq. (7.10) is also a Gaussian, in the Fourier transform variable k. This is a hallmark of a Gaussian distribution. See Eq. (D.6) of Appendix D on the Gaussian distribution, and also Eq. (K.3) of Appendix K on stable distributions.

[8] The non-existence of an equilibrium distribution in x is obvious from the structure of the diffusion equation itself. If such an equilibrium PDF $\mathsf{p}^{\text{eq}}(x)$ existed, it would have to satisfy the ordinary differential equation $D\,d^2\mathsf{p}^{\text{eq}}/dx^2 = 0$. The general solution to this simple equation is $c_1\,x + c_2$, a linear function of x. However, when we impose the condition that the solution be normalizable in $(-\infty\,,\,\infty)$, it is evident that both c_1 and c_2 must vanish identically.

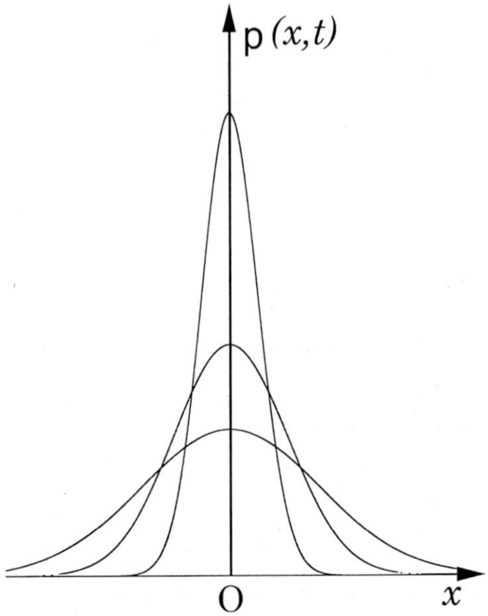

Figure 7.1: The broadening, as time increases, of the fundamental Gaussian solution to the diffusion equation (the normalized PDF of the displacement as given by Eq. (7.11)). The width of the Gaussian increases like the square root of the time elapsed, while the height of the peak decreases like $t^{-1/2}$. The area under the curve remains equal to unity for all t. All the moments of the displacement are finite at all times. There is no nontrivial limiting distribution as $t \to \infty$, because $\mathsf{p}(x,t) \to 0$ as $t \to \infty$ for all x.

Now that we have $\mathsf{p}(x,t)$ explicitly, all the moments of $x(t)$ can be calculated directly. The PDF is an even function of x. Hence all the odd moments of the displacement vanish identically. The even moments are easily determined, using the definite integral given in Eq. (2.8) (or, more simply, by direct analogy with Eq. (2.9)). We have

$$\overline{x^{2l}(t)} = \int_{-\infty}^{\infty} dx\, x^{2l}\, \mathsf{p}(x,t) = \frac{2(2l-1)!}{(l-1)!}\,(Dt)^l, \quad l = 1,\, 2,\, \ldots \tag{7.12}$$

Hence we have, analogous to Eq. (2.9),

$$\overline{x^{2l}(t)} = \frac{(2l-1)!}{2^{l-1}(l-1)!}\left(\overline{x^2(t)}\right)^l, \tag{7.13}$$

illustrating once again a characteristic property of the Gaussian distribution.

It is useful to record also the fundamental solution above in a slightly more general form: Let the starting time be any arbitrary instant t_0, and let the starting point be an arbitrary point x_0. Then the PDF $\mathsf{p}(x,t)$ satisfies the diffusion equation (7.1) and the initial condition $\mathsf{p}(x,t_0) = \delta(x-x_0)$. Reverting to the appropriate notation for the conditional PDF, the solution is

$$\mathsf{p}(x\,,\,t\,|\,x_0\,,\,t_0) = \frac{1}{\sqrt{4\pi D(t-t_0)}}\exp\left(-\frac{(x-x_0)^2}{4D(t-t_0)}\right), \quad t > t_0\,. \tag{7.14}$$

Finally, it is important to understand that

- the linear growth of $\overline{x^2(t)}$ with time for *all* $t \geq 0$ is a direct consequence of the diffusion equation.

The Langevin equation, however, yields a different result, as we shall see in Ch. 11. It will turn out that $\overline{x^2(t)} \sim t$ only for sufficiently *long* times—specifically, when $t \gg \gamma^{-1}$, the correlation time of the velocity. Therefore, anticipating the results to be established in Ch. 11, we state here that:

- The Langevin model provides a time scale γ^{-1} (the velocity correlation time or the Smoluchowski time), based on which we can identify the regime of validity of the diffusion equation (7.1).

- The diffusion equation for the probability density of the position of a diffusing particle is valid at times when $t \gg \gamma^{-1}$. We shall refer to this as the **diffusion regime** (or, more loosely, as the 'diffusion limit').

7.4 Diffusion in three dimensions

Let us now consider all the three Cartesian components of the position of the diffusing particle. The three-dimensional diffusion equation for the PDF $\mathsf{p}(\mathbf{r},t)$ of the position $\mathbf{r}$ of the particle is

$$\frac{\partial \mathsf{p}(\mathbf{r},t)}{\partial t} = D\,\nabla^2 \mathsf{p}(\mathbf{r},t)\,. \tag{7.15}$$

As in the one-dimensional case, we can show directly from this equation that, for a particle starting from the origin at $t=0$,

$$\overline{\mathbf{r}(t)} = 0,\ \overline{r^2(t)} \equiv \overline{x^2(t)} + \overline{y^2(t)} + \overline{z^2(t)} = 6Dt. \tag{7.16}$$

We ask for the normalized fundamental solution to Eq. (7.15), corresponding to the initial condition

$$\mathsf{p}(\mathbf{r},0) = \delta^{(3)}(\mathbf{r}) \equiv \delta(x)\,\delta(y)\,\delta(z) \tag{7.17}$$

and natural boundary conditions

$$\mathsf{p}(\mathbf{r},t) \to 0 \quad \text{as} \quad |\mathbf{r}| \equiv r \to \infty. \tag{7.18}$$

This solution is easily written down. The Laplacian operator ∇^2 is separable in Cartesian coordinates, *and* so are the initial and boundary conditions specified above. It follows that the fundamental solution is

$$\mathsf{p}(\mathbf{r},t) = \frac{1}{(4\pi Dt)^{3/2}} \exp\left(-\frac{x^2+y^2+z^2}{4Dt}\right). \tag{7.19}$$

$\mathsf{p}(\mathbf{r},t)\,d^3r$ is the probability that the diffusing particle will be in the infinitesimal volume element d^3r located at the position vector $\mathbf{r}$ at time t. Re-writing Eq. (7.19) in spherical polar coordinates,

$$\mathsf{p}(\mathbf{r},t) = \frac{1}{(4\pi Dt)^{3/2}} \exp\left(-\frac{r^2}{4Dt}\right). \tag{7.20}$$

Observe that the three-dimensional PDF in Eq. (7.20) is actually independent of the angular coordinates[9] θ and φ, and depends only on the radial coordinate

[9]In this book we do not consider the problem of **rotational diffusion**. Several variants of this phenomenon occur in molecular physics. An example is the diffusive motion of the *orientation* of a unit vector in space. It is clear that this problem is related to the diffusion of a particle restricted to remain on the surface of a unit sphere. It is the angular parts of the Laplacian operator that are relevant in this context.

r. This is a consequence of the rotational invariance of *all* the factors that determine the solution: (i) the differential equation (specifically, the Laplacian operator ∇^2), (ii) the initial condition, Eq. (7.17), *and* (iii) the boundary conditions, Eq. (7.18).

Let us write any spherically symmetric solution of Eq. (7.15) as $\mathsf{p}(r,t)$, instead of $\mathsf{p}(\mathbf{r},t)$, in suggestive notation. The fundamental solution (7.20) is an example of such a PDF. Since only the radial part of the ∇^2 operator contributes in this case, $\mathsf{p}(r,t)$ satisfies the equation

$$\frac{\partial \mathsf{p}(r,t)}{\partial t} = \frac{D}{r^2}\frac{\partial}{\partial r}\left(r^2 \frac{\partial}{\partial r}\mathsf{p}(r,t)\right) = D\left(\frac{\partial^2}{\partial r^2} + \frac{2}{r}\frac{\partial}{\partial r}\right)\mathsf{p}(r,t). \tag{7.21}$$

Now set $\mathsf{p}(r,t) = u(r,t)/r$, so as to eliminate the term involving the first derivative with respect to r in the last equation above[10]. We then find that the function $u(r,t)$ satisfies the equation

$$\frac{\partial u(r,t)}{\partial t} = D\frac{\partial^2 u(r,t)}{\partial r^2}. \tag{7.22}$$

But this is precisely of the same form as the diffusion equation in *one* dimension! This circumstance is often exploited to derive a number of useful results. We must remember, though, that the physical region in r is $0 \le r < \infty$, rather than $(-\infty, \infty)$. Moreover, neither $\mathsf{p}(r,t)$ itself, nor $u(r,t)$, is the normalized PDF of the radial *distance* r from the origin. This PDF, which we shall denote by $\mathsf{p}_{\text{rad}}(r,t)$, is obtained by integrating $\mathsf{p}(r,t)\,d^3r$ over the angular variables θ and φ. Thus

$$\mathsf{p}_{\text{rad}}(r,t) = 4\pi r^2\,\mathsf{p}(r,t) = 4\pi r\,u(r,t). \tag{7.23}$$

In the case of the fundamental solution (7.20), we have

$$\mathsf{p}_{\text{rad}}(r,t) = \frac{1}{(4\pi D t)^{3/2}}\,4\pi r^2 \,\exp\left(-\,r^2/4Dt\right). \tag{7.24}$$

Figure 7.2 shows the radial PDF $\mathsf{p}_{\text{rad}}(r,t)$ as a function of r, for a given value of t (> 0).

[10]This device is perhaps familiar to you from its use in other contexts involving ∇^2 in spherical polar coordinates, such as the solution of Laplace's equation in electrostatics, or the radial Schrödinger equation for a particle moving in a central potential. The trick only works in three dimensions, and not for any dimensionality $d \neq 3$. In Ch. 9, Sec. 9.5, we consider an aspect of the diffusion equation for general d.

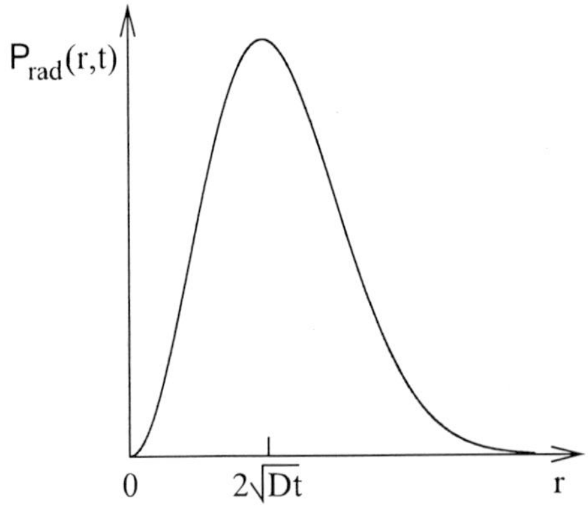

Figure 7.2: The probability density function $p_{rad}(r, t)$ of the radial distance from the origin for diffusion in three dimensions, at some instant $t > 0$. The PDF (given by Eq. (7.24)) increases like r^2 at very short distances, and peaks at a distance $2(Dt)^{1/2}$. This is the most probable value of the radial distance. The PDF falls off very rapidly at large distances, proportional to $r^2 \exp(-r^2/4Dt)$. The mean distance is equal to $4(Dt/\pi)^{1/2}$, which is located to the right of the most probable value.

7.5 Exercises

7.5.1 Derivation of the fundamental solution

Starting from the diffusion equation (7.1), obtain Eq. (7.8) for its transform $\widetilde{p}(k\,,\,s)$. Invert the Laplace transform, and then the Fourier transform, to arrive at the fundamental solution given by Eq. (7.11).

7.5.2 Green function for the diffusion operator

The fact that the solution in Eq. (7.11) corresponds to a sharp initial condition $p(x, 0) = \delta(x)$ enables us to write down the solution corresponding to any arbitrary prescribed initial distribution in the position x. (Recall the analogous case of the Fokker-Planck equation considered in Ch. 6, Sec. 6.5.2.) The solution to the Green function equation

$$\left(\frac{\partial}{\partial t} - D\,\frac{\partial^2}{\partial x^2}\right) G(x\,,\,t\,;\,x_0\,,\,t_0) = \delta(x - x_0)\,\delta(t - t_0) \qquad (7.25)$$

with natural boundary conditions $G \to 0$ as $x \to \pm\infty$ is

$$G(x\,,\,t\,;\,x_0\,,\,t_0) = \frac{1}{\sqrt{4\pi D(t-t_0)}}\,\exp\left(-\frac{(x-x_0)^2}{4D(t-t_0)}\right),\quad t > t_0. \tag{7.26}$$

Show that, if $\mathsf{p}(x,0) = \mathsf{p}_{\text{init}}(x)$ (a given initial PDF of the position), then

$$\mathsf{p}(x\,,\,t) = \int_{-\infty}^{\infty} dx_0\; G(x\,,\,t\,;\,x_0\,,\,0)\,\mathsf{p}_{\text{init}}(x_0) = \int_{-\infty}^{\infty} dx_0\;\frac{e^{-(x-x_0)^2/4Dt}}{\sqrt{4\pi Dt}}\,\mathsf{p}_{\text{init}}(x_0)\,. \tag{7.27}$$

Verify explicitly that, if $\mathsf{p}_{\text{init}}(x_0)$ is an even function of x_0, then so is $\mathsf{p}(x\,,\,t)$ for all t: that is, $\mathsf{p}(-x\,,\,t) = \mathsf{p}(x\,,\,t)$, as we would expect intuitively.

A connection with quantum mechanics: Observe that the diffusion equation (7.1) for the PDF $\mathsf{p}(x,t)$ is very similar in structure to the Schrödinger equation for the position space wave function $\psi(x,t)$ of a nonrelativistic free particle of mass m,

$$\frac{\partial\psi(x,t)}{\partial t} = \frac{i\hbar}{2m}\,\frac{\partial^2\psi(x,t)}{\partial x^2}\,. \tag{7.28}$$

The only difference[11] is that the diffusion constant D (a real number) is replaced by the pure imaginary quantity $i\hbar/(2m)$. Formally, therefore, the counterpart (in the quantum mechanical context) of the Green function in Eq. (7.26) is given by

$$K(x\,,\,t\,;\,x_0\,,\,t_0) = \left[\frac{m}{2\pi i\hbar(t-t_0)}\right]^{1/2}\,\exp\left(\frac{im(x-x_0)^2}{2\hbar(t-t_0)}\right),\quad t > t_0. \tag{7.29}$$

This is, of course, nothing but the free-particle **Feynman propagator**, i. e.,

$$K(x\,,\,t\,;\,x_0\,,\,t_0) = \left\langle x\left|e^{-iH(t-t_0)/\hbar}\right|x_0\right\rangle, \tag{7.30}$$

where H is the free-particle Hamiltonian $p^2/(2m)$. The formal similarity between the diffusion equation and the Schrödinger equation[12] has several ramifications, among which are path integrals, the so-called stochastic approach to quantum mechanics, and so on.

---

[11]But this is a major difference. A nontrivial analytic continuation is involved here, but we shall not go into this question.

[12]This similarity can be extended to the three-dimensional case.

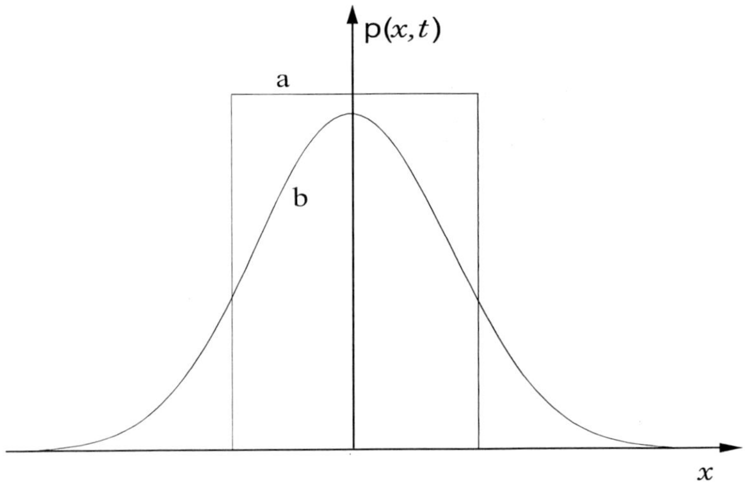

Figure 7.3: (a) An initial rectangular distribution (or uniform distribution in a finite interval) of the position of the diffusing particle; (b) the distribution of the position at a later instant of time. The analytic form of the PDF involves the difference of two error functions, and is given by Eq. (7.32). For values of $|x|$ much larger than the width of the initial distribution, the PDF falls off like the fundamental Gaussian solution, as shown by Eq. (7.33).

7.5.3 Solution for a rectangular initial distribution

Suppose the initial spread in the position is in the form of a rectangular distribution, i. e.,

$$\mathsf{p}_{\text{init}}(x_0) = \begin{cases} 1/(2a) & \text{for} \quad -a \le x_0 \le a, \\ 0 & \text{for} \quad |x_0| > a, \end{cases} \tag{7.31}$$

where a is a positive constant. Figure 7.3(a) shows this initial distribution. It implies that, at $t = 0$, the particle is likely to be found with equal probability in any infinitesimal interval dx in the region $[-a\,,\,a]$. (In terms of the concentration of a diffusing species, the initial concentration is uniform in the region $[-a\,,\,a]$, and zero outside that interval.)

(a) Use Eq. (7.31) in Eq. (7.27) and simplify the resulting expression, to show that

$$\mathsf{p}(x\,,\,t) = \frac{1}{4a}\left[\operatorname{erf}\left(\frac{x+a}{\sqrt{4Dt}}\right) - \operatorname{erf}\left(\frac{x-a}{\sqrt{4Dt}}\right)\right], \tag{7.32}$$

where erf (x) is the error function[13]. Verify that $\mathsf{p}(x\,,\,t)$ as given by Eq. (7.32) is an even function of x. Figure 7.3(b) shows what the solution looks like, for a typical $t > 0$.

(b) The initial distribution under consideration is characterized by the length scale a. Consider any fixed $t > 0$. When $|x| \gg a$, we may expect the solution above to behave, in the leading approximation, like the fundamental Gaussian solution corresponding to an initial δ-function distribution. It is instructive to consider this aspect in some detail. Let $|x| \gg a$. Expand the solution given in Eq. (7.32) in a Taylor series in powers of a. Show that $\mathsf{p}(x\,,\,t)$ can be written as

$$\begin{aligned}\mathsf{p}(x\,,\,t) &= \frac{e^{-x^2/4Dt}}{\sqrt{4\pi Dt}}\left\{1 + \frac{a^2}{12Dt}\left(\frac{x^2}{2Dt} - 1\right)\right.\\ &\left.\quad + \frac{a^4}{160(Dt)^2}\left(\frac{x^4}{12(Dt)^2} - \frac{x^2}{Dt} + 1\right) + \dots\right\}. \qquad (7.33)\end{aligned}$$

Observe that the over-all factor outside the curly brackets is precisely the normalized fundamental Gaussian solution corresponding to an initial δ-function distribution at the origin in x. Only even powers of a occur in the expansion above. The coefficient of a^{2n} is an even polynomial of order $2n$ in x. It is interesting to note that, when integrated over all values of x, the contribution to the integral from each of the terms proportional to a^{2n} $(n \geq 1)$ vanishes[14]. Thus, although the PDF in the present case differs from the fundamental Gaussian form, the 'corrections' to the Gaussian arising from the finite spread of the initial condition do not contribute to the normalization of the PDF.

7.5.4 Distributions involving two independent particles

Suppose we consider *two* tagged particles, each of which starts at the origin and diffuses independently on the x-axis, without any interaction with the other tagged particle. (We also assume that they can 'pass through' each other without any collision.) Let x_1 and x_2 denote the respective positions of the two particles. The positional PDFs of the two particles are $\mathsf{p}(x_1\,,\,t)$ and $\mathsf{p}(x_2\,,\,t)$, where $\mathsf{p}(x\,,\,t)$ is the Gaussian solution given in Eq. (7.11).

(a) Let $x_{\mathrm{cm}} = \frac{1}{2}(x_1 + x_2)$ denote the centre of mass of the two particles. It is clear

---

[13] Appendix A, Sec. A.2. Recall that erf $(x) = (2/\sqrt{\pi})\int_0^x du\, e^{-u^2}$.

[14] Since $\overline{x^2(t)} = 2Dt$ and $\overline{x^4(t)} = 12(Dt)^2$, this is at once seen to be valid for the terms that have been written out explicitly in Eq. (7.33).

that the normalized PDF of x_{cm} is given by[15]

$$\mathsf{p}_{\text{cm}}(x_{\text{cm}}\,,\,t) = \int_{-\infty}^{\infty} dx_1 \int_{-\infty}^{\infty} dx_2\; \mathsf{p}(x_1\,,\,t)\,\mathsf{p}(x_2\,,\,t)\,\delta\left(x_{\text{cm}} - \frac{x_1+x_2}{2}\right). \tag{7.34}$$

Simplify this expression[16] and show that

$$\mathsf{p}_{\text{cm}}(x_{\text{cm}}\,,\,t) = \frac{1}{\sqrt{2\pi Dt}}\,\exp\left(-\frac{x_{\text{cm}}^2}{2Dt}\right). \tag{7.35}$$

The effective diffusion constant for the centre of mass is therefore $\frac{1}{2}D$.

(b) Similarly, let $\xi = x_1 - x_2$ denote the separation between the particles. The normalized PDF of ξ is given by

$$\mathsf{p}_{\text{sep}}(\xi, t) = \int_{-\infty}^{\infty} dx_1 \int_{-\infty}^{\infty} dx_2\; \mathsf{p}(x_1\,,\,t)\,\mathsf{p}(x_2\,,\,t)\,\delta\big(\xi - (x_1 - x_2)\big). \tag{7.36}$$

Show that

$$\mathsf{p}_{\text{sep}}(\xi, t) = \frac{1}{\sqrt{8\pi Dt}}\,\exp\left(-\frac{\xi^2}{8Dt}\right). \tag{7.37}$$

The effective diffusion constant for the separation between the particles is therefore $2D$.

(c) More generally, show that the PDF of the linear combination ax_1+bx_2 (where a and b are real constants) is again a Gaussian, with a variance given by $2(a^2+b^2)Dt$. Once again, this result verifies that the PDF of a linear combination of independent Gaussian random variables is also a Gaussian[17].

7.5.5 Moments of the radial distance

(a) From the expression in Eq. (7.24) for $\mathsf{p}_{\text{rad}}(r, t)$, show that the moments of the radial distance are given by

$$\overline{r^l(t)} = \frac{2^{l+1}}{\sqrt{\pi}}\,(Dt)^{l/2}\,\Gamma\left(\tfrac{1}{2}(l+3)\right). \tag{7.38}$$

---

[15] Recall the form of Eq. (2.30) in Sec. 2.4.3 for the PDF of the relative velocity of two tagged particles, and the comments made there on the use of the Dirac δ-function to find the PDF of a combination of independent random variables.

[16] As in all such cases, you will need to use the properties $\delta(x) = \delta(-x)$ and $\delta(ax) = (1/|a|)\,\delta(x)$, where x is the variable of integration.

[17] See Appendix D, Sec. D.4.

For even and odd values of l, respectively, verify that Eq. (7.38) simplifies to

$$\overline{r^l(t)} = \begin{cases} \left[(l+1)!/(\frac{1}{2}l)!\right] (Dt)^{l/2} & \text{for even } l \\ \left[2^{l+1}/\sqrt{\pi}\right] \left(\frac{1}{2}(l+1)\right)! \, (Dt)^{l/2} & \text{for odd } l. \end{cases} \tag{7.39}$$

(b) Owing to the presence of the factor r^2 arising from the volume element in three dimensions, the first two *negative* moments of the radial distance also exist. Show that

$$\overline{r^{-1}(t)} = (\pi Dt)^{-1/2} \quad \text{and} \quad \overline{r^{-2}(t)} = (2Dt)^{-1}. \tag{7.40}$$

7.5.6 Stable distributions related to the Gaussian

The fundamental Gaussian solution for $\mathsf{p}(x,t)$ in Eq. (7.11) leads to other well-known forms of probability distributions, when we consider suitable functions of the random variable x. Here are two interesting examples.

(i) Consider, for any fixed value of $t\,(>0)$, the dimensionless random variable $\xi = (2Dt)/x^2$. As x increases from $-\infty$ through 0 to ∞, ξ increases from 0 to ∞, and then decreases once again to 0. In order to make the change of variables one-to-one, let us consider just the region $0 \le x < \infty$. The PDF of x, normalized to unity in this region, is clearly given by $2\mathsf{p}(x,t) = (\pi Dt)^{-1/2} \exp\left[-x^2/(4Dt)\right]$. Show that the normalized PDF of ξ is given by

$$p(\xi) = \frac{1}{(2\pi\xi^3)^{1/2}} \exp\left(-\frac{1}{2\xi}\right), \quad 0 \le \xi < \infty. \tag{7.41}$$

The form of the PDF above implies that the random variable ξ has a **Lévy distribution**, or, to give it its full name, a Lévy skew alpha-stable distribution[18] with exponent $\frac{1}{2}$. The Lévy distribution will be encountered once again when we consider the distribution of the first-passage time for diffusion, in Ch. 10.

(ii) Consider two tagged particles, each starting from the origin at $t = 0$, diffusing on the x-axis. The particles are assumed to have no interaction with each other (and to be able to 'pass through' each other). Let x_1 and x_2 be the positions of the particles, and let D_1 and D_2 be their respective diffusion constants (the particles may have different masses and/or radii, for instance). The probability densities of x_1 and x_2 are given by the fundamental Gaussian solution to the diffusion equation.

[18] See Appendix K.

Let $\xi = x_1/x_2$ be the ratio of the positions of the two particles. It is clear that $\xi \in (-\infty, \infty)$. The PDF of ξ is given by

$$p(\xi\,,\,t) = \int_{-\infty}^{\infty} dx_1 \int_{-\infty}^{\infty} dx_2\; \mathsf{p}(x_1\,,\,t)\,\mathsf{p}(x_2\,,\,t)\,\delta\left(\xi - \frac{x_1}{x_2}\right) \tag{7.42}$$

where $\mathsf{p}(x_1\,,\,t)$ (respectively, $\mathsf{p}(x_2\,,\,t)$) is the fundamental Gaussian solution with diffusion constant D_1 (respectively, D_2). Use the δ-function to eliminate the integration over x_1, and then carry out the integration over x_2. Show that Eq. (7.42) yields

$$p(\xi) = \frac{\sqrt{D_1\,D_2}}{\pi\,(D_1 + D_2\,\xi^2)} \quad \text{for all } t > 0. \tag{7.43}$$

Note that the t-dependence actually disappears from the PDF of ξ. The functional form of $p(\xi)$ corresponds to another member of the family of stable distributions—the **Cauchy distribution**, which is a Lévy alpha-stable distribution with exponent 1 (see Appendix K, Sec. K.3). The shape of the PDF in Eq. (7.43) is the familiar **Lorentzian**.

In Ch. 10, Sec. 10.3.3, we shall see how this result has an interesting complement in the time domain, in connection with first-passage times. It will be shown there that the ratio of the square roots of two random variables, each of which has a Lévy distribution, also has a Cauchy distribution.

(iii) Following up on the result just derived, it is natural to ask what happens if we consider n non-interacting, independently diffusing particles with positions $x_1\,, \ldots\,, x_n$, all of them starting from the origin at $t = 0$. Let us assume, for simplicity, that D is the same for all the particles. Scale out the positions of the particles by that of any one of them, say x_1. That is, define the ratios

$$\xi_1 = \frac{x_2}{x_1}\,,\; \xi_2 = \frac{x_3}{x_1}\,,\; \ldots\,,\; \xi_{n-1} = \frac{x_n}{x_1}\,. \tag{7.44}$$

Each ξ_i can take on values in $(-\infty, \infty)$. What is the joint distribution of the $(n-1)$ random variables $\xi_1\,, \xi_2\,, \ldots\,, \xi_{n-1}$ at any time t? Show that the joint PDF of these random variables is also time-independent, and is given by

$$p(\xi_1\,, \ldots\,, \xi_{n-1}) = \frac{\Gamma(n/2)}{\pi^{n/2}\,\left(1 + \xi_1^2 + \ldots + \xi_{n-1}^2\right)^{n/2}}\,, \quad (n \geq 2) \tag{7.45}$$

for all $t > 0$. Verify that the PDF above is correctly normalized to unity. [Hint: Change to 'spherical' polar coordinates in $(n-1)$ dimensions!]

Chapter 8

Diffusion in a finite region

We consider the one-dimensional diffusion equation in a finite region, with reflecting boundary conditions and with absorbing boundary conditions. Solutions using the method of separation of variables are presented. The survival probability in a finite region in the case of absorbing boundaries is determined. An alternative approach to the solution of the diffusion equation, based on the method of images, is also described. The remarkable relationship between the representations for the PDF $\mathsf{p}(x, t)$ *obtained by the two methods is clarified using Poisson's summation formula.*

8.1 Diffusion on a line with reflecting boundaries

In the preceding chapter, we have derived and analyzed the fundamental Gaussian solution to the diffusion equation. The range of the spatial variable x was $(-\infty\,,\,\infty)$ (or all points in space, in the three-dimensional case). The boundary condition used was simply the requirement that the PDF vanish at spatial infinity, a necessary condition for its normalizability.

In physical applications of diffusion, we are frequently concerned with the case in which the position of the diffusing particle is restricted to a finite region. Depending on the boundary conditions (based, in turn, on the physical situation at hand), several possibilities can occur. It turns out that the solution to the diffusion equation is strongly dependent on the boundary conditions imposed.

We begin with diffusion in one dimension, in the region $-a \le x \le a$, where $a > 0$. We assume that the 'walls' or 'barriers' at the two ends $x = -a$ and $x = a$

are impenetrable, and perfectly reflecting. The problem is to find the solution of the diffusion equation $\partial \mathsf{p}(x,t)/\partial t = D\,\partial^2 \mathsf{p}(x,t)/\partial x^2$ in this region, with the initial condition $\mathsf{p}(x,0) = \delta(x)$. The boundary conditions at $x = \pm a$ are dictated by the fact that the walls are neither permeable (there is no 'leakage' through them), nor 'sticky' (the particle does not stick to the wall once it reaches a boundary). These conditions imply that the diffusion *current* vanishes at the end-points, i. e., the *flux* through these boundaries is zero. In the absence of any external force, i. e., for 'free diffusion', this current is $-D\partial \mathsf{p}/\partial x$ (recall Fick's first law). The boundary conditions are therefore

$$\left.\frac{\partial \mathsf{p}}{\partial x}\right|_{x=\pm a} = 0 \quad \text{for all } t \geq 0 \quad \text{(reflecting boundary conditions).} \tag{8.1}$$

The problem is eminently suited to the application of the method of separation of variables. The calculation is further simplified if we keep the following in mind. Note that the diffusion equation is unchanged under the transformation $x \to -x$. In the present instance, the initial condition as well as the boundary conditions are *also* invariant under this transformation. We can then assert that the solution $\mathsf{p}(x,t)$ itself must be an even function of x at all times. This observation helps us select the correct solution, which turns out to be

$$\mathsf{p}(x,t) = \frac{1}{2a} + \frac{1}{a}\sum_{n=1}^{\infty} \cos\left(\frac{n\pi x}{a}\right) \exp\left(-\frac{n^2\pi^2 Dt}{a^2}\right) \quad \text{(reflecting b.c.).} \tag{8.2}$$

It is trivially verified that

$$\int_{-a}^{a} dx\; \mathsf{p}(x\,,\,t) = 1 \quad \text{for all } t \geq 0. \tag{8.3}$$

Hence the total probability that the diffusing particle is in the region $[-a\,,\,a]$ is unity for all t, as we would expect on physical grounds: the initial PDF $\delta(x)$ is normalized to unity, and there is no 'leakage of probability' through the perfectly reflecting barriers at the two end-points. Moreover, a non-vanishing equilibrium PDF does exist in this situation. Not surprisingly, it is simply the uniform PDF in the region $[-a\,,\,a]$:

$$\mathsf{p}^{\text{eq}}(x) = \lim_{t\to\infty} \mathsf{p}(x\,,\,t) = \frac{1}{2a}\,, \quad |x| \leq a. \tag{8.4}$$

Infinite series similar to that in Eq. (8.2) appear very frequently in solutions to the diffusion problem with finite boundaries. We shall come across a number of examples in the sequel. Typically, these series involve exponentials of *quadratic* functions[1] of the summation index n.

8.2 Diffusion on a line with absorbing boundaries

The second important case is that of diffusion with absorbing boundary conditions. If the end-points $x = \pm a$ are perfect *absorbers*, then the diffusing particle is absorbed (or annihilated) when it hits either of the two end-points for the first time, and the process of diffusion terminates. It is plausible that a particle starting at the origin at $t = 0$ will surely hit one of the end-points, given sufficient time[2]. We may then expect the total probability $\int_{-a}^{a} dx\, \mathsf{p}(x,t)$ to decrease from its initial value of unity, as t increases, and eventually vanish. These conjectures are indeed borne out, as we shall see.

The appropriate boundary conditions that take into account perfectly absorbing barriers at $x = \pm a$ are

$$\mathsf{p}(x,t)\big|_{x=\pm a} = 0 \quad \text{for all } t \geq 0 \quad \text{(absorbing boundary conditions).} \tag{8.5}$$

Once again, we may use the method of separation of variables to obtain the solution to the diffusion equation. The symmetry property $\mathsf{p}(x,t) = \mathsf{p}(-x,t)$ remains valid in this case too, and helps simplify the calculation. The solution corresponding to the initial condition $\mathsf{p}(x,0) = \delta(x)$ is now given by

$$\mathsf{p}(x,t) = \frac{1}{a}\sum_{n=0}^{\infty} \cos\left(\frac{(2n+1)\pi x}{2a}\right) \exp\left(-\frac{(2n+1)^2\pi^2 Dt}{4a^2}\right) \quad \text{(absorbing b.c.).} \tag{8.6}$$

This expression should be compared with that obtained for reflecting boundary conditions, Eq. (8.2). It is immediately evident that $\mathsf{p}(x,t)$ now vanishes as $t \to \infty$, in marked contrast to the previous case.

---

[1]Such sums cannot be expressed in closed form in terms of elementary algebraic or transcendental functions, such as trigonometric, hyperbolic or logarithmic functions. However, they can be written in closed form in terms of the so-called **Jacobi theta functions** and their derivatives.

[2]While this is plausible, one must not jump to the conclusion that it is *obviously* true! For instance, it is conceivable that the diffusing particle simply moves back and forth forever at random, crossing the origin repeatedly, without ever reaching either $-a$ or a on the x-axis. However, this does not happen. The particle *will* reach either $-a$ or a with probability 1, given sufficient time, no matter how large a is. This is proved in Ch. 10, where we discuss these aspects at greater length.

The total probability at any $t > 0$, namely,

$$\int_{-a}^{a} dx\, \mathsf{p}(x,t) \equiv S(t\,,\,\pm a\,|\,0\,), \tag{8.7}$$

has a simple interpretation. It is precisely the **survival probability** that a diffusing particle, starting from $x = 0$ at $t = 0$, remains in the region $(-a\,,\,a)$ without having hit either of the boundaries till time t. Using the solution (8.6) for $\mathsf{p}(x,t)$ in Eq. (8.7), we get the important result

$$S(t\,,\,\pm a\,|\,0\,) = \frac{4}{\pi}\sum_{n=0}^{\infty}\frac{(-1)^n}{(2n+1)}\exp\left(-\frac{(2n+1)^2\pi^2 Dt}{4a^2}\right). \tag{8.8}$$

We already know that $\int_{-a}^{a} dx\, \mathsf{p}(x,0) = \int_{-a}^{a} dx\, \delta(x) = 1$. We may verify that Eq. (8.8) matches this result at $t = 0$, because the series[3]

$$\sum_{n=0}^{\infty}\frac{(-1)^n}{(2n+1)} = 1 - \frac{1}{3} + \frac{1}{5} - \ldots = \frac{\pi}{4}. \tag{8.9}$$

Now, each exponential factor in the sum in Eq. (8.8) is less than unity for any $t > 0$, and moreover decreases sharply in magnitude with increasing n. This suggests that $S(t\,,\,\pm a\,|\,0) < 1$ for all $t > 0$, and that it is a monotonically decreasing function of time. In the limit $t \to \infty$, $S(t\,,\,\pm a\,|\,0)$ vanishes, indicating that absorption at a boundary is a 'sure' event—that is, it will occur with certainty (or probability one), given sufficient time. Figure 8.1 depicts the survival probability as a function of t. We will have occasion to use the expression obtained above for $S(t\,,\,\pm a\,|\,0\,)$ in Ch. 10, Sec. 10.3.2.

8.3 Solution by the method of images

The solutions in the two cases above, corresponding to reflecting and absorbing boundary conditions, respectively, may also be obtained by using the method of images. This method is perhaps more familiar to you in the context of electrostatics. It is used there to find the solution to Poisson's equation in the presence of specific boundary conditions, provided the problem has a convenient symmetry. But the method is applicable in other instances as well. In broad terms,

[3] A little history of mathematics is not out of place here. The infinite sum in Eq. (8.9) is the well-known Leibniz-Gregory series. Its value ($\frac{1}{4}\pi$) was first deduced by the father of mathematical analysis, the Indian mathematician Madhava of Sangamagrama (1350-1425), as a special case of the more general infinite series for $\tan^{-1} x$ discovered by him. The latter was rediscovered a few centuries later by Gregory (1638-1675), and is now known as the Madhava-Gregory series.

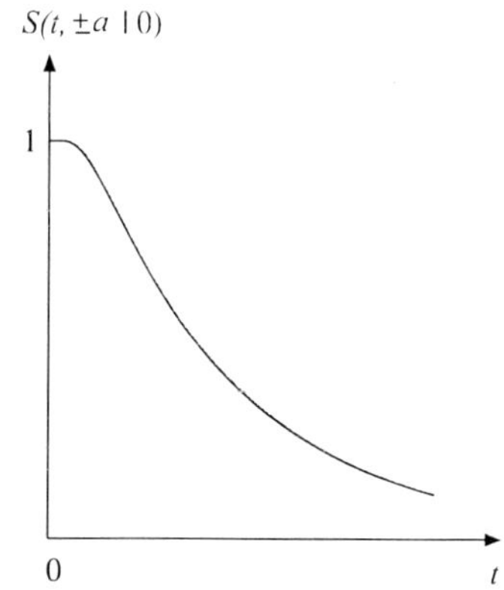

Figure 8.1: The survival probability $S(t\,,\,\pm a\,|\,0)$ as a function of the time. This is the probability that the diffusing particle, starting at $t=0$ from the origin on the x-axis, has not yet hit either of the points $x=-a$ or $x=a$ until time t. The probability S is a superposition of an infinite number of decaying exponentials with characteristic time constants $4a^2/[(2n+1)^2\pi^2 D]$, where $n=0,1,\ldots$. Thus S decays monotonically to zero, with a leading asymptotic behavior $\sim \exp(-\pi^2 Dt/4a^2)$.

it is a technique to find the Green function of a linear differential operator with specific boundary conditions, provided the problem has some symmetry that can be exploited. It hinges on the fact that the differential equation concerned, taken together with appropriate initial and boundary conditions, has a *unique* solution. Our purpose in applying it to the problem at hand is two-fold: first, to bring out the fact that the method is a general one, applicable to time-dependent problems as well; second, to show how powerful the method can be, when the conditions are right. We shall use the method of images once again in Ch. 10, Secs. 10.2 and 10.3.1.

For simplicity and subsequent comparison with the solutions in Eqs. (8.2) and (8.6), we assume once again that the diffusing particle starts at the origin at $t=0$. Imagine a mirror to be placed at each boundary, i. e., at the points a and $-a$ on the x-axis. The image of any point $x \in (-a\,,\,a)$ in the mirror at a is of course located at the point $2a-x$, while the image of the point x in the mirror at $-a$ is located $-2a-x$. As these two mirrors face each other, an initial source point x has an infinite number of images. For instance, the image of the point $2a-x$ in the mirror at $-a$ is located at $-2a-(2a-x)=-4a+x$, while the image of the point $-2a-x$ in the mirror at a is located at $2a-(-2a-x)=4a+x$, and so on, *ad infinitum*. Figure 8.2 shows the location of a source point at x and the first few

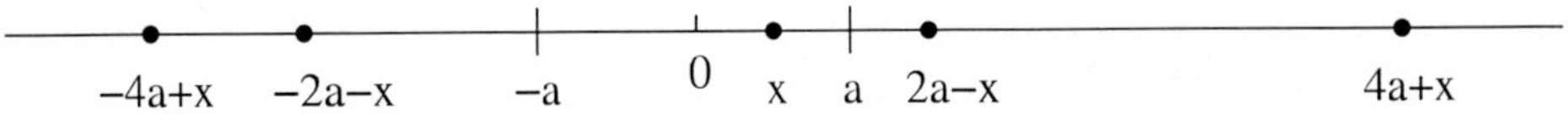

Figure 8.2: Locations of an arbitrary point $x \in (-a\,,\,a)$ and its first few images for the case of two boundaries, or 'mirrors', at a and $-a$ respectively. The number of image points is infinite in this case. The contribution of each image to the probability density function $\mathsf{p}(x,t)$ must be incorporated with an appropriate sign. For reflecting boundary conditions at $\pm a$, all the signs are positive. For absorbing boundary conditions at $\pm a$, the sign is $(-1)^n$ for the contribution from an image point that arises as the result of n reflections.

images (of the infinite set of images) in the pair of mirrors at a and $-a$, respectively.

There is a contribution to the PDF $\mathsf{p}(x,t)$ associated with the point x and each of its infinite set of images. Each contribution is just the fundamental Gaussian solution to the diffusion equation (with the coordinate of the corresponding image point taking the place of x). The probability density $\mathsf{p}(x,t)$ in the physical region between the two boundary points is a linear combination of the contributions from all the images, with coefficients that are $+1$ or -1. *The boundary conditions are incorporated* by means of the following simple sign rule:

- For reflecting boundary conditions at $\pm a$, all the coefficients are equal to $+1$.
- For absorbing boundary conditions at $\pm a$, the coefficient multiplying the contribution from an image point arising from n reflections is $(-1)^n$.

Consider, first, the case of reflecting boundaries at both ends. After a bit of simplification, this procedure yields the solution

$$\mathsf{p}(x,t) = \frac{1}{\sqrt{4\pi Dt}} \sum_{n=-\infty}^{\infty} \exp\left(-\frac{(x+2na)^2}{4Dt}\right) \quad \text{(reflecting b.c.).} \tag{8.10}$$

It is not difficult to verify that the expression in Eq. (8.10) satisfies the boundary conditions

$$[\partial \mathsf{p}/\partial x]_{x=\pm a} = 0.$$

The result in Eq. (8.10) may be written in the alternative form

$$\mathsf{p}(x,t) = \frac{e^{-x^2/4Dt}}{\sqrt{4\pi Dt}} \left\{1 + 2\sum_{n=1}^{\infty} e^{-n^2a^2/Dt} \cosh\left(\frac{nax}{Dt}\right)\right\} \quad \text{(reflecting b.c.).} \tag{8.11}$$

This representation brings out clearly the role of the finite boundaries. The leading term is just the fundamental Gaussian solution to the diffusion equation on the infinite line, in the absence of any boundaries. The rest of the terms may be regarded as 'corrections' that arise from the presence of the reflecting boundaries at $\pm a$.

The solution when there are absorbing boundaries present at both ends is found similarly—but we have to keep track of the sign of each term in the sum. The contribution from the original source point x occurs with a plus sign; those from the images at $2a - x$ and $-2a - x$, with minus signs; those from the images (at $4a + x$ and $-4a + x$) of these images, with plus signs; and so on. The final result is

$$\mathsf{p}(x,t) = \frac{1}{\sqrt{4\pi Dt}} \sum_{n=-\infty}^{\infty} (-1)^n \exp\left(-\frac{(x+2na)^2}{4Dt}\right) \quad \text{(absorbing b.c.).} \qquad (8.12)$$

It is easily verified that the solution above satisfies the boundary conditions

$$[\mathsf{p}(x\,,\,t)]_{x=\pm a} = 0.$$

As in the previous case, Eq. (8.12) may be re-written in the suggestive form

$$\mathsf{p}(x,t) = \frac{e^{-x^2/4Dt}}{\sqrt{4\pi Dt}} \left\{1 + 2\sum_{n=1}^{\infty}(-1)^n\, e^{-n^2a^2/Dt} \cosh\left(\frac{nax}{Dt}\right)\right\} \quad \text{(absorbing b.c.).} \qquad (8.13)$$

Again, this representation of the solution exhibits the amendments to the fundamental Gaussian that arise from the presence of the absorbing boundaries at $\pm a$.

- Compare the expressions in Eqs. (8.10) and (8.12) for the solutions in the presence of reflecting and absorbing boundaries, respectively. It is remarkable that the *only* difference between them is the extra factor $(-1)^n$ in the summand of the latter equation—but this is sufficient to alter completely the behavior of the PDF $\mathsf{p}(x,t)$!

An interesting question arises now. Observe that the factor t appears in the *numerator* in the exponents occurring in the solutions obtained by the method of separation of variables, Eqs. (8.2) and (8.6). These expressions thus involve damped exponentials in time. Hence they may be expected to be particularly useful in finding the asymptotic or *large-t* behavior of $\mathsf{p}(x,t)$. In marked contrast to this, the t-dependence is in the *denominator* of the exponent both in the fundamental Gaussian solution $\exp(-x^2/4Dt)$, as well as in the exponents in the summands of Eqs. (8.10) and (8.12).[4] Thus there are two distinct representations for $\mathsf{p}(x,t)$ in each case, apparently with very different kinds of t-dependence. How are these to be reconciled with each other? How do we establish their equality?

The answer lies in one of those far-reaching, almost magical, formulas of mathematics—the **Poisson summation formula**. The details are described in one of the exercises that follow.

8.4 Exercises

8.4.1 Solution by separation of variables

(a) Using the method of separation of variables, derive the solution given in Eq. (8.2) to the diffusion equation in the presence of perfectly reflecting barriers at $\pm a$.

(b) Similarly, derive the solution given in Eq. (8.6) to the diffusion equation in the presence of perfectly absorbing barriers at $\pm a$.

You will need the following representation of the initial PDF $\delta(x)$, in order to determine the unknown coefficients that appear in the course of the derivation. Start with the identity

$$\delta(\xi) = \sum_{n=-\infty}^{\infty} e^{2\pi n i \xi} \tag{8.14}$$

for a dimensionless variable ξ. Set $\xi = x/(2a)$, so that

$$\delta(x/2a) = \sum_{n=-\infty}^{\infty} e^{\pi n i x/a}. \tag{8.15}$$

For the purpose at hand, it is best to re-write Eq. (8.15) as

$$\delta(x) = \frac{1}{2a} + \frac{1}{a}\sum_{n=1}^{\infty} \cos\left(\frac{n\pi x}{a}\right). \tag{8.16}$$

---

[4]For large values of t, these exponents tend to zero. Since $e^0 = 1$, one might wonder about the convergence of the various infinite series concerned in the long-time limit. We shall not enter here into this interesting, but essentially mathematical, question—except to point out that extracting the $t \to \infty$ behavior of $\mathsf{p}(x,t)$ from Eqs. (8.10) and (8.12) is a more delicate task than doing so from Eqs. (8.2) and (8.6).

8.4.2 Diffusion on a semi-infinite line

We have obtained solutions to the diffusion equation in the presence of two boundaries at $\pm a$ on the x-axis. A much simpler exercise is to find the solution in the presence of a single boundary. Consider diffusion on the semi-infinite line $(-\infty\,,\,a]$ where a is any real number. Let $x_0 < a$ be the starting point of the diffusing particle at $t = 0$. Use the method of images to write down the solution to the diffusion equation with a natural boundary condition at $-\infty$, and (i) a reflecting boundary at a; (ii) an absorbing boundary at a. The expressions sought are:

$$\mathsf{p}(x,t) = \frac{1}{\sqrt{4\pi Dt}}\left[\exp\left(-\frac{(x-x_0)^2}{4Dt}\right) \pm \exp\left(-\frac{(2a-x-x_0)^2}{4Dt}\right)\right], \qquad (8.17)$$

where the plus and minus signs correspond, respectively, to a reflecting boundary at a and an absorbing boundary at a. Note that these expressions are *not* functions of the difference $(x - x_0)$ alone. The presence of the boundary at the finite point a breaks the translational invariance that obtains in the case of the infinite line.

8.4.3 Application of Poisson's summation formula

The task here is to show that the expressions in Eqs. (8.10) and (8.2) for the solution in the presence of reflecting barriers at $\pm a$ are, in fact, identically equal to each other; and, similarly, that the expressions in Eqs. (8.12) and (8.6) for the solution in the presence of absorbing barriers at $\pm a$ are identically equal to each other. As mentioned in the text, this is best done with the help of Poisson's summation formula.

Let q and k be real variables in $(-\infty\,,\,\infty)$, and let $f(q)$ and $\widetilde{f}(k)$ be a Fourier transform pair, i. e.,

$$f(q) = \frac{1}{2\pi}\int_{-\infty}^{\infty} dk\, e^{ikq}\, \widetilde{f}(k), \quad \text{so that} \quad \widetilde{f}(k) = \int_{-\infty}^{\infty} dq\, e^{-ikq}\, f(q). \qquad (8.18)$$

Then the Poisson summation formula states that

$$\sum_{n=-\infty}^{\infty} f(\lambda n) = \frac{1}{\lambda}\sum_{n=-\infty}^{\infty} \widetilde{f}\left(\frac{2\pi n}{\lambda}\right), \qquad (8.19)$$

where λ is any real number. This means that

- if a function $f(q)$ is *sampled* at all the integer multiples of some number λ, and the values so obtained are added up, the outcome is the sum obtained by sampling its Fourier transform $\widetilde{f}(k)$ at all the integer multiples of $2\pi/\lambda$, apart from an over-all factor of $1/\lambda$.

- The important thing to note is that if the sampling interval λ in the variable q is small, the required sampling interval $2\pi/\lambda$ in the conjugate variable k is large, and *vice versa*. This is a manifestation of a deep duality that has profound implications in many fields.

(a) Use the Poisson summation formula to prove that the right-hand sides of Eqs. (8.10) and (8.2) are identically equal to each other, as follows. First set $\lambda = a/\sqrt{Dt}$. If the right-hand side of Eq. (8.10) is identified with $\sum_{-\infty}^{\infty} f(\lambda n)$, it follows that

$$f(q) = \frac{1}{\sqrt{4\pi Dt}} \exp\left\{ -\left(q + \frac{x}{\sqrt{4Dt}}\right)^2 \right\}. \tag{8.20}$$

Work out the Fourier transform (with respect to q) of this function[5], to obtain

$$\widetilde{f}(k) = \frac{1}{\sqrt{4Dt}} \exp\left(-\frac{1}{4}k^2 + \frac{ikx}{\sqrt{4Dt}} \right). \tag{8.21}$$

Hence show that $\lambda^{-1} \sum_{-\infty}^{\infty} \widetilde{f}(2\pi n/\lambda)$ reduces precisely to the right-hand side of Eq. (8.2), establishing the identity desired.

(b) Similarly, use the summation formula to prove that the right-hand sides of Eqs. (8.12) and (8.6) are identically equal to each other. This is a little less straightforward than the preceding case. If we want to identify the right-hand side of Eq. (8.12) with $\sum_{-\infty}^{\infty} f(\lambda n)$, where $\lambda = a/\sqrt{Dt}$, then what is $f(q)$? What should one do with the factor $(-1)^n$ in the summand, when n is not an integer?

Show that the Fourier transform of $f(q)$ is, in this case,

$$\widetilde{f}(k) = \frac{1}{\sqrt{4Dt}} \exp\left\{ -\frac{1}{4}\left(k - \frac{\pi\sqrt{Dt}}{a}\right)^2 + \frac{ix}{2\sqrt{Dt}}\left(k - \frac{\pi\sqrt{Dt}}{a}\right) \right\}. \tag{8.22}$$

Hence show that $\lambda^{-1} \sum_{-\infty}^{\infty} \widetilde{f}(2\pi n/\lambda)$ reduces precisely to the expression in Eq. (8.6).

[5] Once again, the basic integral required for this purpose is the Gaussian integral given in Eq. (2.31) or Eq. (A.4).

Chapter 9

Brownian motion

We consider the stochastic process represented by the position of a diffusing particle, when its PDF satisfies the diffusion equation. This is the Wiener process, or Brownian motion in the mathematical sense. Several salient features of this random process are described, including the continuous but non-smooth nature of its sample paths; its precise short-time and long-time behavior; the recurrence properties of Brownian motion in one and more dimensions; and the stochastic differential equations for some functions of Brownian motion. Some brief remarks are also made on the Itô calculus.

9.1 The Wiener process (Standard Brownian motion)

In the preceding chapter, we studied the diffusion equation satisfied by the PDF of the position of the tagged particle. It was pointed out that the diffusion equation

$$\frac{\partial \mathsf{p}(x,t)}{\partial t} = D\,\frac{\partial^2 \mathsf{p}(x,t)}{\partial x^2} \tag{9.1}$$

is actually an approximation that is valid on time scales much larger than the velocity correlation time γ^{-1}. In this chapter, we continue to work in the diffusion regime, and consider another aspect of the motion of a diffusing particle: namely, the nature and properties of the stochastic process represented by the position $x(t)$ of the particle.

We begin with the observation that Eq. (9.1) itself appears to have the structure of a Fokker-Planck equation (whose general form is given by Eq. (6.7)), in which the drift f is absent, and the function g is just a constant. It is therefore

natural to pose the inverse question: Is there a stochastic differential equation satisfied by the position variable x, such that its PDF is simply the diffusion equation? We have said that the FPE (6.7) follows from the SDE (6.6). Based on this relationship, it follows that—*in the diffusion regime*—the position variable x satisfies the rather simple-looking SDE
indexstochastic differential equation

$$\dot{x} = \sqrt{2D}\,\zeta(t), \tag{9.2}$$

where $\zeta(t)$ is a stationary, Gaussian, δ-correlated, Markov process (a Gaussian white noise). Repeating Eqs. (6.3), its mean and autocorrelation are given by

$$\langle \zeta(t) \rangle = 0 \quad \text{and} \quad \big\langle \zeta(t)\,\zeta(t') \big\rangle = \delta(t - t'). \tag{9.3}$$

Hence:

- In the diffusion regime or approximation, the position x is 'the integral of a Gaussian white noise'. It is a *non*-stationary Gaussian Markov process, called **Brownian motion** in the mathematical literature[1].

In mathematical treatments of stochastic processes, it is customary to choose units such that $2D = 1$, and to refer to the resulting process as **standard Brownian motion**. However, we shall retain the scale factor $2D$, as we are interested in the physical process of diffusion, with a specific diffusion constant that will be related to other physical parameters. On the other hand, it will be very useful to have a separate notation for standard Brownian motion itself. Now, the Gaussian white noise $\zeta(t)$ that we have used rather casually is really a very singular object, from the rigorous mathematical point of view. It turns out to be more suitable to work with standard Brownian motion, which is often denoted by $w(t)$, and whose increment is given by the stochastic equation

$$dw(t) = \zeta(t)\,dt. \tag{9.4}$$

Standard Brownian motion is also called the **Wiener process**. The process $x(t)$ that we are dealing with in this chapter is of course just a constant multiple of $w(t)$, namely,

$$x(t) = \sqrt{2D}\,w(t). \tag{9.5}$$

[1]It will become evident that, in this chapter, a slightly higher degree of mathematical sophistication is required on the part of the reader, than elsewhere in this book. This is a reflection of the nature of the subject matter itself. But our attempt will be, as always, to phrase even technical details in as simple a language as possible.

It has already been mentioned that this was essentially Einstein's original model for the physical phenomenon of Brownian motion. We see now that this model is tantamount to assuming that the velocity of a Brownian particle is a Gaussian white noise with zero correlation time. Although we understand that this is at best a rough approximation as far as the *physical* phenomenon of Brownian motion is concerned, the mathematical properties of the random process are of great interest in their own right.

- The Wiener process, or standard Brownian motion, is a basic paradigm in the study of random processes. It is the continuum limit (i. e., the limit in which the step length in space, and the time step, both tend to zero) of a simple **random walk**. The latter, in turn, is a basic model for random processes in discrete space and time[2].

- It is also used as a model in a large and diverse number of applications, ranging all the way from economics to quantum field theory.

It is therefore worth getting acquainted with some of its numerous remarkable properties[3]. We shall merely list these properties here, with no attempt to prove them formally. Throughout this chapter, by Brownian motion we mean the mathematical stochastic process $x(t)$ defined by Eq. (9.5).

There are actually several equivalent ways of defining Brownian motion mathematically. A convenient one is as follows. For definiteness (and notational simplicity), let us choose the origin on the x-axis to be the starting point of the random process $x(t)$ at $t = 0$. Then,

- Brownian motion may be *defined* as a continuous random process, such that the family of random variables $\{x(t)\}_{t\geq 0}$ is Gaussian (that is, all their joint probability distributions are Gaussian), with zero mean, namely,

$$\overline{x(t)} = 0 \quad \text{for all } t, \tag{9.6}$$

and autocorrelation function

$$\overline{x(t)\,x(t')} = 2D \min(t\,,\,t'). \tag{9.7}$$

---

[2]I have therefore devoted Appendix E to showing how a simple random walk goes over into diffusion in the continuum limit.

[3]The mathematical literature on the subject and its ramifications is truly vast. However, this is neither the topic of this book, nor is it our intention here to get into the mathematics of such stochastic processes at anything approaching a mathematically rigorous level.

Remarkably enough, all the other properties of Brownian motion follow from this definition. In particular, $x(t)$ can be shown to be a Markov process. It follows from Eq. (9.7) that the expectation value of the square of the difference $x(t) - x(t')$ is given by

$$\overline{\big(x(t) - x(t')\big)^2} = 2D\,|t - t'|. \tag{9.8}$$

Brownian motion perpetually renews itself, or 'starts afresh', in the following sense:

- For all $t > t'$, the difference $x(t) - x(t')$ is a random process that is statistically independent of $x(t')$, and has the same probability distribution as $x(t')$.
- Hence, for a *given* value of $x(t')$, the *conditional* average value of $x(t)$ is precisely $x(t')$ itself, for all $t > t'$.

What is the origin of the specific form of the autocorrelation function in Eq. (9.7)? As introduced above, it is part of the very definition of Brownian motion in the mathematical sense, but where does this particular functional form come from? In Ch. 7, Sec. 7.2, the mean squared value of the position (or the displacement) of the diffusing particle was calculated. We did not calculate the autocorrelation function of its position. To do so, we need the joint two-time PDF of $x(t)$. If we make use of the Markov property of Brownian motion, then we can write down this joint PDF in terms of the conditional PDF $\mathsf{p}(x\,,\,t\,|\,x_0\,,\,t_0\,)$. The autocorrelation function of $x(t)$ may then be calculated. The result turns out to be precisely that given in Eq. (9.7) (see the exercises at the end of this chapter). Alternatively, the autocorrelation function concerned can be determined exactly from the Langevin equation, as we shall show in Ch. 11. It will then become clear that Eq. (9.7) is an approximation, valid in the diffusion regime, to the exact expression for this quantity—just as we would expect it to be.

9.2 Properties of Brownian paths

The possible paths or trajectories of a particle undergoing Brownian motion, or, equivalently, the **sample paths** of a Wiener process $w(t)$, are continuous. The kind of continuity possessed by the sample paths of Brownian motion has a precise mathematical characterization, which is as follows.

In general, a curve $x(t)$ that is parametrized by the continuous variable t is said to be **Hölder continuous** of order β (where β is a positive number), if there exists a positive constant C such that

$$|x(t+\epsilon) - x(t)| \leq C\,|\epsilon|^{\beta}\,, \tag{9.9}$$

for sufficiently small ϵ. If β is equal to 1, it is clear that the curve is differentiable[4]. On the other hand, if $\beta < 1$, the curve is continuous but not differentiable. In the case of Brownian motion it turns out that, at almost all points, the exponent β can be taken to be *as close to* $\frac{1}{2}$ *from below* as we like. However, it cannot be set *equal* to $\frac{1}{2}$. In fact, the set of points at which a sample path has Hölder continuity with $\beta = \frac{1}{2}$ is a set of measure zero!

An immediate consequence of the foregoing is that the sample paths of Brownian motion are *non-differentiable* almost everywhere. The local maxima and minima of $x(t)$ (i. e., the peaks and troughs) form a dense set. Figure 9.1 is a (very) schematic depiction of such a path. In actuality, the kinks in the graph occur almost everywhere, and on all scales—and this feature persists under arbitrarily high magnification of any segment of the curve. If any portion of the curve is selected, and the abscissa (time) is re-scaled by a factor α, while the ordinate (position) is re-scaled by $\sqrt{\alpha}$, the new curve would be indistinguishable, statistically, from the original one. More precisely:

- If $x(t)$ $(t > 0)$ is a Brownian motion (that starts at 0 at $t = 0$), so is the random process $\sqrt{\alpha}\, x(t/\alpha)$, where α is any positive number.

There is no interval of time, however small, in which $x(t)$ is of bounded variation! The curve is a self-similar **random fractal**. In every realization of the random process, $x(t)$ is a wildly oscillating function. Almost surely, it crosses the starting point $x = 0$ repeatedly in any time interval t.

The instants marking these crossings (referred to as the **zero set** of Brownian motion) themselves form a dense set, and is uncountable. This zero set has no isolated points, but it does not include any interval, and its Lebesgue measure (i. e., its length) is zero. The **fractal dimension**[5] of the zero set is equal to $\frac{1}{2}$.

---

[4]Hölder continuity of order 1 is also called **Lipschitz continuity**.

[5]This is yet another concept with which I have assumed some familiarity on the part of the reader. Although fractals have become part of the educational landscape today (in more senses than one!), a short explanation is perhaps in order. The standard example is that of a curve that is 'kinky' or 'jagged' at almost all points on the curve. As we refine our scale of resolution, i. e., the least count l of the measuring rod, the kinks and bends get taken into account more and more accurately. Hence the over-all measured length of the curve actually increases in absolute magnitude (in any given units) as l becomes smaller and smaller. For a true fractal curve, the length of the curve would *diverge* in the limit $l \to 0$, like some negative power l^{-d_H}. The number d_H is called the fractal dimension of the curve. This is somewhat loose terminology, although it has become established in use. (It is actually the so-called box-counting or **box dimension** of the curve. There is also another, related, concept called the **Hausdorff dimension**. These two

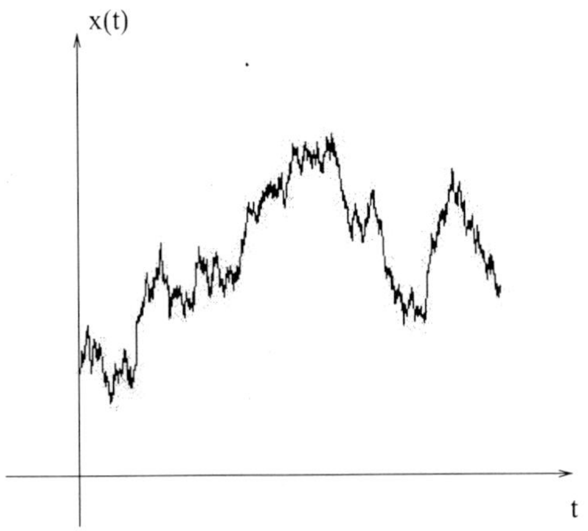

Figure 9.1: A rough schematic of a sample path of Brownian motion. The actual path is kinky and non-differentiable at almost all points, although it is continuous. $x(t)$ is also unbounded, and crosses any specified value infinitely often from below and from above. The curve is a random fractal. However, it has numerous well-defined statistical and probabilistic properties. Some of these are described in the text.

It is clear from the foregoing that Brownian motion is *extremely* irregular motion. The velocity of a particle executing Brownian motion is unbounded. Given this, is it possible to have, nevertheless, quantitative answers—albeit at a probabilistic level—to the following questions: In a time interval t, how far could the particle travel? Where, typically, would it be at a time t, if it started from a given origin at $t = 0$? How often would it cross its starting point? What would be the mean time between successive crossings? In a given time interval, what is the average value of the fraction of the total time that the particle spends on either side of the x-axis? What is the probability distribution of this fraction? Remarkably enough, precise answers exist for all these questions (and several others as well), in terms of probabilities. Let us mention some of these.

dimensionalities are not necessarily the same in all cases. It would be out of place here to discuss the subtle difference, in certain cases, between the box dimension and the Hausdorff dimension.) The fractal dimensions of other geometrical objects such sets of points, surfaces, volumes, etc. are defined analogously. Fractal dimensions need not be integers—indeed, this is the reason for the very name.

9.3 Khinchin's law of the iterated logarithm

We know that the mean value $\overline{x(t)} = 0$ for all $t > 0$, while the mean square $\overline{x^2(t)} = 2Dt$. (Recall that we have chosen $x(0) = 0$.) These are average values, the averages being taken over all realizations of the driving noise $\zeta(t)$. In any individual realization, however, $x(t)$ can be arbitrarily large for any $t > 0$ (however small), because the instantaneous velocity is unbounded. But there exists a rigorous result, Khinchin's **law of the iterated logarithm**, that tells us what happens almost always. We define the function

$$h(t) = [2Dt \log \log (\tau/t)]^{1/2}, \tag{9.10}$$

where τ is an arbitrary positive constant with the physical dimensions of time, that has been introduced merely for dimensional reasons. Then, for arbitrarily small positive ϵ it can be shown that, as $t \to 0$ from the positive side,

$$\text{Prob}\,\big\{x(t) < (1+\epsilon)h(t)\big\} = 1, \tag{9.11}$$

while

$$\text{Prob}\,\big\{x(t) < (1-\epsilon)h(t)\big\} = 0. \tag{9.12}$$

Together, these equations imply that the least upper bound on $x(t)$, as $t \to 0$, must essentially be $h(t)$. The precise statement is

$$\text{Prob}\left\{ \limsup_{t\downarrow 0} \frac{x(t)}{h(t)} = 1\right\} = 1. \tag{9.13}$$

The Wiener process enjoys several symmetry properties. If $x(t)$ is such a process (that starts at 0 at $t = 0$), so is $-x(t)$. This helps establish that the greatest lower bound on $x(t)$ as $t \to 0$ is essentially $-h(t)$, in the sense that

$$\text{Prob}\left\{ \liminf_{t\downarrow 0} \frac{x(t)}{h(t)} = -1\right\} = 1. \tag{9.14}$$

Figure 9.2 helps us understand the import of the law of the iterated logarithm. The Brownian particle is located infinitely often in the shaded region. Hence it also crosses the origin an arbitrarily large number of times. And, as we have already mentioned, the instants of zero crossings form a random 'dust' of fractal dimension $\frac{1}{2}$. We have already stated that the Wiener process incessantly renews itself and 'starts all over again'. The law of the iterated logarithm just stated is therefore applicable at any point, with an appropriate shift of the origin. That is, if the origin in Fig. 9.2 corresponds to any instant of time t_0, and to any value $x(t_0) = x_0$ of the random process, then the figure applies to the process $x(t) - x_0$

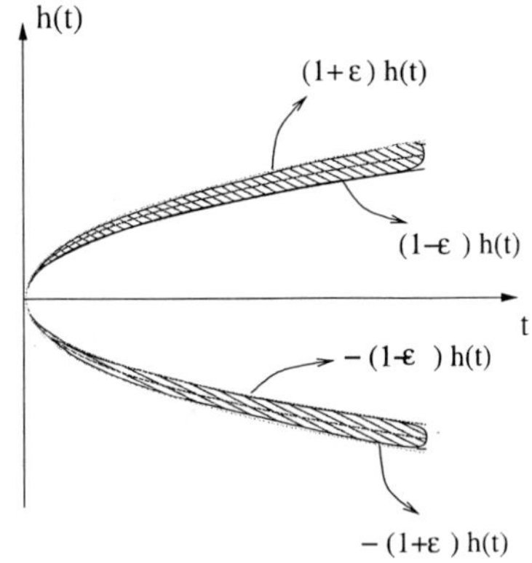

Figure 9.2: The function $h(t)$ that appears in the short-time law of the iterated logarithm for Brownian motion. For sufficiently small values of t, the Brownian particle (which is taken to start at $x = 0$ at $t = 0$) almost never goes beyond the outermost curves above and below the t-axis. It also crosses the t-axis an arbitrarily large number of times. With an appropriate shift of the origin in t and x, this behavior applies at any instant of time, and at any threshold value of x, as described in the text.

as $t \to t_0$ from above. The corresponding zero set is then the set of crossings of the value x_0. Thus, the law of the iterated logarithm essentially specifies the local behavior of Brownian motion at all times.

Mathematical Brownian motion has an interesting **time inversion property** that helps us relate its short-time behavior to its long-time behavior.

- If $x(t)$ $(t > 0)$ is a Brownian motion (that starts at 0 at $t = 0$), it can be shown that so is $(t/\tau)\, x(\tau^2/t)$, and that these two processes have exactly the same distribution.

This duality between processes as $t \to 0$ and $t \to \infty$ enables us to deduce that the long-time behavior of Brownian motion is also given by an iterated-logarithm law. Define the function

$$H(t) = [2Dt \log \log (t/\tau)]^{1/2}. \tag{9.15}$$

Then, as $t \to \infty$,

$$\text{Prob}\left\{ \limsup_{t\to\infty} \frac{x(t)}{H(t)} = 1 \right\} = 1 \tag{9.16}$$

and

$$\text{Prob}\left\{ \liminf_{t\to\infty} \frac{x(t)}{H(t)} = -1 \right\} = 1\,. \tag{9.17}$$

These results tell us a good deal about the behavior of the trajectory of a particle executing Brownian motion, except at points that form a set of measure zero. It says, very roughly, that the distance the particle travels on either side of the origin is typically of the order of $t^{1/2}$, but the foregoing are very much more precise statements than this rough version. In particular, the fine 'logarithmic correction' to the general $t^{1/2}$ power law is spelt out precisely.

The law of the iterated logarithm and the properties of the zero set of Brownian motion (in particular, the fact that it is dense and uncountable) give us a glimpse into the strange properties of this motion. A third aspect of its peculiarities concerns an interesting 'persistence' property. Given a Brownian motion that starts at the origin ($x = 0$) at $t = 0$, and a specified time interval t, we may ask for the probability distribution of the total time t_+ that the particle spends in the region $x > 0$, which may be termed the **occupation time distribution**. (Equivalently, we may consider the total time t_- that it spends in the region $x < 0$.) The PDF of the random variable t_+ (whose sample space is obviously the interval $[0, t]$) can be shown to be given by

$$f(t_+ \,|\, t) = \frac{1}{\pi\sqrt{t_+(t-t_+)}}\,, \qquad 0 \le t_+ \le t. \tag{9.18}$$

Interestingly enough, this PDF, which is symmetric about the mid-point $\frac{1}{2}t$, has a minimum at that point. Hence t_+ is least likely to have a value in an infinitesimal neighborhood of $\frac{1}{2}t$, although its average value is given by $\langle t_+\rangle = \frac{1}{2}t$. The PDF diverges at the end-points 0 and t. This implies that the particle is overwhelmingly likely to spend most of the time (for any given value of the total time t) either in the region $x < 0$ or in the region $x > 0$, in spite of the properties of the zero set of the motion. The cumulative distribution function of t_+ is given by

$$F(t_+ \,|\, t) = \int_0^{t_+} dt'_+\, f(t'_+ \,|\, t) = \frac{2}{\pi}\,\sin^{-1}\sqrt{\frac{t_+}{t}}\,. \tag{9.19}$$

This is the famous **Lévy arcsine law** for the total probability that the fraction of the time spent by the particle in the region $x > 0$ is less than any given value t_+/t, for a specified value of t.

9.4 Brownian trails in d dimensions: recurrence properties

Brownian motion on a line has the following properties that will be established in Ch. 10. Given sufficient time, a Brownian particle starting at the origin is sure to visit every point on the x-axis eventually: that is, the process $x(t)$ will take on any real value with probability 1. The random process is then said to be **recurrent**. Owing to the fact that the Wiener process perpetually starts afresh (as we have already emphasized), all points of the entire x-axis are repeatedly visited by the particle—an infinite number of times, as $t \to \infty$. However, the mean time between successive visits diverges, i. e., is infinite. The process is then said to be **null-recurrent**.

It is of considerable interest (and importance, given the relevance of the Wiener process in numerous contexts) to examine the properties of Brownian motion in spaces of higher dimensions, $d \geq 2$. We now list a few of these properties in brief. For our present purposes, we restrict ourselves to the usual Euclidean spaces. Of necessity, several technical terms have been used in the short, qualitative description that follows. Explaining each of them at the level of detail that the matter deserves would take us too far afield[6].

For all $d \geq 2$, a Brownian trail is a self-similar random fractal curve whose fractal dimension is 2. A more precise statement is that the Hausdorff dimension of a Brownian trail in a space of dimension $d \geq 2$ is arbitrarily close to 2 from below.

In $d = 2$, i. e., on an infinite plane, a Brownian trail is space-filling. The probability of a return to any starting point is 1. The mean time between successive returns is, however, infinite. Brownian motion is null-recurrent in $d \leq 2$. The number of double points, i e., the number of points visited twice by the particle, is infinite. So is the number of points visited n times for *any* higher value of $n\,(= 3, 4, \ldots)$.

In $d = 3$, almost every Brownian trail again has an infinite number of double points. In fact, any point on a Brownian trail is, with probability 1, the limit point of double points. However, the probability of a return to a given point is

[6]Therefore I have not attempted to do so. Another reason is to encourage the student reader to hunt through the literature for detailed explanations—and, perhaps, to be inspired to enter the arena of research in modern statistical physics and its applications.

less than 1, i. e., the process is **transient** (as opposed to recurrent). And almost every Brownian trail has *no* triple or higher-order point.

In $d \geq 4$, almost every Brownian trail has *no* double points[7]. Heuristically (*very* heuristically!), we may understand this as follows. We may interpret a double point on the trail as the meeting point of two independent Brownian trails. As each trail has a fractal dimensionality of 2, generically there will be no intersection of two such objects in a space of dimensionality ≥ 4. This is essentially the reason why the **self-avoiding random walk** and the ordinary random walk turn out to have similar properties in $d \geq 4$. It is also clear (at least intuitively) that $d = 4$ is a marginal dimensionality in this sense[8], and hence that random walks in $d = 4$ will behave slightly differently from those in $d > 4$.

9.5 The radial distance in d dimensions

In Ch. 7, Sec. 7.4, we considered the diffusion equation in three dimensions. In particular, we obtained the PDF of the radial distance from the origin and analyzed its moments. Let us now consider the stochastic properties of the radial distance from the origin, $r(t)$, for diffusion in a (Euclidean) space of more than one dimension. The random variable $r(t)$ takes values in $[0\,,\,\infty)$. It is convenient to consider the general case of Brownian motion in an arbitrary number of dimensions $d\,(= 2, 3, 4, \ldots)$. The results for any specific d can then be recovered as a special case.

One of our objectives is to find the correct stochastic differential equation obeyed by $r(t) = \big(x_1^2 + \ldots + x_d^2\big)^{1/2}$, where each Cartesian component x_i is an independent Brownian motion (with the same diffusion constant D). That is,

[7]The student who is not yet familiar with the theory of phase transitions and critical phenomena may wonder why we speak of $d > 3$ at all. A very short explanation is as follows. It has been found that a proper understanding of the complicated behavior of systems in the vicinity of phase transitions is only possible if the number of spatial dimensions, d, is regarded as a variable parameter. The behavior of various physical quantities must be studied as *functions* of d as well as other parameters (specifically, the dimensionality of the order parameter that distinguishes the ordered phase from the disordered one). It turns out to be absolutely crucial, in this regard, to consider values of d greater than 3.

[8]In the language of critical phenomena, $d = 4$ is the **upper critical dimension** for the random walk problem. One would therefore expect 'logarithmic corrections to scaling' to occur precisely at $d = 4$. In the same vein, $d = 3$ is the upper critical dimension for the intersection of three independent random walks. For $d \geq 3$, the probability of such an intersection is zero. For $d < 3$, nontrivial intersection properties exist.

each $x_i(t)$ is a zero-mean Gaussian Markov process, and

$$\overline{x_i(t)\, x_j(t')} = 2D\,\delta_{ij}\, \min\,(t\,,\,t'), \quad 1 \leq i,j \leq d. \tag{9.20}$$

It will turn out that $r(t)$ is itself a diffusion process, satisfying a stochastic differential equation in which both a drift term and a diffusion term are present. $r(t)$ is also known as the **Bessel process**.

The rigorous way of deriving the SDE we seek, as well as the other SDEs that we shall consider in the exercises accompanying this chapter, is by applying what is known as the Itô calculus to various functions of the Wiener process. In this book, however, we do not discuss this mathematical topic[9]. Instead, we shall use the following strategy to attain our goal. We know that the diffusion equation (9.1) and the stochastic differential equation (9.5) are related by the SDE↔FPE correspondence. Given the PDF of $x(t)$, we can determine the PDF of any function of x by a straightforward transformation of variables. Similarly, we can determine the master equation satisfied by this PDF by changing variables appropriately in the diffusion equation. We then use the SDE↔FPE correspondence backwards to write down the correct SDE satisfied by the new random variable, without having to use the Itô rules explicitly.

We start with the fundamental Gaussian solution of the diffusion equation in d dimensions, and work out the corresponding PDF of the radial distance $r(t)$. The fundamental solution to the diffusion equation (7.15) in d-dimensional space, with an initial condition

$$\mathsf{p}(\mathbf{r},0) = \delta^{(d)}(\mathbf{r}) \tag{9.21}$$

and natural boundary conditions, is a straightforward generalization of the result in the three-dimensional case, Eq. (7.20). It is given by

$$\mathsf{p}(\mathbf{r},t) = \frac{1}{(4\pi Dt)^{d/2}}\, \exp\left(-\frac{r^2}{4Dt}\right). \tag{9.22}$$

As the Gaussian in r factors into a product of Gaussians in each of the Cartesian coordinates $x_i\,,\ 1 \leq i \leq d$, it is easy to see that the mean squared displacement is

$$\overline{r^2(t)} = \sum_{i=1}^{d} \overline{x_i^2(t)} = 2dDt. \tag{9.23}$$

[9]However, a word on what this calculus is all about is not out of place here. We have already stated that the sample paths of a Wiener process are non-differentiable almost everywhere. This fact is taken into account in the Itô calculus by an amendment of the rules of differentiation when applied to the Wiener process and its functions and functionals. See the remarks in Sec. 9.9, at the end of this chapter.

- The fact that $\overline{r^2(t)}$ remains linearly proportional to t in *any* number of dimensions is of great significance in many other contexts involving the Green function of the diffusion operator.

Let us again write the PDF $\mathsf{p}(\mathbf{r},t)$ as $\mathsf{p}(r,t)$ to indicate its dependence on r rather than $\mathbf{r}$. As only the radial part of the ∇^2 operator acts on it, we find that $\mathsf{p}(\mathbf{r},t)$ satisfies the differential equation

$$\frac{\partial \mathsf{p}(r,t)}{\partial t} = \frac{D}{r^{d-1}}\frac{\partial}{\partial r}\left(r^{d-1}\frac{\partial}{\partial r}\mathsf{p}(r,t)\right) = D\left(\frac{\partial^2}{\partial r^2} + \frac{(d-1)}{r}\frac{\partial}{\partial r}\right)\mathsf{p}(r,t). \tag{9.24}$$

This is the d-dimensional counterpart of the three-dimensional case, Eq. (7.21).[10] In order to identify the corresponding PDF of the radial distance, $\mathsf{p}_{\rm rad}(r,t)$, we must multiply $\mathsf{p}(r,t)$ by the volume element in d-dimensional space written out in 'spherical' polar coordinates[11], and integrate over all the angular variables. The result is

$$\mathsf{p}_{\rm rad}(r,t) = \frac{2\pi^{d/2}}{\Gamma(\frac{1}{2}d)}\, r^{d-1}\, \mathsf{p}(r,t), \tag{9.25}$$

where the constant factor on the right-hand side is the generalization of the total solid angle 4π that obtains in 3-dimensional space. It is just the $(d-1)$-dimensional surface of a sphere of unit radius embedded in d-dimensional Euclidean space[12]. Thus, it is respectively equal to 2π, 4π, $2\pi^2$, ... for $d = 2, 3, 4, \ldots$. Combining Eqs. (9.25) and (9.22), we get

$$\mathsf{p}_{\rm rad}(r,t) = \frac{r^{d-1}}{2^{d-1}\Gamma(\frac{1}{2}d)(Dt)^{d/2}} \exp\left(-\frac{r^2}{4Dt}\right). \tag{9.26}$$

We have already written down the PDF of the radial distance in the three-dimensional case, in Eq. (7.24). In two dimensions, Eq. (9.26) reduces to

$$\mathsf{p}_{\rm rad}(r,t) = \frac{r}{2Dt} \exp\left(-\frac{r^2}{4Dt}\right). \tag{9.27}$$

---

[10]When $d = 3$, a further change of dependent variables eliminates the term proportional to $\partial \mathsf{p}/\partial r$ without introducing any new terms, as we have seen in Eq. (7.22). This is possible only when $d = 3$, as mentioned in Sec. 7.4.

[11]That is, plane polar coordinates for $d = 2$, the usual spherical polar coordinates for $d = 3$, and hyperspherical polar coordinates for $d > 3$.

[12]Incidentally, setting $d = 1$ in Eq. (9.25) yields $\mathsf{p}_{\rm rad}(r,t) = 2\mathsf{p}(x,t)$ in this case. You should understand the origin of the factor 2. On the one hand, it is the 'surface' of a line segment (the two end points). On the other, it is the normalization factor required to ensure that $\int_0^\infty dr\, \mathsf{p}_{\rm rad}(r,t)$ is normalized to unity, since r is just $|x|$ in one dimension.

The probability distribution corresponding to a PDF of the form in Eq. (9.27) is called a **Rayleigh distribution**.

Equation (9.25) shows that $\mathsf{p}_{\text{rad}}(r,t)$ is proportional to $r^{d-1}\,\mathsf{p}(r,t)$. We may use this fact in the first of Eqs. (9.24) to deduce the master equation satisfied by $\mathsf{p}_{\text{rad}}(r,t)$ itself. A bit of algebra leads to the partial differential equation

$$\frac{\partial}{\partial t}\,\mathsf{p}_{\text{rad}}(r,t) = -\frac{\partial}{\partial r}\left[\frac{D(d-1)}{r}\mathsf{p}_{\text{rad}}(r,t)\right] + D\,\frac{\partial^2}{\partial r^2}\,\mathsf{p}_{\text{rad}}(r,t). \tag{9.28}$$

But Eq. (9.28) is in the form of a Fokker-Planck equation, Eq. (6.7), in which both a drift term as well as a diffusion term are present. We have already stated that such a Fokker-Planck equation follows from the stochastic differential equation (6.6) for a general diffusion process. For ready reference (both here and in the sequel), let us recall the SDE↔FPE correspondence between the stochastic differential equation satisfied by a diffusion process $\xi(t)$ that is driven by the Gaussian white noise $\zeta(t)$, and the Fokker-Planck equation for its conditional PDF $p(\xi,t)$. We have already pointed out that the mathematically appropriate way to write the stochastic differential equation (6.6) is in terms of the Wiener process $w(t)$ (or rather, its increment $dw(t)$). Thus, the SDE for a general diffusion process is

$$d\xi(t) = f(\xi,t) + g(\xi,t)\,dw(t). \tag{9.29}$$

This is known as the **Itô equation** . According to the SDE↔FPE correspondence, Eq. (9.29) implies that the PDF of $\xi(t)$ satisfies the Fokker-Planck equation

$$\frac{\partial p}{\partial t} = -\frac{\partial}{\partial \xi}\,(fp) + \frac{1}{2}\frac{\partial^2}{\partial \xi^2}\,(g^2\,p)\,, \tag{9.30}$$

and *vice versa.* Working backwards from the master equation (9.28), we can read off the quantities f and g in the case at hand. Hence, we conclude that the stochastic differential equation satisfied by the radial distance $r(t)$ is given by

$$dr(t) = \frac{D(d-1)}{r}\,dt + \sqrt{2D}\,dw(t). \tag{9.31}$$

It is interesting to note that a drift velocity proportional to r^{-1} automatically emerges in the above, even though each Cartesian component $x_i(t)$ undergoes unbiased Brownian motion such that its mean value is zero at all times. When the Brownian particle is at the origin, it is easy to see that, no matter in which direction the particle moves out of the origin, $r(t)$ always increases, in every realization of the process. This acts as a sort of drift velocity pushing the particle away from the origin. What is noteworthy is that a similar argument can be given at all subsequent times, for *any* instantaneous position of the particle: the probability that an incremental change in r *increases* the radial distance is greater than the probability that it *decreases* this distance[13].

When the starting point of the Brownian motion is not the origin, but any other point, the distribution of the radial distance $r(t)$ can be obtained using the backward Kolmogorov equation. The result is, naturally, a more complicated formula than the one given in Eq. (9.26). It involves modified Bessel functions multiplied by a Gaussian factor.

9.6 Sample paths of diffusion processes

We have seen that the sample paths of Brownian motion are highly irregular, far-from-smooth curves. A similar remark applies, in fact, to *all* diffusion processes, i. e., all those continuous Markov processes described by the general stochastic equation (6.6) or (9.29). All these processes are driven by the Gaussian white noise $\zeta(t)$. It is ultimately the highly irregular nature of this noise that leads to the extremely jagged, continuous-yet-non-differentiable nature of Brownian paths. Not surprisingly, this property is retained by *all* diffusion processes:

- The sample paths of any diffusion process $\xi(t)$, by which we mean a (Markov) process whose increment is given by the Itô stochastic differential equation $d\xi(t) = f(\xi, t) + g(\xi, t)\, dw(t)$, are non-differentiable almost everywhere.

- In particular, this applies to the sample paths of the Ornstein-Uhlenbeck velocity process as well. The SDE in this case is, of course, the Langevin equation.

As the OU process is of primary interest to us in this book, the next question that arises is the following:

- Is there an analog of the law of the iterated logarithm for the OU process?

---

[13]In order to visualize this argument, consider the two-dimensional case. Suppose the instantaneous position of the particle is on the x-axis, at $(r\,,\,0)$, say. Draw a circle of radius r centered at the origin, and the vertical tangent to the circle at $(r\,,\,0)$. In the next instant, the x-coordinate of the particle is equally likely to increase or to decrease—that is, the particle is equally likely to be to the left or to the right of the tangent. Points to the right of the tangent certainly correspond to an increase of the radial distance. But even points to the left of the tangent, that lie outside the arc of the circle, contribute to the probability measure of an increase of the radial distance.

The answer is 'yes'. At *short* times, the behavior of this process is similar to that of the Wiener process or Brownian motion. That is, the function $[t \ln \ln (1/t)]^{1/2}$ (the counterpart of the function $h(t)$ defined in Eq. (9.10)) essentially controls the behavior of the process. This is to be expected on physical grounds because, at very short times, the drift term $-\gamma v$ in the Langevin equation 'has not had time' to make its presence felt. At very long times, however, it is no longer the function $[t \ln \ln t]^{1/2}$ (the counterpart of the function $H(t)$ defined in Eq. (9.15)) that decides how far the process extends. Instead, it is replaced by the function $(\ln t)^{1/2}$ (see Sec. 9.7 below). Once again, we know that the OU velocity process tends to a nontrivial stationary distribution, the Maxwellian $p^{\text{eq}}(v)$, and so there is no long-range diffusion in velocity space—hence the counterpart of $H(t)$ cannot involve a leading power-law behavior proportional to $t^{1/2}$. There is no subdiffusive behavior, either, and so a power-law behavior in t, with an exponent less than $\frac{1}{2}$, is also not possible. As it happens, there is only a weak logarithmic growth of the span of the process with time.

9.7 Relationship between the OU and Wiener processes

The statements just made can be established rigorously by exploiting a certain transformation of the independent variable t that enables the OU process to be mapped, in a mathematical sense, to a Wiener process whose argument is a function of t. A proper discussion of this transformation is outside the scope of this book. However, for the sake of completeness, let us quote the result. We know that the velocity $v(t)$ of the tagged particle satisfies the Langevin equation. Its conditional mean and autocorrelation function are given by

$$\overline{v(t)} = v_0 \, e^{-\gamma t} \quad \text{and} \quad \overline{\delta v(t) \, \delta v(t')} = \frac{k_B T}{m} \, e^{-\gamma |t-t'|}, \tag{9.32}$$

where $\delta v(t) = v(t) - \overline{v(t)}$. The conditional PDF of $v(t)$ is given by the PDF of an OU process, Eq. (6.14). Now let $w(t)$ be a Wiener process—that is, a Gaussian Markov process with mean and autocorrelation given by with

$$\overline{w(t)} = 0 \quad \text{and} \quad \overline{w(t) \, w(t')} = \min(t \, , \, t') \, . \tag{9.33}$$

Then

$$\delta v(t) \stackrel{d}{=} \left(\frac{k_B T}{m}\right)^{1/2} e^{-\gamma t} \, w\left(e^{2\gamma t}\right), \tag{9.34}$$

where $\stackrel{d}{=}$ indicates that the processes on the two sides of it are *equal in distribution* (i. e., they have identical sets of probability distributions). Loosely speaking, we may say, "In logarithmic time, the velocity process given by the Langevin equation is essentially a (mathematical) Brownian motion". It is a simple matter to check that Eqs. (9.32) are recovered on substituting Eq. (9.34) in Eqs. (9.33).[14] Similarly, the combination $(t \ln \ln t)^{1/2}$ representing the function $H(t)$ in the case of $w(t)$ becomes the combination $e^{\gamma t} (\ln 2\gamma t)^{1/2}$ in the case of $w(e^{2\gamma t})$, or $(\ln 2\gamma t)^{1/2}$ in the case of $e^{-\gamma t}\, w(e^{2\gamma t})$. Hence the large-$t$ behavior of $v(t)$ in the counterpart of the law of the iterated logarithm is specified by just $(\ln t)^{1/2}$, as stated in the preceding section.

We shall encounter another instance of an OU process in Ch. 13. It will turn out that, in the long-time regime, the *position* of a harmonically bound particle is also an OU process.

Going just a bit further, here is another reason why the Wiener process is of such significance:

- Essentially *all* Gaussian Markov processes can be related to the Wiener process along the lines just described. That is, any such process $\xi(t)$ can be written as

$$\xi(t) \stackrel{d}{=} S(t)\, w\big(T(t)\big), \tag{9.35}$$

  where the function $S(t)$ may be regarded as a change of scale, and the function $T(t)$ as a re-parametrization of the time.

We now begin to appreciate the special nature of the OU process, and how Doob's Theorem, stated in Ch. 6, Sec. 6.1, comes about. It is only when $S(t)$ and $T(t)$ are appropriate *exponential* functions of t, that there is a possibility of converting the nonstationary expression $\min(t\,,\,t\,')$ to a function of the difference $(t - t\,')$. This is how the OU process turns out to be the *only* Gaussian Markov process that is also *stationary*.

9.8 Exercises

9.8.1 $r^2(t)$ in d-dimensional Brownian motion

It can be shown that

- in the case of Brownian motion in d dimensions, the square of the radial distance, $R(t) \equiv r^2(t) = x_1^2(t) + \ldots + x_d^2(t)$, is also a diffusion process.

[14]Observe, in particular, how the autocorrelation of $v(t)$ becomes a function of $(t - t\,')$, even though that of $w(t)$ is not a function of this difference alone. See the remarks following Eq. (9.35) below.

We would like to verify this statement by establishing the stochastic differential equation satisfied by this process. Naively, one would think that all one needs to do is to multiply both sides of the stochastic differential equation (9.31) for $r(t)$ by r, and use the fact that $dR = 2r\,dr$. The result is

$$dR = 2D(d-1)dt + 2\sqrt{2DR}\,dw(t).$$

However, *this is incorrect.*

This example serves to illustrate, once again, the need to apply the Itô calculus to navigate without error between stochastic differential equations. As before, we use an alternative route to arrive at the correct answer: we shall use the SDE↔FPE correspondence summarized in Eqs. (9.29) and (9.30), as this follows rigorously from the Itô calculus.

We know the Fokker-Planck equation satisfied by $\mathsf{p}_{\text{rad}}(r,t)$, namely, Eq. (9.28). Let $\mathsf{p}_{\text{sqr}}(R,t)$ be the (conditional) PDF of $R(t)$. It follows by definition that

$$\mathsf{p}_{\text{sqr}}(R,t)\,dR \equiv \mathsf{p}_{\text{rad}}(r,t)\,dr, \tag{9.36}$$

so that

$$\mathsf{p}_{\text{rad}}(r,t) = (dR/dr)\,\mathsf{p}_{\text{sqr}}(R,t) = 2\sqrt{R}\,\mathsf{p}_{\text{sqr}}(R,t). \tag{9.37}$$

Use this relation in the Fokker-Planck equation (9.28) to show that the Fokker-Planck equation satisfied by $\mathsf{p}_{\text{sqr}}(R,t)$ is

$$\frac{\partial}{\partial t}\,\mathsf{p}_{\text{sqr}}(R,t) = -2Dd\,\frac{\partial}{\partial R}\,\mathsf{p}_{\text{sqr}}(R,t) + 4D\,\frac{\partial^2}{\partial R^2}\left[R\,\mathsf{p}_{\text{sqr}}(R,t)\right]. \tag{9.38}$$

It follows from the SDE↔FPE correspondence that the stochastic differential equation satisfied by the process $R(t)$ is

$$dR(t) = 2dD\,dt + 2\sqrt{2DR(t)}\,dw(t)\,. \tag{9.39}$$

Observe that the drift is a constant. However, the noise is now *multiplicative*, because the factor multiplying $dw(t)$ depends on the state variable $R(t)$.

Setting $d = 1$ in Eq. (9.39), we get the stochastic differential equation

$$dR(t) = 2D\,dt + 2\sqrt{2DR(t)}\;dw(t) \tag{9.40}$$

for the square of a one-dimensional Brownian motion $x(t)$. (Convince yourself that we *can* set $d = 1$ in Eq. (9.39) to get the correct SDE for $x^2(t)$.) Hence

- the square of a one-dimensional Brownian motion is also a diffusion process, with a constant drift and a multiplicative noise.

9.8.2 The nth power of Brownian motion

In exactly the same manner as in the preceding problem, we can show that the nth power of Brownian motion is also a diffusion process. Set $x_n(t) = \big(x(t)\big)^n$ where n is any positive integer. Show that the stochastic differential equation satisfied by the process $x_n(t)$ is

$$dx_n(t) = Dn(n-1)\, x_n^{(n-2)/n}\, dt + n\sqrt{2D}\, x_n^{(n-1)/n}\, dw(t). \tag{9.41}$$

9.8.3 Geometric Brownian motion

Consider the random process $\xi(t) = \exp\big(x(t)/L\big)$, where $x(t)$ is a Brownian motion, and L is an arbitrary (and trivial) length scale introduced to make the exponent dimensionless. Then:

- The exponential of a Brownian motion is also a diffusion process. It is a special case of a process known as **geometric Brownian motion**[15].

We want to find the stochastic differential equation satisfied by $\xi(t)$. Now, ξ is a monotonically increasing function of x, with a range $(0\,,\infty)$ corresponding to the range $(-\infty\,,\infty)$ of x. Once again, suppose we use $x(t) = L\ln\xi(t)$ to write $dx = L\,d\xi/\xi$. We might then be tempted to start with the stochastic differential equation $dx(t) = \sqrt{2D}\,dw(t)$, and simply transform it to $d\xi = (\sqrt{2D}/L)\,\xi\,dw(t)$ for $\xi(t)$. However, this is not the correct Itô equation for $\xi(t)$.

As in the previous example, we work backwards from the diffusion equation for the PDF $\mathsf{p}(x,t)$ of $x(t)$. If $\mathsf{p}_{\text{geom}}(\xi,t)$ is the (conditional) PDF of ξ, then

$$\mathsf{p}(x,t)\,dx \equiv \mathsf{p}_{\text{geom}}(\xi,t)\,d\xi \Longrightarrow \mathsf{p}(x,t) = \frac{\xi}{L}\,\mathsf{p}_{\text{geom}}(\xi,t). \tag{9.42}$$

Show that the (one-dimensional) diffusion equation for $\mathsf{p}(x,t)$ leads to the Fokker-Planck equation

$$\frac{\partial}{\partial t}\,\mathsf{p}_{\text{geom}}(\xi,t) = -\frac{D}{L^2}\frac{\partial}{\partial\xi}\,[\xi\,\mathsf{p}_{\text{geom}}(\xi,t)] + \frac{D}{L^2}\frac{\partial^2}{\partial\xi^2}\,[\xi^2\,\mathsf{p}_{\text{geom}}(\xi,t)] \tag{9.43}$$

for $\mathsf{p}_{\text{geom}}(\xi,t)$. Hence read off the stochastic differential equation satisfied by geometric Brownian motion,

$$d\xi(t) = \frac{D}{L^2}\,\xi(t)\,dt + \frac{\sqrt{2D}}{L}\,\xi(t)\,dw(t). \tag{9.44}$$

[15] It is also called **exponential Brownian motion**.

Once again, the noise is multiplicative. The noteworthy feature is that both the drift and diffusion coefficients are linear functions of ξ in this case. Geometric Brownian motion is the generalization of Eq. (9.44) to

$$d\xi(t) = \alpha\, \xi(t)\, dt + \beta\, \xi(t)\, dw(t), \tag{9.45}$$

where α and β are constants. This process is related to the exponential of a Wiener process according to

$$\xi\big(w(t), t\big) = \exp\,\big[\beta\, w(t) + \big(\alpha - \tfrac{1}{2}\beta^2\big)\, t\big]\,. \tag{9.46}$$

The notation on the left-hand side is meant to show that ξ is not only a function of $w(t)$, but also has an explicit dependence on t. Geometric Brownian motion is of great importance in certain applications—notably, in **econophysics**: the **Black-Scholes model** is based on this process.

9.9 Brief remarks on the Itô calculus

Before this chapter is concluded, some brief remarks are perhaps in order, so as to de-mystify the cryptic references made to the Itô calculus. You must understand *why* the SDEs we have derived for various functions of Brownian motion differ from what one would obtain by naive differentiation with respect to t—in other words, exactly why the Itô calculus (which we have skirted around) differs from standard calculus. The answer lies in the Hölder continuity of Brownian motion, as given by Eq. (9.9), and the fact that the sample paths of this process are not differentiable. Owing to the fact that $|w(t+\epsilon) - w(t)| \le C\, |\epsilon|^{\beta}$ where $\beta < \frac{1}{2}$, it is evident that $|w(t+\epsilon) - w(t)|/\epsilon \to \infty$ as $\epsilon \to 0$. Hence we cannot write $dw(t)$ as $[dw(t)/dt]\, dt$, as we normally would in the case of a differentiable function of t. But a clever 'amendment' of the rules of calculus enables us to handle differentials such as $dw(t)$. Equation (9.8) suggests strongly that, in some sense, the *square* $(dw(t))^2$ of the infinitesimal $dw(t)$ acts like dt itself. In fact, this is a precise statement in the Itô calculus, where one writes $(dw(t))^2 = dt$. Thus the infinitesimal $dw(t)$ may be regarded as the 'square root' of the infinitesimal dt. Objects such as $(dt)^2$ and $(dw)(dt)$ are therefore higher order infinitesimals (and hence negligible to first order in dt), while $(dw)^2$ is not. It is this idea that is made precise and built upon in the Itô calculus. Once this mathematical technique is in place, we have a rigorous way of handling functions of a Wiener process $w(t)$, as well as its *functionals* that are nonlocal in time (the so-called **Brownian functionals**[16]).

[16] We mention here that the **Feynman-Kac formula** is of special importance in this regard, from the point of view of applications in nonequilibrium statistical mechanics. I apologize for the omission of this important topic from this book. A proper treatment of this aspect would involve path integrals (functional integration), and would take us too far afield.

The basic rules of the Itô calculus are deduced from the defining properties of the Wiener process. Let $[t_1\,,\,t_2]$ and $[t_3\,,\,t_4]$ be any two intervals of time. It is easily shown that the mean value of the product

$$\big(w(t_2) - w(t_1)\big)\,\big(w(t_4) - w(t_3)\big)$$

is simply the length of the interval representing the intersection (if any) of the two intervals. It follows that $\overline{dw(t)\,dw(t)} = dt$. Next, consider the integral $\int_0^t w(t')\,dw(t')$. Breaking up the interval $[0, t]$ into successive infinitesimal subintervals marked by the instants $t_0 = 0\,,\, t_1\,,\, t_2\,,\,\ldots\,,\, t_n\,,\, t_{n+1} = t$, the integral is interpreted as

$$\int_0^t w(t')\,dw(t') = \lim_{n\to\infty} \sum_{i=0}^{n} [w(t_i)]\,[w(t_{i+1}) - w(t_i)]. \tag{9.47}$$

Since the Wiener process continually renews itself, each of the two factors in square brackets in the summand is statistically independent of the other. Hence the mean value of the integral is actually zero. *This would not be so if the time argument of the first factor in the sum above is* t_i *rather than* t_{i+1} *, or even any combination of* t_i *and* t_{i+1} *such as* $\frac{1}{2}(t_i + t_{i+1})$.

The Itô calculus also leads to the following basic formula: If $F\big(w(t)\big)$ is a function of the Wiener process,

$$\int_{t_1}^{t_2} F'\big(w(t)\big)\,dw(t) = F\big(w(t_2)\big) - F\big(w(t_1)\big) - \tfrac{1}{2}\int_{t_1}^{t_2} F''\big(w(t)\big)\,dt, \tag{9.48}$$

where $F'(w) = dF/dw$ and $F''(w) = d^2F/dw^2$. The extra term on the right-hand side of Eq. (9.48) implies that the differential of a function of a Wiener process is given by

$$\begin{aligned} dF(w(t)) &= F'\big(w(t)\big)\,dw(t) + \tfrac{1}{2}F''\big(w(t)\big)\,dw(t)\,dw(t) \\ &= F'\big(w(t)\big)\,dw(t) + \tfrac{1}{2}F''\big(w(t)\big)\,dt. \end{aligned} \tag{9.49}$$

In the case of an explicitly time-dependent function $F\big(w(t), t\big)$, we have

$$dF\big(w(t), t\big) = F'\big(w(t), t\big)\,dw(t) + \left[\frac{1}{2}F''\big(w(t), t\big) + \frac{\partial F\big(w(t), t\big)}{\partial t}\right] dt. \tag{9.50}$$

As an example, apply this formula to verify that the function in Eq. (9.46) is indeed the solution to the stochastic differential equation (9.45).

Equations (9.48) and (9.50) are, respectively, the integral and differential forms of the fundamental rule of the Itô calculus.

The specific prescription for integrals over a Wiener process implied by the sum in Eq. (9.47) also underlines the Itô interpretation of the SDE (9.29), $d\xi(t) = f(\xi, t) + g(\xi, t)\, dw(t)$. The time argument of the random process ξ in the SDE is t in both $f(\xi, t)$ and $g(\xi, t)$, while $dw(t)$ stands for the *forward* increment $w(t + dt) - w(t)$. As we have stressed more than once, the Fokker-Planck equation that is consistent with this SDE, Eq. (9.30), is based on this interpretation. It is this form of the SDE↔FPE correspondence that has been used throughout this book.

We have already stated that Brownian motion has the following property. For any $t > t'$, the *conditional* average of $x(t)$, for a *given* value of $x(t')$, is $x(t')$ itself. A stochastic process that satisfies this condition is called a **martingale**: Brownian motion is the prime continuous-time example of a martingale. There exists an extensively developed theory of martingales. This theory provides a powerful technique—perhaps the most powerful one— to establish rigorously a large number of results connected with mathematical Brownian motion. We make these remarks here because of the importance of these ideas in many applications of stochastic processes, and the increasingly prominent role played by them in nonequilibrium statistical physics.

Chapter 10

First-passage time

We consider the first-passage time problem for Brownian motion on a line. We derive an expression for the PDF of the time when the diffusing particle, starting from some arbitrary initial point x_0 at $t = 0$, reaches any other given point x for the first time. Two different methods are used for this purpose. The first relies on a renewal equation for Brownian motion. The second determines the PDF required from a certain survival probability. The PDF is shown to correspond to a stable distribution with characteristic exponent $\frac{1}{2}$. The mean and higher moments of this distribution are infinite. We discuss the nature of this divergence. The distribution of the time to reach a given distance from the starting point for the first time is also determined.

10.1 First-passage time distribution from a renewal equation

In this chapter, we turn to another important aspect of Brownian motion—the **first-passage time distribution**. In Ch. 9, Sec. 9.4, it was stated that a Brownian particle that starts at the origin and moves on the x-axis is sure to visit every point on the x-axis with probability 1. The instant of time when the diffusing particle, starting from some arbitrary initial point x_0 at $t = 0$, reaches the point $x\,(\neq x_0)$ for the *first* time is called the **first-passage time**[1] to go from x_0 to x. This interval of time is clearly a random variable. A knowledge of its PDF (often called the first-passage time distribution) is of importance in numerous applications of stochastic processes. In particular, the mean and variance of

[1]It is also called the **hitting time** or the **exit time**, depending on the context.

the first-passage time are required in many situations—for instance, in chemical reaction theory, an important area of application of these aspects of stochastic processes.

In what follows, we shall derive the PDF of the first-passage time for one-dimensional Brownian motion. There is more than one way of finding this PDF. We shall describe two of these methods, as they not only help us understand the nature of the Wiener process in greater depth, but are also instructive in terms of possible extensions to other random processes.

Let $\mathsf{q}(t\,,\,x\,|\,0\,,\,x_0)$ denote the probability density function in t for the particle to reach the point x for the *first* time, given that it started from x_0 at $t = 0$. The first method of finding $\mathsf{q}(t\,,\,x\,|\,0\,,\,x_0)$ is based on a special chain condition that is valid for Markov processes[2]. Suppose $x > x_0$, for definiteness. Let x_1 be any point to the right of x on the x-axis. Clearly, to reach the point x_1 at some time t_1, the diffusing particle (which starts at x_0) must necessarily pass through the point x *for the first time* at some instant of time $t \le t_1$. Summing over all possible intermediate events of this kind, we have the relation

$$\mathsf{p}(x_1\,,\,t_1\,|\,x_0\,,\,0) = \int_0^{t_1} dt\,\mathsf{p}(x_1\,,\,t_1\,|\,x\,,\,t)\,\mathsf{q}(t\,,\,x\,|\,0\,,\,x_0). \qquad (10.1)$$

Equation (10.1) is an example of a **renewal equation**. It must be pondered over. It is important to realize that *first* passage to any given point x at an instant t_1, and *first* passage to this point at any *other* instant t_2, are mutually exclusive events. This is why it is permissible, as well as necessary, to sum over these possibilities, i. e., to integrate over all physically admissible values of the intermediate instant of time, t.

We know that the conditional PDF $\mathsf{p}(x\,,\,t\,|\,x_0\,,\,t_0)$ is, in fact, a function of the time difference $(t - t_0)$. (Its explicit form is given by the Gaussian in Eq. (7.14).) We may therefore write Eq. (10.1) in the more suggestive form

$$\mathsf{p}(x_1\,,\,t_1\,|\,x_0) = \int_0^{t_1} dt\,\mathsf{p}(x_1\,,\,t_1 - t\,|\,x)\,\mathsf{q}(t\,,\,x\,|\,x_0). \qquad (10.2)$$

But the right-hand side of this equation is in the form of a convolution of two functions of time, so that it is natural to work with Laplace transforms. By the convolution theorem, the transform of the right-hand side is just the product of

[2]It is actually valid in somewhat greater generality, but this is not relevant here.

the Laplace transforms of the two functions in the integrand. Using a tilde to denote the Laplace transform with respect to the time, we therefore get

$$\widetilde{\mathsf{q}}(s\,,\,x\,|\,x_0) = \frac{\widetilde{\mathsf{p}}(x_1\,,\,s\,|\,x_0)}{\widetilde{\mathsf{p}}(x_1\,,\,s\,|\,x)}\,. \tag{10.3}$$

Now, the Laplace transform of the fundamental Gaussian solution

$$\mathsf{p}(x\,,\,t\,|\,x_0) = \frac{1}{\sqrt{4\pi Dt}}\,\exp\left(-\frac{(x-x_0)^2}{4Dt}\right) \tag{10.4}$$

is given by[3]

$$\widetilde{\mathsf{p}}(x\,,\,s\,|\,x_0) = \frac{1}{\sqrt{4Ds}}\,e^{-|x-x_0|\,\sqrt{(s/D)}}\,. \tag{10.5}$$

Inserting this expression for $\widetilde{\mathsf{p}}$ in Eq. (10.3), and using the fact that $x_0 < x < x_1$, we get

$$\widetilde{\mathsf{q}}(s\,,\,x\,|\,x_0) = \frac{e^{-(x_1-x_0)\,\sqrt{(s/D)}}}{e^{-(x_1-x)\,\sqrt{(s/D)}}} = e^{-(x-x_0)\,\sqrt{(s/D)}}\,. \tag{10.6}$$

Note how the dependence on the arbitrary point x_1 cancels out neatly[4] in the ratio involved in Eq. (10.6), leaving behind a function of x and x_0 (and of course s), as required. The inverse Laplace transform of Eq. (10.6) (see the second transform pair quoted in the footnote) immediately yields

$$\mathsf{q}(t\,,\,x\,|\,x_0) = \frac{(x-x_0)}{\sqrt{4\pi Dt^3}}\,\exp\left(-\frac{(x-x_0)^2}{4Dt}\right), \qquad x > x_0\,. \tag{10.7}$$

The restriction $x > x_0$ is easily removed. When x is to the left of x_0, the first factor in the numerator becomes $(x_0 - x)$. Hence we have, for all x_0 and x,

$$\mathsf{q}(t\,,\,x\,|\,x_0) = \frac{|x-x_0|}{\sqrt{4\pi Dt^3}}\,\exp\left(-\frac{(x-x_0)^2}{4Dt}\right). \tag{10.8}$$

---

[3]See, e. g., F. Oberhettinger and L. Badii, *Tables of Laplace Transforms*, Springer-Verlag, New York, 1973. The basic Laplace transforms we need for our present purposes are the following. Defining the Laplace transform of a function $f(t)$ $(t \geq 0)$ in the usual manner as $\widetilde{f}(s) = \int_0^\infty dt\, f(t)\, \exp(-st)$, we have :

$$f(t) = t^{-1/2}\,\exp(-\alpha/4t) \Longleftrightarrow \widetilde{f}(s) = (\pi/s)^{1/2}\,\exp(-\sqrt{\alpha s}\,)$$

and

$$f(t) = t^{-3/2}\,\exp(-\alpha/4t) \Longleftrightarrow \widetilde{f}(s) = 2(\pi/\alpha)^{1/2}\,\exp(-\sqrt{\alpha s}\,),$$

for Re $\alpha \geq 0$, Re $s > 0$.

[4]This cancellation is a general feature arising from the Markovian nature of the random process under consideration.

We emphasize once again that this is a PDF in t: the quantity $\mathsf{q}(t\,,\,x\,|\,x_0)\,dt$ is the probability that, starting at x_0 at $t = 0$, the first passage to x occurs in the time interval $(t\,,\,t + dt)$. We will discuss the nature of this PDF shortly.

We have claimed that first passage from x_0 to x is a sure event, i.e., that it occurs with probability one. This is equivalent to asserting that the first-passage time density $\mathsf{q}(t\,,\,x\,|\,x_0)$ is normalized to unity according to[5]

$$\int_0^\infty dt\,\mathsf{q}(t\,,\,x\,|\,x_0) = 1. \tag{10.9}$$

We can verify that this relation is satisfied by carrying out the integration involved (or by referring to a table of definite integrals, perhaps). But there is an easier way! We have merely to observe that the left-hand side of Eq. (10.9) is nothing but the Laplace transform $\widetilde{\mathsf{q}}(s\,,\,x\,|\,x_0)$ at $s = 0$. Setting s equal to zero in Eq. (10.6), we find at once that $\widetilde{\mathsf{q}}(0\,,\,x\,|\,x_0) = 1$. This confirms that

- first passage from any x_0 to any x on the infinite line occurs with probability 1. Hence the corresponding first-passage time is indeed a 'proper' random variable.

10.2 Survival probability and first passage

The result obtained above for the first-passage time density $\mathsf{q}(t\,,\,x\,|\,x_0)$ may also be derived in the following alternative way. Let us again choose $x > x_0$ for definiteness. We now imagine that there is an *absorbing* boundary at the point x. We first calculate the survival probability $S(t\,,\,x\,|\,x_0)$ that the diffusing particle, starting from x_0 at $t = 0$, continues to remain in the region $(-\infty\,,\,x)$ till time t without having reached x, i. e., without having been absorbed at that boundary. For this purpose, we first need the probability density that a particle, starting from x_0 at $t = 0$, is at any arbitrary position x' (where $-\infty < x' < x$) at time t, *in the presence of the absorbing boundary* at x. Let us denote this PDF by $\mathsf{p}(x'\,,\,t\,|\,x_0)\big|_{\text{abs. at } x}$. We can easily write down an expression for it[6] using the method of images. The image of a source point at x' in the mirror at x is located

[5]Once again, this follows because first passage at different times are mutually exclusive events.

[6]The alert reader will note that this has already been done! In Sec. 8.4.2, Eq. (8.17), we have found the PDF of the position for diffusion on the semi-infinite line $(-\infty\,,\,a]$ in the presence of an absorbing barrier at a (the minus sign between the two terms on the right-hand side corresponds to an absorbing barrier). All you have to do is to replace x by x' and a by x in that equation, to obtain $\mathsf{p}(x'\,,\,t\,|\,x_0)\big|_{\text{abs. at } x}$.

at $(2x - x')$. We must incorporate a minus sign in the contribution from the image, because the absorbing boundary condition at x requires the vanishing of $\mathsf{p}(x', t \,|\, x_0)\big|_{\text{abs. at } x}$ at $x' = x$ for all t. Thus

$$\mathsf{p}(x', t \,|\, x_0)\big|_{\text{abs. at } x} = \mathsf{p}(x', t \,|\, x_0) - \mathsf{p}(2x - x', t \,|\, x_0), \tag{10.10}$$

where $\mathsf{p}(x', t \,|\, x_0)$ is of course the familiar fundamental Gaussian solution to the diffusion equation that we have written down several times in the foregoing.

The survival probability we seek is then equal to the total probability, at time t, of finding the particle in the region $(-\infty, x)$. Thus

$$S(t, x \,|\, x_0) = \int_{-\infty}^{x} dx' \, \mathsf{p}(x', t \,|\, x_0)\big|_{\text{abs. at } x} \tag{10.11}$$

A moment's thought shows that

- the first-passage time density for reaching x at time t is nothing but the negative of the rate of change of the survival probability $S(t, x \,|\, x_0)$.

That is,

$$\begin{aligned} \mathsf{q}(t, x \,|\, x_0) &= -\frac{\partial}{\partial t} S(t, x \,|\, x_0) \\ &= -\int_{-\infty}^{x} dx' \Big\{ \frac{\partial}{\partial t} \mathsf{p}(x', t \,|\, x_0) - \frac{\partial}{\partial t} \mathsf{p}(2x - x', t \,|\, x_0) \Big\}, \end{aligned} \tag{10.12}$$

where we have used Eq. (10.10) for $\mathsf{p}(x', t \,|\, x_0)\big|_{\text{abs. at } x}$. Substituting from the diffusion equation for the basic PDF p in the equation above,

$$\mathsf{q}(t, x \,|\, x_0) = -D \int_{-\infty}^{x} dx' \Big\{ \frac{\partial^2}{\partial x'^2} \mathsf{p}(x', t \,|\, x_0) - \frac{\partial^2}{\partial x'^2} \mathsf{p}(2x - x', t \,|\, x_0) \Big\}. \tag{10.13}$$

Carrying out the integration over x',

$$\mathsf{q}(t, x \,|\, x_0) = -D \Big\{ \frac{\partial}{\partial x'} \mathsf{p}(x', t \,|\, x_0)\Big|_{x'=-\infty}^{x} - \frac{\partial}{\partial x'} \mathsf{p}(2x - x', t \,|\, x_0)\Big|_{x'=-\infty}^{x} \Big\}. \tag{10.14}$$

We have now only to substitute the fundamental solution

$$\mathsf{p}(x', t \,|\, x_0) = \frac{1}{\sqrt{4\pi D t}} \exp\left(-\frac{(x' - x_0)^2}{4Dt} \right), \tag{10.15}$$

and the corresponding expression for $\mathsf{p}(2x - x', t \,|\, x_0)$, in the right-hand side of Eq. (10.14). Simplifying, we recover precisely the solution for $\mathsf{q}(t, x \,|\, x_0)$ quoted earlier, in Eq. (10.7). The case $x < x_0$ is handled similarly. The general expression is that given in Eq. (10.8).

The method just discussed is based on the identification of the first passage time density as the magnitude of the rate of decrease of the survival probability $S(t\,,\,x\,|\,x_0)$ in the region $(-\infty\,,\,x)$. This interpretation is obvious, on physical grounds. Yet another, related, interpretation can be given for this quantity. The probability current or flux at any point $x' \in (-\infty\,,\,x)$ is just

$$j(x')\big|_{\text{abs. at } x} = -D\frac{\partial}{\partial x'}\,\mathsf{p}(x'\,,\,t\,|\,x_0)\big|_{\text{abs. at } x}\,. \tag{10.16}$$

It is evident that the expression on the right-hand side of Eq. (10.14) is nothing but the current or flux through the point x, since the flux at $-\infty$ vanishes. Hence

- the PDF of the time of first passage to x is therefore just the probability current at x, evaluated *as if* there was an absorbing boundary at that point.

Having obtained an explicit expression for the first-passage time density in Eq. (10.8), let us consider its form and its consequences. We note that the first-passage time $t(x_0 \to x)$ is a continuous random variable that takes values in $[0\,,\,\infty)$. Referring to Appendix K on stable distributions, we recognize that the PDF in Eq. (10.8) corresponds to a member of the family of stable distributions—namely, a Lévy skew alpha-stable distribution with index or characteristic exponent equal to $\frac{1}{2}$, also known as a Lévy distribution[7]. Figure 10.1 depicts this PDF.

Interestingly, the cumulative probability distribution corresponding to the density function $\mathsf{q}(t\,,\,x\,|\,x_0)$ can also be written in closed form. This quantity represents the probability that the point x is reached for the first time within a given time t, for Brownian motion starting at x_0 at $t = 0$. It is more useful to consider the complement of this quantity, namely, the probability that first passage from x_0 to x has *not* occurred *till* time t. But this is precisely the survival probability[8] $S(t\,,\,x\,|\,x_0)$! A little simplification yields

$$1 - \int_0^t dt'\,\mathsf{q}(t'\,,\,x\,|\,x_0) = S(t\,,\,x\,|\,x_0) = \operatorname{erf}\left(\frac{|x - x_0|}{\sqrt{4Dt}}\right), \tag{10.17}$$

where erf denotes the error function (see Appendix A). As $\operatorname{erf}(\infty) = 1$ and $\operatorname{erf}(0) = 0$, the foregoing probability starts at unity for $t = 0$ and tends to zero as $t \to \infty$, as expected.

[7] Recall that we have already come across this distribution, in another context, in Sec. 7.5.6, Eq. (7.41).

[8] This is quite obvious. But we can also show this formally by integrating the first equation in (10.12) from 0 to t with respect to time, and using the fact that $S(0\,,\,x\,|\,x_0) = 1$ for $x \neq x_0$.

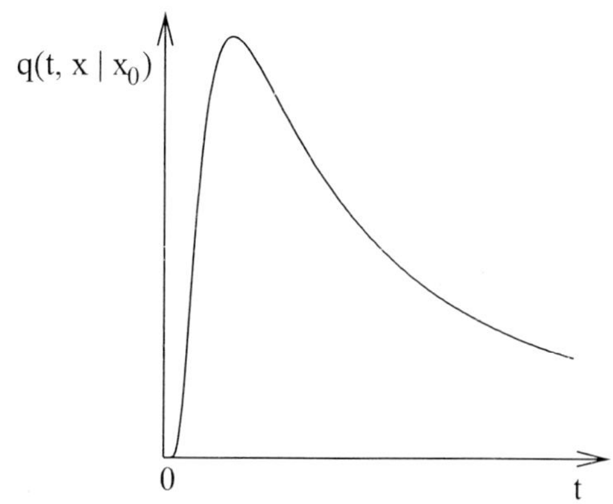

Figure 10.1: The probability density in time, $q(t, x \mid x_0)$, for first passage from any point x_0 to any other point $x > x_0$, for Brownian motion on a line of infinite extent to the left of x_0. The PDF, given by Eq. (10.8), corresponds to a stable distribution with index $\frac{1}{2}$. The density is extremely flat at the origin: it vanishes at $t = 0$, as do all its (right) derivatives. The peak of the PDF occurs at $t = (x - x_0)^2/6D$. The density falls off at long times like $t^{-3/2}$. Owing to this slow decay, the mean value and all the higher moments of the first-passage time are infinite.

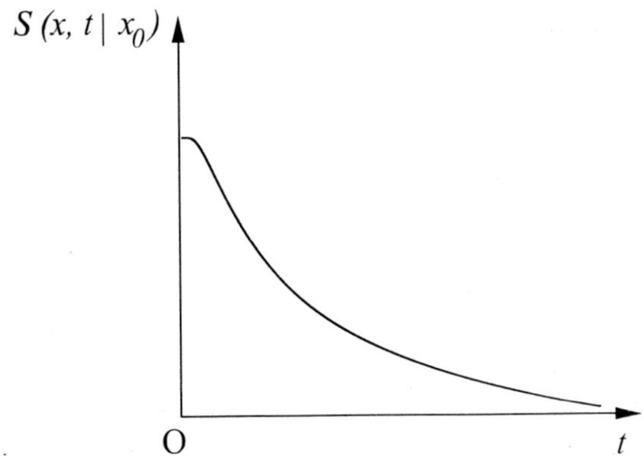

Figure 10.2: The survival probability $S(t\,,\,x\,|\,x_0)$ that a diffusing particle starting at a point $x_0\,(< x)$ remains in the region $(-\infty\,,\,x)$ for a time interval t without hitting the upper boundary at x. [Alternatively, $S(t\,,\,x\,|\,x_0)$ is the survival probability that the particle starts at a point $x_0\,(> x)$ and remains in the region $(x\,,\,\infty)$ for a time interval t without hitting the lower boundary at x.] The exact expression for $S(t\,,\,x\,|\,x_0)$ is an error function, given in Eq. (10.17). The variation of $S(t\,,\,x\,|\,x_0)$ with time appears to be qualitatively similar to that of the survival probability $S(t\,,\,\pm a\,|\,0)$ shown in Fig. 8.1. However, the time dependences in the two cases are quite different from each other. $S(t\,,\,\pm a\,|\,0)$ falls off exponentially with time, as shown in Ch. 8. In contrast, the asymptotic fall-off of $S(t\,,\,x\,|\,x_0)$ is a much slower power law, $\sim t^{-1/2}$.

- Equation (10.17) is a basic result pertaining to diffusion in one dimension.

The leading asymptotic behavior of the survival probability for very large values of t, that is, for $t \gg (x - x_0)^2/D$, also follows immediately: $S(t\,,\,x\,|\,x_0) \sim t^{-1/2}$. We conclude that:

- Given any arbitrary 'threshold' x, unbiased Brownian motion that starts at any point in the region $(-\infty\,,\,x)$ [or in the region $(x\,,\,\infty)$] persists in that region, without hitting the threshold value x, with a probability that decays with time like $t^{-1/2}$.

Figure 10.2 shows $S(t\,,\,x\,|\,x_0)$ as a function of t for a fixed value of $(x - x_0)$.

An even more interesting result now follows. As we are assured that first passage from any x_0 to any x is a sure event, we may ask for the **mean first-passage time**. This is given formally by

$$\overline{t(x_0 \to x)} = \int_0^\infty dt\; t\; \mathsf{q}(t\,,\,x\,|\,x_0). \tag{10.18}$$

However, since $\mathsf{q}(t\,,\,x\,|\,0) \sim t^{-3/2}$ at long times, it follows that the mean first-passage time is actually infinite—the integral in Eq. (10.18) diverges, because

$$\int^\infty dt\; t\; \mathsf{q}(t\,,\,x\,|\,x_0) \sim \int^\infty dt\, t^{-1/2} \to \infty.$$

This means that:

- Even though first passage from any x_0 to any other x is a sure event for Brownian motion on the infinite line, the mean time for such a first passage is infinite. The higher moments of the first-passage time are also infinite, of course[9].

The cause of this divergence is the infinite extent of the line on the side of x_0 *away* from x (i. e., to the left of x_0 if x is to its right, and *vice versa*). The summation in Eq. (10.18) is dominated by those realizations of the random process in which the particle spends an arbitrarily large amount of time exploring this infinite expanse before hitting x for the first time. As a consequence, the integral diverges.

We may understand this unbounded mean first-passage time in another way. Imagine placing a reflecting boundary at some distant point $-L$ on the left, thus

[9]It is clear that the *fractional* moments $\overline{t^\alpha(x_0 \to x)}$ of the first-passge time are finite as long as $\alpha < \frac{1}{2}$.

bounding the position of the particle from below. The mean first-passage time from 0 to any $x > 0$ now becomes finite. It turns out to be a linear function of the parameter L, as we shall see in an exercise below. Hence the mean first-passage time diverges *linearly* in the cut-off L as the latter is allowed to tend to infinity[10].

10.3 Exercises

10.3.1 Divergence of the mean first-passage time

Consider Brownian motion on the portion $[-L\,,\,\infty)$ of the x-axis, with a reflecting boundary at $-L$. We want to calculate the first-passage time density to go from any initial point $x_0\,(> -L)$ to any point x, where x is to the right of x_0, and then to find the corresponding mean first-passage time. Let us use the renewal equation method for this purpose.

Let $\mathsf{p}(x\,,\,t\,|\,x_0)\big|_{\text{refl. at }-L}$ be the probability density of the position in the presence of the reflector at $-L$. Using the method of images, this is just

$$\mathsf{p}(x\,,\,t\,|\,x_0)\big|_{\text{refl. at }-L} = \mathsf{p}(x\,,\,t\,|\,x_0) + \mathsf{p}(-2L - x\,,\,t\,|\,x_0), \tag{10.19}$$

where $\mathsf{p}(x\,,\,t\,|\,x_0)$ is of course the fundamental Gaussian solution to the diffusion equation on the infinite line[11]. Observe that the plus sign preceding the contribution from the image point at $-(2L + x)$ takes care of the reflecting boundary condition at $-L$.

(a) Let x_1 be an arbitrary point to the right of x. Use the formula obtained in Eq. (10.3) and the expression for the Laplace transform of the fundamental Gaussian solution in Eq. (10.5) to show that the Laplace transform of the first-passage time density in the presence of the reflecting boundary is

$$\widetilde{\mathsf{q}}(s\,,\,x\,|\,x_0)\big|_{\text{refl. at }-L} = \frac{\cosh\left\{(L + x_0)\sqrt{(s/D)}\right\}}{\cosh\left\{(L + x)\sqrt{(s/D)}\right\}}\,. \tag{10.20}$$

Since this function is equal to unity at $s = 0$, it is obvious that the first-passage

[10] This statement is only valid in the case of *unbiased* diffusion or free Brownian motion. It is altered drastically in the case of diffusion in the presence of an external force or in an external potential. This aspect, important in physical applications, will be considered further in Ch. 13.

[11] We have written this solution down several times so far, most recently in Eq. (10.15).

time density is normalized to unity, as required[12].

(b) The mean first-passage time is therefore

$$\begin{aligned}\overline{t(x_0 \to x)} &= \int_0^\infty dt\; t\; \mathsf{q}(t\,,\, x\,|\, x_0)\big|_{\text{refl. at } -L} \\ &= \left[-\frac{\partial}{\partial s}\widetilde{\mathsf{q}}(s\,,\, x\,|\, x_0)\big|_{\text{refl. at } -L}\right]_{s=0}, \end{aligned} \tag{10.21}$$

in terms of the transform of the density. (The second line in the above represents a useful formula for the mean first-passage time.) Show that this mean value is given by

$$\overline{t(x_0 \to x)} = \frac{1}{2D}(2L + x + x_0)(x - x_0). \tag{10.22}$$

Hence we conclude that the mean first-passage time increases like L as $L \to \infty$, and is infinite on the infinite line.

10.3.2 Distribution of the time to reach a specified distance

We have seen that the law of the iterated logarithm tells us how almost all sample paths of Brownian motion behave near the origin of the process. The results we have already derived also enable us to answer the following questions: if the process starts at the origin at $t = 0$, what is the first-passage time density for the particle to reach a given distance a on *either* side of the origin? What is the mean time taken to reach this distance for the first time?

(a) In Sec. 8.2, Eq. (8.8), we have found the survival probability $S(t\,,\, \pm a\,|\, 0)$ that, at time t, the particle remains in the region $-a < x < a$ in the case of diffusion with absorbing barriers at both a and $-a$. For ready reference, let us write down this expression once again[13]:

$$S(t\,,\, \pm a\,|\, 0\,) = \frac{4}{\pi}\sum_{n=0}^{\infty}\frac{(-1)^n}{(2n+1)}\exp\left(-\frac{(2n+1)^2\pi^2 Dt}{4a^2}\right). \tag{10.23}$$

---

[12]The Laplace transform given by Eq. (10.20) can be inverted, and the first-passage time density itself can be written in closed form in terms of a derivative of a certain Jacobi theta function.

[13]Compare the fairly complicated expression for $S(t\,,\, \pm a\,|\, 0)$ in Eq. (10.23) with the relatively simple one for the survival probability $S(t\,,\, x\,|\, x_0)$ given in Eq. (10.17). The former refers to the case of *two* absorbing boundaries at $\pm a$, while the latter refers to a *single* absorbing boundary at x, the other end $(-\infty)$ being left free. This is another reminder of the complications that arise from the presence of finite boundaries. In the caption of Fig. 10.2, we have pointed out how the asymptotic behavior of $S(t\,,\, \pm a\,|\, 0)$ differs drastically from that of $S(t\,,\, x\,|\, x_0)$.

Use this result to show that the first-passage time density we seek is simply

$$\mathsf{q}(t\,,\,\pm a\,|\,0) = \frac{\pi D}{a^2}\sum_{n=0}^{\infty}(-1)^n(2n+1)\exp\left(-\frac{(2n+1)^2\pi^2 Dt}{4a^2}\right). \tag{10.24}$$

(b) The corresponding *mean* time to hit either one of the boundaries at $\pm a$ can now be calculated. This may be done directly from the definition of the mean first-passage time as the first moment of $\mathsf{q}(t\,,\,\pm a\,|\,0)$. Alternatively: since the Laplace transform of e^{-ct} (where $c > 0$) is just $(s+c)^{-1}$, write down the Laplace transform $\widetilde{\mathsf{q}}\,(s\,,\,\pm a\,|\,0)$ from Eq. (10.24). Now use the formula

$$\overline{t(0 \rightarrow \pm a)} = \left[-\frac{\partial}{\partial s}\,\widetilde{\mathsf{q}}\,(s\,,\,\pm a\,|\,0)\right]_{s=0} \tag{10.25}$$

to show that, for a Brownian particle starting at $t = 0$ from the origin, the mean time to reach a distance a from the origin for the first time is just[14]

$$\overline{t(0 \rightarrow \pm a)} = \frac{a^2}{2D}\,. \tag{10.26}$$

(c) Note how neatly the results

$$\overline{x^2(t)} = 2Dt \quad \text{and} \quad 2D\,\overline{t(0 \rightarrow \pm x)} = x^2 \tag{10.27}$$

complement each other! The mean squared distance travelled in a given time t is $2Dt$, while the mean time to travel a given distance is the square of that distance divided by $2D$.[15] But one should not forget the highly irregular nature of Brownian motion in time as well as space. It is important to keep in mind that the results above refer to *mean* values, over all realizations of the random process.

The following exercise will help you appreciate this a little better. Calculate the mean squared value of the first-passage time to $\pm a$. It is helpful to exploit the fact that this quantity is just the second derivative of $\widetilde{\mathsf{q}}\,(s\,,\,\pm a\,|\,0)$ at $s = 0$.[16]

---

[14] You will need the sum $\sum_{n=0}^{\infty}(-1)^n/(2n+1)^3 = \pi^3/32\,.$

[15] It is important to remember that $\overline{t(0 \rightarrow \pm x)}$ is the mean time to reach *either* the point x *or* the point $-x$, starting at the origin. It is this mean time that is finite. Individually, the mean first passage times $\overline{t(0 \rightarrow x)}$ and $\overline{t(0 \rightarrow -x)}$ are infinite for free (or unbiased) diffusion on an infinite line, as we have seen.

[16] In this case you need the sum $\sum_{n=0}^{\infty}(-1)^n/(2n+1)^5 = 5\pi^5/1536\,.$

Hence find the standard deviation of the first-passage time, $\Delta t = [\overline{t^2} - (\,\overline{t}\,)^2]^{1/2}$. Show that the relative fluctuation in the first-passage time is given by

$$\Delta t(0 \to \pm a) \Big/ \overline{t(0 \to \pm a)} = \sqrt{2/3}\,. \tag{10.28}$$

This is a pure number, independent of a and D. The fact that it is so large (it is of the order of unity) is again an indication of how irregular the Wiener process can be.

10.3.3 Yet another aspect of the $x^2 \sim t$ scaling

It should be clear by now that the fundamental property of diffusive motion is that, broadly speaking, '*the distance scales like the square root of the time*'. We have, of course, obtained several exact formulas whose ultimate origin lies in this general property. Here is one more variation on this theme. Incidentally, it also provides another example of certain relationships between different stable distributions. Recall that we have touched upon such relationships in Ch. 7, Sec. 7.5.6.

Consider the first-passage time for a diffusing particle starting at the origin on the infinite line to reach any point x_1 for the first time. Call this random variable t_1. The PDF of t_1 is given by Eq. (10.8), with x_0 set equal to zero (there is no loss of generality in doing so). We have

$$\mathsf{q}(t_1\,,\, x_1\,|\,0) = \frac{|x_1|}{\sqrt{4\pi D t_1^3}}\, \exp\left(-\frac{x_1^2}{4Dt_1}\right). \tag{10.29}$$

Similarly, consider the time of first passage from the origin to any *other* point x_2, again on the infinite line. Call this random variable t_2. The PDF of t_2 is of course again given by Eq. (10.29), with t_1 simply replaced by t_2 and x_1 by x_2.

Now consider the *ratio* of the two random variables above, $u = t_1/t_2$. It is obvious that u, like t_1 and t_2, cannot take on any negative values. The (normalized) PDF $f(u)$ of u is given by

$$f(u) = \int_0^\infty dt_1 \int_0^\infty dt_2\, \mathsf{q}(t_1\,,\, x_1\,|\,0)\, \mathsf{q}(t_2\,,\, x_2\,|\,0)\, \delta\left(u - \frac{t_1}{t_2}\right). \tag{10.30}$$

Show that this formula yields

$$f(u) = \frac{2|x_1\, x_2|}{\pi\sqrt{u}\,(x_1^2 + x_2^2 u)}\,. \tag{10.31}$$

Define the random variable

$$\tau = \sqrt{u} \equiv \sqrt{t_1/t_2}\,. \tag{10.32}$$

Hence show that the PDF of τ is again a Lorentzian (or rather, 'half' a Lorentzian, since τ cannot be negative), given by

$$q(\tau) = \frac{2|x_1\,x_2|}{\pi\,(x_1^2 + x_2^2\,\tau^2)}\,. \tag{10.33}$$

In particular, if $x_2 = -x_1$, then $q(\tau)$ reduces to $2/[\pi\,(1+\tau^2)]$.

- In Sec. 7.5.6 we found that, if x_1 and x_2 are the positions of two independently diffusing particles starting from the origin at $t = 0$, then each of these random variables has a Gaussian distribution, with a variance that increases linearly with t. Their ratio x_1/x_2 has a Cauchy distribution for all $t > 0$, with a PDF given by Eq. (7.43).

- We now have a complement to the foregoing result. If t_1 and t_2 are the first-passage times to reach two different points x_1 and x_2, starting from the origin at $t = 0$, then the ratio $\sqrt{t_1/t_2}$ has a Cauchy distribution for all $x_1 \neq x_2$, with a PDF given by Eq. (10.33).

Chapter 11

The displacement of the tagged particle

We find the exact expression for the mean squared displacement of a tagged particle in the Langevin model. The long-time limit of this result is compared with that obtained from the diffusion equation. This leads to a fundamental relation between the diffusion coefficient and the dissipation parameter occurring in the Langevin equation. The autocorrelation function of the displacement is calculated, showing explicitly how it depends on both its time arguments. The cross-correlation between the velocity and the position of the tagged particle is also determined.

11.1 Mean squared displacement in equilibrium

In this chapter, we return to the Langevin model for the motion of the tagged particle. Our focus now is on the properties of the displacement of the tagged particle that follow directly from the Langevin equation for its velocity. We could start with the formal solution for this displacement, and proceed as we have done in the case of the velocity of the particle in Ch. 4. We shall follow this path subsequently. But it is instructive to see right away how the mean squared displacement in thermal equilibrium, as predicted by the Langevin model, differs from the simple expression $2Dt$ that follows from the diffusion equation. The desired result follows in a few short steps from the expression already derived for the equilibrium velocity autocorrelation function.

The diffusion equation leads to a linear increase with time of the mean squared displacement of a tagged particle, for *all* $t > 0$. However, as mentioned at the end of Sec. 7.3 in Ch. 7, this may be expected to be valid only at sufficiently long times. An exact expression for the mean squared displacement of a tagged particle in thermal equilibrium follows from the Langevin equation for its velocity. How is this expression to be deduced, in the absence of a limiting or equilibrium distribution for the displacement? The equilibrium *velocity* autocorrelation function suffices for this purpose, as follows.

For brevity, let us write

$$X(t) = x(t) - x(0) = x(t) - x_0 \tag{11.1}$$

for the displacement in a time interval t. It is obvious that working with $X(t)$ rather than $x(t)$ enables us to eliminate any explicit dependence on the arbitrary initial value x_0 of the position of the particle. We begin with the identity

$$X(t) = \int_0^t dt_1\, v(t_1). \tag{11.2}$$

Squaring both sides and taking the (complete) average of the right-hand side, we get—*by definition*—the mean squared displacement of the tagged particle in a state of thermal equilibrium:

$$\left\langle X^2(t) \right\rangle_{\text{eq}} = \int_0^t dt_1 \int_0^t dt_2\, \langle v(t_1)\, v(t_2) \rangle_{\text{eq}}\,. \tag{11.3}$$

But the equilibrium autocorrelation of the velocity is a function of the difference $(t_1 - t_2)$, and also a symmetric function of its argument, i. e., it is a function of $|t_1 - t_2|$ (see Eqs. (4.8)). Hence the integrand can be doubled, while the region of integration (originally a square in the $(t_1\,,\, t_2)$ plane) is halved, to run over the triangle in which $t_2 < t_1$. Thus

$$\left\langle X^2(t) \right\rangle_{\text{eq}} = 2 \int_0^t dt_1 \int_0^{t_1} dt_2\, \langle v(t_1)\, v(t_2) \rangle_{\text{eq}}\,. \tag{11.4}$$

(Note the change in the upper limit of the integration over t_2.) Equation (11.4) is actually a *general* formula for the mean squared displacement, and we will return to it later on. If we now insert the expression found in Eq. (4.4) for the velocity autocorrelation, we get

$$\left\langle X^2(t) \right\rangle_{\text{eq}} = \frac{2k_BT}{m} \int_0^t dt_1 \int_0^{t_1} dt_2\, e^{-\gamma(t_1 - t_2)} = \frac{2k_BT}{m\gamma^2} \left(\gamma t - 1 + e^{-\gamma t}\right). \tag{11.5}$$

Figure 11.1 shows the time variation of the mean squared displacement.

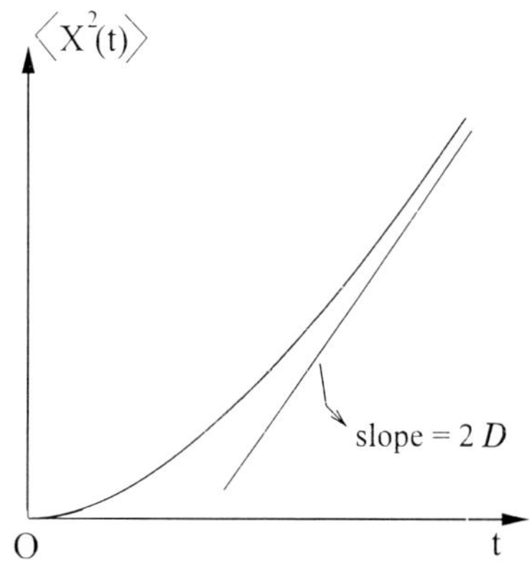

Figure 11.1: The mean squared displacement (Eq. (11.5)) of a tagged particle satisfying the Langevin equation, as a function of time. At very short times ($\gamma t \ll 1$), the leading behavior of the mean squared displacement is proportional to t^2. At very long times ($\gamma t \gg 1$), the curve tends to its asymptote, given by the straight line with ordinate $2D(t - \gamma^{-1})$.

When $\gamma t \gg 1$, $\langle X^2(t)\rangle_{\text{eq}}$ tends to its asymptotic form $2k_BTt/(m\gamma)$. This expression is linear in t, corroborating what was stated in Ch. 7, Sec. 7.3. Compare this asymptotic form with the result $2Dt$ for the mean squared displacement (see Eq. (7.6)) that follows from the diffusion equation.

- A very important relationship between the diffusion constant and the friction coefficient γ then becomes evident. We are led to conclude that the diffusion coefficient must be given by

$$D = \frac{k_BT}{m\gamma}\,. \tag{11.6}$$

This is yet another form of the FD relation that we have already found in earlier chapters (in the form $\Gamma = 2m\gamma k_BT$). It relates the strength of the fluctuations in the position of the tagged particle, as measured by the diffusion coefficient D, with the dissipation as measured by the friction coefficient γ.

If we assume a spherical shape of radius a for the tagged particle, the magnitude of the viscous drag on it is given by the Stokes formula $F_{\text{drag}} = 6\pi\eta a v$, where η is the viscosity of the fluid[1] in which it moves. Equating this to $m\gamma v$, we have

[1]To repeat the footnote in Sec. 4.1, this η should not be confused with the random force $\eta(t)$.

$m\gamma = 6\pi\eta a$. Using this in Eq. (11.6), we get

$$D = \frac{k_B T}{6\pi\eta a}, \tag{11.7}$$

relating the diffusion coefficient to the viscosity and the temperature. This relation, derived independently in 1905 by Sutherland and by Einstein, is properly known as the Stokes-Einstein-Sutherland relation[2]. It must be noted that η itself is strongly dependent on the temperature, so that the primary T-dependence of the diffusion coefficient comes from that of the viscosity η rather than from the numerator in Eq. (11.7). Typically, the dependence of η on T is given by an Arrhenius formula, $\eta \propto \exp(Q/k_B T)$, where Q is some characteristic 'activation energy'. Hence $D \propto T \exp(-Q/k_B T)$, showing that D is very sensitively dependent on the temperature: typically, D decreases very rapidly as T is lowered[3].

11.2 Time scales in the Langevin model

We have mentioned in Ch. 4, at the end of Sec. 4.1, that time scales are of crucial importance in the Langevin model. We are now in a position to discuss this aspect in a little more detail. As in Sec. 4.1, we consider a micron-sized tagged particle (radius $a \sim 1\,\mu$m to $10\,\mu$m, mass $m \sim 10^{-15}$ kg to 10^{-12} kg). We found that $\gamma^{-1} \sim 10^{-7}$ s to 10^{-5} s for a tagged particle moving in water at room temperature. Now, the r.m.s. velocity in equilibrium, $\langle v^2 \rangle_{\text{eq}}^{1/2} = (k_B T/m)^{1/2}$, represents a natural scale of speed for the particle. In the present instance, we have, at room temperature, $(k_B T/m)^{1/2} \sim 10^{-3}$ to 10^{-5} m/s. We may combine the r.m.s. velocity with the velocity correlation time to obtain a length scale—the distance that would be travelled by the tagged particle in the Smoluchowski

[2]The manner in which Einstein himself arrived at this relationship in 1905 is interesting, as the Langevin model was not extant at the time. He made clever use of the condition for a steady state in the presence of a constant external force. We shall not go into this here. Diffusion in a constant force field is discussed in Ch. 13, Sec. 13.8.3.

[3]The expression $T \exp(-Q/k_B T)$ also suggests that D vanishes only at $T = 0$. In actuality, the fluid would solidify well before the temperature reached absolute zero. Diffusion does take place (albeit relatively slowly) in the solid state, too, but the mechanisms involved in that case are different from that of the simple Langevin model with which we are concerned here.

There is another interesting aspect to the Arrhenius-like, or $\exp(1/T)$, dependence of the viscosity on the temperature. In **glassy dynamics**, when a liquid solidifies into an amorphous solid or a glass, a divergence of the viscosity occurs at a *nonzero* temperature T_0 according to $\eta \sim \exp[1/(T - T_0)]$ as T is decreased. This is called the **Vogel-Fulcher law**, and is an important aspect of the glass transition. The physical mechanisms leading to a form like $\exp[1/(T - T_0)]$, which has an an essential singularity at T_0, are far from obvious, and are the subject of many investigations.

time if it moved at the r.m.s. velocity. This is given by $\langle v^2\rangle_{\text{eq}}^{1/2}/\gamma$. But we can arrive at this length scale in an alternative way. The diffusion constant D has the physical dimensions (length)^2/time. It can therefore be combined with the velocity correlation time to define a length scale $(D/\gamma)^{1/2}$ characterizing the motion of the tagged particle. Using the relation $D = k_B T/(m\gamma)$ (Eq. (11.6)), we see that this is precisely the characteristic length scale identified above, namely,

$$\frac{\langle v^2\rangle_{\text{eq}}^{1/2}}{\gamma} = \left(\frac{k_B T}{m\gamma^2}\right)^{1/2} = \left(\frac{D}{\gamma}\right)^{1/2}. \tag{11.8}$$

The fact that the two length scales are actually identical to each other is again a consequence of the consistency condition imposed on the entire formalism by the FD relation.

Putting in the numerical values we have for our tagged particle, we find that its diffusion constant ranges from $D \sim 10^{-13}\,\text{m}^2/\text{s}$ to $10^{-14}\,\text{m}^2/\text{s}$. The distance scale given by Eq. (11.8) is then seen to be $\sim 10^{-10}\,\text{m}$. But this is of atomic dimensions, and is much smaller than the size of the tagged particle itself! What this means is that the time scale on which the thermalization or equilibration of the velocity of the tagged particle takes place is really practically instantaneous as far as the physical displacement of the tagged particle is concerned.

Let us consider the different time scales involved in the problem at hand, from first principles. The molecules of the fluid interact with each other via forces that are, ultimately, electromagnetic in origin. The time scale associated with these interactions are extremely small, $\lesssim 10^{-15}\,\text{s}$. A rough estimate of the time between successive interactions (or collisions) is easily made. The density $1\,\text{kg/m}^3$ together with Avogadro's number and the molecular weight of water (= 18) gives a mean inter-particle distance $\sim 3 \times 10^{-10}\,\text{m}$, which is in the right 'ball park'. The thermal speed $(k_B T/m_{\text{mol}})^{1/2}$ then yields a mean time between collisions that is of the order of a fraction of a picosecond[4]. It is natural to associate this inter-collision time scale with the correlation time of the noise $\eta(t)$ that drives the fluctuations in the velocity of the tagged particle in the Langevin model. We have also identified a much longer time scale—the Smoluchowski time, which is in the microsecond range—that is associated with the motion of the tagged particle itself. The white noise assumption made for the random force $\eta(t)$ is thus justified. On physical grounds, therefore, the Langevin equation for our tagged particle may be expected

[4]Thus, the inter-collision time is already several orders of magnitude larger than the interaction time scale. This is why, in kinetic theory, one generally assumes the collisions to be essentially instantaneous.

to be a satisfactory description on time scales much longer than (but certainly not on those comparable to, or smaller than) the molecular collision time scale of picoseconds. And on time scales $\gg \gamma^{-1}$, we go over into the diffusion regime, in which the velocity of the tagged particle may itself be assumed to be essentially δ-correlated. Thus, we have a rather broad temporal regime in which the Langevin equation provides a good description of the motion of the tagged particle.

It is this sort of clear separation of different time scales that makes the Langevin description tenable. Some further remarks related to the time scales considered here are made in Ch. 12, Sec. 12.5, and in Ch. 13, Sec. 13.5.

11.3 Equilibrium autocorrelation function of the displacement

Equation (11.5) gives the exact expression that follows from the Langevin equation for the mean squared displacement of the tagged particle in a state of thermal equilibrium. It is easy to extend the derivation to find the autocorrelation function of the displacement, again in thermal equilibrium. Consider any two instants t, t' such that $0 < t' < t$, without loss of generality. Then

$$X(t) = \int_0^t dt_1\, v(t_1) \quad \text{and} \quad X(t') = \int_0^{t'} dt_2\, v(t_2). \tag{11.9}$$

Therefore

$$\langle X(t)\, X(t')\rangle_{\text{eq}} = \int_0^t dt_1 \int_0^{t'} dt_2\, \langle v(t_1)\, v(t_2)\rangle_{\text{eq}}. \tag{11.10}$$

Although the integrand (a function of $|t_1 - t_2|$) is unaltered under the interchange of t_1 and t_2, the *region* of integration is no longer symmetric. It is a rectangle rather than a square in the $(t_1\,,\, t_2)$ plane. Putting in the known expression for $\langle v(t_1)\, v(t_2)\rangle_{\text{eq}}$, a short calculation gives

$$\langle X(t)\, X(t')\rangle_{\text{eq}} = \frac{k_B T}{m\gamma^2}\left(2\gamma t' - 1 + e^{-\gamma t'} + e^{-\gamma t} - e^{-\gamma(t-t')}\right) \quad (t > t'). \tag{11.11}$$

For $t < t'$, the roles of t and t' are interchanged in the foregoing expression. Further, when $t' = t$, we recover Eq. (11.5) for the mean squared displacement $\left\langle (X(t))^2 \right\rangle_{\text{eq}}$. It is important to note that the latter is explicitly dependent on t. Correspondingly, the autocorrelation $\langle X(t)\, X(t')\rangle_{\text{eq}}$ of the displacement is a function of *both* t and t', rather than a function of the difference $(t - t')$ alone. We therefore reiterate that,

- in a state of thermal equilibrium, the position (or the displacement) of the tagged particle is *not* a *stationary* random process, in contrast to the velocity of the particle.

In the diffusion regime in which $t \gg \gamma^{-1}$, $t' \gg \gamma^{-1}$, Eq. (11.11) yields the leading behavior $\langle X(t)\, X(t')\rangle_{\text{eq}} \to 2k_BTt'/(m\gamma)$. Therefore, identifying $k_BT/(m\gamma)$ with D, we have in general

$$\big\langle X(t)\, X(t')\big\rangle_{\text{eq}} \to 2D\, \min\,(t\,,\,t') = 2D\,\big[\,t\,\theta(t-t') + t'\,\theta(t'-t)\,\big] \tag{11.12}$$

in the diffusion regime. Even in this regime, the autocorrelation function continues to be a function of both its time arguments. As we have already seen in Ch. 9, the displacement is well approximated in the diffusion regime by Brownian motion (a Wiener process).

11.4 Conditional autocorrelation function

Next, let us calculate *ab initio* the conditional autocorrelation function of the position of the tagged particle, from the solution of the Langevin equation. We recall that the average involved here is conditioned upon a given initial velocity v_0 and a given initial position[5] x_0. This calculation will also help us see how the results deduced in the preceding sections indeed represent the *equilibrium* averages of the quantities concerned.

In the absence of an external force, the formal solution for the velocity of the tagged particle is given by Eq. (3.7). We begin by writing this solution as

$$v(t_1) = v_0\, e^{-\gamma t_1} + \frac{1}{m}\int_0^{t_1} dt_2\, e^{-\gamma(t_1-t_2)}\, \eta(t_2). \tag{11.13}$$

Integrate this expression once again with respect to time. We then get, for the displacement at any time t,

$$\begin{aligned} X(t) \;&\equiv\; x(t) - x_0 = \int_0^t dt_1\, v(t_1) \\ &=\; \frac{v_0}{\gamma}(1-e^{-\gamma t}) + \frac{1}{m}\int_0^t dt_1 \int_0^{t_1} dt_2\, e^{-\gamma(t_1-t_2)}\, \eta(t_2). \end{aligned} \tag{11.14}$$

[5]However, as we have already seen, all dependence on x_0 is eliminated by considering the displacement rather than the position itself.

Interchanging the order of integration in the second term on the right-hand side of Eq. (11.14), we have

$$\int_0^t dt_1 \int_0^{t_1} dt_2\, e^{-\gamma(t_1-t_2)}\, \eta(t_2) = \int_0^t dt_2 \int_{t_2}^t dt_1\, e^{-\gamma(t_1-t_2)}\, \eta(t_2). \tag{11.15}$$

But the integration over t_1 can now be carried out explicitly. Hence the displacement at time t is given by

$$X(t) = \frac{v_0}{\gamma}(1 - e^{-\gamma t}) + \frac{1}{m\gamma}\int_0^t dt_2\, \big(1 - e^{-\gamma(t-t_2)}\big)\, \eta(t_2). \tag{11.16}$$

The conditional mean value is therefore

$$\overline{X(t)} = \frac{v_0}{\gamma}(1 - e^{-\gamma t}), \tag{11.17}$$

once again using the fact that the noise $\eta(t)$ has zero mean at all times. For completeness (and for ready reference when needed in the sequel), we also record the trivial extensions of Eqs. (11.16) and (11.17), namely,

$$x(t) = x_0 + \frac{v_0}{\gamma}(1 - e^{-\gamma t}) + \frac{1}{m\gamma}\int_0^t dt_2\, \big(1 - e^{-\gamma(t-t_2)}\big)\, \eta(t_2) \tag{11.18}$$

and

$$\overline{x(t)} = x_0 + \frac{v_0}{\gamma}(1 - e^{-\gamma t}). \tag{11.19}$$

The conditional autocorrelation of the displacement may now be worked out in a straightforward manner, similar to the calculation of the conditional autocorrelation function of the velocity leading to Eq. (4.3). As before, we use the relation $\overline{\eta(t_1)\,\eta(t_2)} = \Gamma\,\delta(t_1 - t_2)$ where $\Gamma = 2m\gamma k_B T$. Taking $t > t'$ for definiteness, we get

$$\begin{aligned}\overline{X(t)\,X(t')} &= \frac{v_0^2}{\gamma^2}\big(1 - e^{-\gamma t} - e^{-\gamma t'} + e^{-\gamma(t+t')}\big) \\ &+ \frac{D}{\gamma}\big(2\gamma t' - 2 + 2e^{-\gamma t'} + 2e^{-\gamma t} - e^{-\gamma(t-t')} - e^{-\gamma(t+t')}\big).\end{aligned} \tag{11.20}$$

Interchanging t and t' gives the expression for the autocorrelation in the case $t < t'$. Observe the explicit dependence on v_0 in Eqs. (11.17) and (11.20). As a check on the foregoing calculation, let us go on to deduce the equilibrium autocorrelation of the displacement from Eq. (11.20). This may be done by averaging the right-hand side over v_0 using the equilibrium Maxwellian PDF $p_{\text{eq}}(v_0)$. The only factor in Eq. (11.20) that requires to be averaged over is v_0^2. Since $\langle v_0^2 \rangle_{\text{eq}} = k_B T/m = D\gamma$, the result is

$$\langle X(t)\, X(t') \rangle_{\text{eq}} = \frac{D}{\gamma} \left(2\gamma\, t' - 1 + e^{-\gamma t'} + e^{-\gamma t} - e^{-\gamma(t-t')}\right) \quad (t > t'). \tag{11.21}$$

But this is precisely the expression found already, in Eq. (11.11).

11.5 Fluctuations in the displacement

Equation (11.20) gives the conditional autocorrelation function of the displacement $X(t)$. As the mean value of $X(t)$ is not zero, this autocorrelation function does not pinpoint exclusively the contribution arising from the fluctuations. In order to do so, we must of course consider the autocorrelation of the deviation $\delta X(t)$. From Eqs. (11.16) and (11.17), the deviation of the displacement from its (conditional) mean value is given by

$$X(t) - \overline{X(t)} \equiv \delta X(t) = \frac{1}{m\gamma} \int_0^t dt_2 \left(1 - e^{-\gamma(t-t_2)}\right) \eta(t_2). \tag{11.22}$$

We note that $\delta X(t)$ is independent of v_0. It is also obvious that

$$\delta X(t) = \delta x(t) \tag{11.23}$$

where

$$\delta x(t) \equiv x(t) - \overline{x(t)}. \tag{11.24}$$

Hence the statistical properties of $\delta X(t)$ and $\delta x(t)$ are identical. From the calculation that yielded Eq. (11.20) we can now read off the result

$$\overline{\delta X(t)\, \delta X(t')} = \frac{D}{\gamma} \left(2\gamma t' - 2 + 2e^{-\gamma t'} + 2e^{-\gamma t} - e^{-\gamma(t-t')} - e^{-\gamma(t+t')}\right), \tag{11.25}$$

for $t \geq t'$. The right-hand side of Eq. (11.25) does not have any dependence at all on x_0 and v_0. It follows that the result represents, as it stands, the autocorrelation of $\delta X(t)$ (or $\delta x(t)$) in a state of thermal equilibrium. Thus

$$\overline{\delta X(t)\, \delta X(t')} = \overline{\delta x(t)\, \delta x(t')} = \langle \delta X(t)\, \delta X(t') \rangle_{\text{eq}} = \langle \delta x(t)\, \delta x(t') \rangle_{\text{eq}}, \tag{11.26}$$

each of these quantities being given by Eq. (11.25). Setting $t = t'$ in this equation yields the conditional variance of the displacement. Once again, the same expression also represents the variance of the displacement (which is also the variance

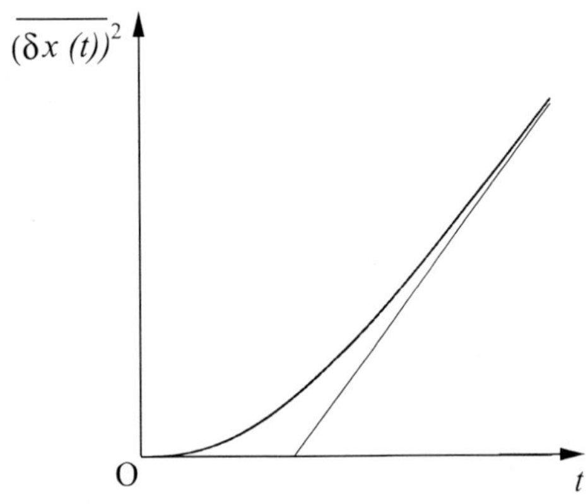

Figure 11.2: The variance of the displacement of the tagged particle as a function of time, as given by Eq. (11.27). This is the exact expression for the variance that follows from the Langevin equation. At short times ($\gamma t \ll 1$), the behavior of the variance is proportional to t^3. In the diffusion regime, the variance tends to $2Dt$, as expected. Hence the slope of the asymptote is equal to $2D$.

of the position) in the state of thermal equilibrium. We have

$$\begin{aligned}\overline{(\delta X(t))^2} &= \overline{(\delta x(t))^2} = \left\langle (\delta X(t))^2 \right\rangle_{\text{eq}} = \left\langle (\delta x(t))^2 \right\rangle_{\text{eq}} \\ &= \frac{D}{\gamma}\big(2\gamma t - 3 + 4e^{-\gamma t} - e^{-2\gamma t}\big). \end{aligned} \tag{11.27}$$

Figure 11.2 shows the time variation of the variance of the displacement. The small-t behavior of this quantity is $\sim \frac{2}{3}\gamma^3 t^3$, while its leading asymptotic behavior is $2Dt$, as expected. We will have occasion, in Ch. 12, Sec. 12.3, to recall the expression above for the variance of $x(t)$.

11.6 Cross-correlation of the velocity and the displacement

Finally, we may also ask precisely how the velocity and the displacement of the tagged particle are correlated with each other. We are now in a position to compute the equal-time correlation function $\overline{\delta x(t)\,\delta v(t)}$ directly. Repeating the relevant equations for ready reference, we have

$$\delta x(t) = \delta X(t) = \frac{1}{m\gamma}\int_0^t dt_2\,\left(1 - e^{-\gamma(t-t_2)}\right)\,\eta(t_2) \tag{11.28}$$

and

$$\delta v(t) = v - v_0\, e^{-\gamma t} = \frac{1}{m}\int_0^t dt_1\, e^{-\gamma(t-t_1)}\, \eta(t_1). \tag{11.29}$$

Using once again the fact that $\overline{\eta(t_1)\,\eta(t_2)} = 2m\gamma k_B T\,\delta(t_1 - t_2)$, a short calculation yields the result

$$\overline{\delta x(t)\,\delta v(t)} = D\,\left(1 - e^{-\gamma t}\right)^2. \tag{11.30}$$

Once again, we will recall Eq. (11.30) in Ch. 12, Sec. 12.3. Observe that this cross-correlation vanishes at $t = 0$. This is to be expected, since the initial conditions on the position and the velocity are, after all, independent of each other.

11.7 Exercises

11.7.1 Verification

Derive Eq. (11.11) using the result

$$\langle v(t_1)\, v(t_2)\rangle_{\text{eq}} = (k_B T/m)\, e^{-\gamma|t_1 - t_2|} \tag{11.31}$$

in Eq. (11.10).

11.7.2 Variance of X from its mean squared value

Show that Eq. (11.27) for the variance $\left\langle \left(\delta X(t)\right)^2\right\rangle_{\text{eq}}$, namely,

$$\left\langle \left(\delta X(t)\right)^2\right\rangle_{\text{eq}} = \frac{D}{\gamma}\left(2\gamma t - 3 + 4e^{-\gamma t} - e^{-2\gamma t}\right), \tag{11.32}$$

follows from Eq. (11.5) for $\langle X^2(t)\rangle_{\text{eq}}$, that is,

$$\langle X^2(t)\rangle_{\text{eq}} == \frac{D}{\gamma}\left(\gamma t - 1 + e^{-\gamma t}\right). \tag{11.33}$$

Use the definition $\delta X(t) = X(t) - \overline{X(t)}$, and hence

$$\left(\delta X(t)\right)^2 = X^2(t) - 2X(t)\,\overline{X(t)} + \left(\,\overline{X(t)}\,\right)^2. \tag{11.34}$$

Now take 'equilibrium averages' on both sides, after using Eq. (11.17) for $\overline{X(t)}$.

11.7.3 Velocity-position equal-time correlation

Here is a simple way to obtain the various versions of the cross-correlation between the position (or displacement) and the velocity. This exercise should also help you gain a better understanding of the differences between various conditional and equilibrium averages, as well as the implications of the nonstationary nature of the random process represented by the position of the tagged particle.

(a) Recall Eq. (11.5), which says that

$$\left\langle (x(t)-x_0)^2 \right\rangle_{\text{eq}} = \frac{2k_BT}{m\gamma^2}\left(\gamma t - 1 + e^{-\gamma t}\right). \tag{11.35}$$

Differentiate both sides[6] with respect to t, to find that

$$\langle x(t)\, v(t)\rangle_{\text{eq}} = D\left(1-e^{-\gamma t}\right). \tag{11.36}$$

This result should remind you once again that the position $x(t)$ of a freely diffusing tagged particle is not a stationary random process: if it had been so, the cross-correlation above would have vanished identically. The general argument is as follows (we use the notation $\langle\cdots\rangle$ for the average over all realizations of the process):

Suppose $\xi(t)$ is a *stationary* random process. Then $\langle \xi(t)\,\xi(t+t')\rangle$ must be a function of t' alone, and not of t. Hence

$$\frac{d}{dt}\langle \xi(t)\,\xi(t+t')\rangle = \langle \dot\xi(t)\,\xi(t+t')\rangle + \langle \xi(t)\,\dot\xi(t+t')\rangle = 0. \tag{11.37}$$

Setting $t'=0$, it follows that

$$\langle \xi(t)\,\dot\xi(t)\rangle = 0. \tag{11.38}$$

That is, the equal-time cross-correlation between the process ξ and its derivative process $\dot\xi$ must vanish for *all* t. In the case at hand, however, the cross-correlation $\langle x(t)\,\dot x(t)\rangle_{\text{eq}} = \langle x(t)\,v(t)\rangle_{\text{eq}}$ does not do so. This is yet another consequence of the nonstationary nature of the position process $x(t)$. In Ch. 13, Sec. 13.4, we shall see how the result in Eq. (11.36) contrasts with that in the case of a harmonically bound particle, the Brownian oscillator (see Eq. (13.23)).

---

[6]It is evident that the operations of differentiation and averaging commute with each other.

(b) Now recall Eq. (11.27), which says that

$$\overline{(\delta x(t))^2} = \left\langle (\delta x(t))^2 \right\rangle_{\text{eq}} = \frac{D}{\gamma}\left(2\gamma t - 3 + 4e^{-\gamma t} - e^{-2\gamma t}\right). \tag{11.39}$$

Differentiate both sides with respect to t, to obtain

$$\overline{\delta x(t)\, \delta v(t)} = \langle \delta x(t)\, \delta v(t)\rangle_{\text{eq}} = D\left(1 - e^{-\gamma t}\right)^2. \tag{11.40}$$

This is precisely the expression found in Eq. (11.30) for $\overline{\delta x(t)\, \delta v(t)}$.

11.7.4 Velocity-position unequal-time correlation

Equation (11.30) or (11.40) gives the equal-time correlation between the position and the velocity of a particle obeying the Langevin equation. We may also ask for the more general *un*equal-time correlation $\overline{\delta x(t)\, \delta v(t\,')}$. Using Eqs. (11.28) and (11.29), we have

$$\overline{\delta x(t)\, \delta v(t\,')} = \frac{\Gamma}{m^2\gamma}\int_0^t dt_2 \int_0^{t\,'} dt_1 \left(1 - e^{-\gamma(t-t_2)}\right) e^{-\gamma(t\,'-t_1)}\, \delta(t_1 - t_2). \tag{11.41}$$

Set $\Gamma = 2m\gamma k_B T$ as usual. Evaluate the double integral in Eq. (11.41) carefully, in the two cases $t > t\,'$ and $t < t\,'$, respectively. Identifying D with $k_B T/(m\gamma)$, show that

$$\overline{\delta x(t)\, \delta v(t\,')} = D\left(2 - 2e^{-\gamma t\,'} - e^{-\gamma(t-t\,')} + e^{-\gamma(t+t\,')}\right) \quad \text{for} \quad t > t\,', \tag{11.42}$$

while

$$\overline{\delta x(t)\, \delta v(t\,')} = D\left(e^{-\gamma(t\,'-t)} - 2e^{-\gamma t\,'} + e^{-\gamma(t+t\,')}\right) \quad \text{for} \quad t < t\,'. \tag{11.43}$$

Each of these expressions reduces correctly to that given by Eq. (11.30) when $t = t\,'$.

Note the asymmetry between the two cases $t > t\,'$ and $t < t\,'$. This is not surprising, as the velocity and the position are certainly not on the same footing. The random force $\eta(t)$ induces an acceleration in the particle, leading to a change in its velocity. In turn, this leads to a change in its position. The velocity at any given time is dependent on the previous acceleration history, while the position at any time is dependent on the velocity history. Suppose we let both t and $t\,'$ become very large. The correlation in Eq. (11.42) (between the velocity at an earlier time and the position at a later time) then tends to a constant, $2D$, with a 'correction' given by the damped exponential $D\exp[-\gamma(t-t')]$. The latter vanishes as the difference $(t-t')$ itself becomes very large. On the other hand, the correlation in Eq. (11.43) (between the position at an earlier time and the velocity at a later time) tends, for very large t and t', to the damped exponential $D\exp[-\gamma(t'-t)]$, and this quantity tends to zero as the difference $(t'-t)$ becomes very large[7].

[7]Based on causality arguments, we might have expected this latter correlation function to vanish identically. In Ch. 17, we shall examine more carefully the validity of the Langevin equation *vis-à-vis* causality.

Chapter 12

The Fokker-Planck equation in phase space

We consider the Fokker-Planck equation for the joint probability density in the velocity and the position of a diffusing tagged particle, i. e., its phase space density. The covariance matrix is defined, and the generalized Gaussian solution to the Fokker-Planck equation is presented. In the diffusion regime, the joint density factorizes into two Gaussians: the equilibrium Maxwellian PDF in the velocity, and the fundamental Gaussian solution of the diffusion equation, in the position.

12.1 Recapitulation

Let us pause for a moment to repeat and take stock of the two different equations we have discussed for the PDFs in the velocity and the position, respectively.

In Ch. 6, we considered the Fokker-Planck equation satisfied by the PDF $p(v\,,t)$ of the velocity of the tagged particle, Eq. (6.9):

$$\frac{\partial p}{\partial t} = \gamma \frac{\partial}{\partial v}(vp) + \frac{\gamma k_B T}{m}\frac{\partial^2 p}{\partial v^2}. \tag{12.1}$$

The fundamental solution to this master equation, corresponding to a sharp initial PDF $\delta(v - v_0)$, is the Ornstein-Uhlenbeck (OU) distribution. Recalling Eq. (6.14), we have

$$p(v\,,t\,|\,v_0) = \left[\frac{m}{2\pi k_B\,T(1-e^{-2\gamma t})}\right]^{\frac{1}{2}} \exp\left[\frac{-m(v - v_0 e^{-\gamma t})^2}{2k_B T(1-e^{-2\gamma t})}\right], \quad t > 0\,. \tag{12.2}$$

The corresponding stochastic differential equation for the random variable v is the Langevin equation,

$$\dot{v} = -\gamma v + \frac{1}{m}\eta(t) = -\gamma v + \frac{\sqrt{\Gamma}}{m}\zeta(t). \tag{12.3}$$

The driving Gaussian white noise $\zeta(t)$ has zero mean and autocorrelation function $\langle\zeta(t)\,\zeta(t')\rangle = \delta(t-t')$. The SDE of Eq. (12.3) has both a drift term and a diffusion term. The noise is additive (its coefficient is a constant). The drift is linear in v, the random variable concerned. The strength of the noise is related to the dissipation via the FD relation $\Gamma = 2m\gamma k_B T$.

In Ch. 7, we considered the diffusion equation for the positional PDF $\mathsf{p}(x,t)$ of the tagged particle, Eq. (7.1):

$$\frac{\partial \mathsf{p}}{\partial t} = D\,\frac{\partial^2 \mathsf{p}}{\partial x^2}\,. \tag{12.4}$$

The fundamental solution to this master equation, corresponding to a sharp initial PDF $\delta(x - x_0)$, is

$$\mathsf{p}(x\,,\,t\,|\,x_0) = \frac{1}{\sqrt{4\pi Dt}}\,\exp\left(-\frac{(x-x_0)^2}{4Dt}\right), \quad t > 0\,. \tag{12.5}$$

The diffusion equation is valid in the diffusion regime $t \gg \gamma^{-1}$. In this regime, the velocity itself is approximated by a Gaussian white noise. The corresponding stochastic differential equation for x, Eq. (9.2), is simply

$$\dot{x} = \sqrt{2D}\,\zeta(t). \tag{12.6}$$

This SDE has no drift term. The noise is additive. The strength of the noise in this case is related to the dissipation via $D = k_BT/(m\gamma)$.

12.2 Two-component Langevin and Fokker-Planck equations

We now ask: Is there a single master equation for the PDF of the particle in *both* the dynamical variables, velocity and position? In other words, is there an FPE for the probability density in *phase space*? The answer is: yes, there is such an equation. We can write it down, if we exploit an extended version of the correspondence between the stochastic differential equation for a diffusion process and the Fokker-Planck equation for its PDF. As we have already used up the symbols p and p for the PDFs of the velocity and position, respectively, let us denote the *joint* or **phase space density**[1] by $\rho(x\,,\,v\,,\,t)$. This is also a convenient point at which to summarize what we have learnt so far, even at the expense of some repetition. As we are concerned with the motion in phase space, it is helpful to combine the position and velocity into a two-component random process

$$\mathbf{x} = \begin{pmatrix} x \\ v \end{pmatrix}. \tag{12.7}$$

In the absence of an external force, the process $\mathbf{x}(t)$ satisfies the Langevin equation

$$\dot{\mathbf{x}} + \Upsilon\,\mathbf{x} = \frac{1}{m}\boldsymbol{\eta}(t), \tag{12.8}$$

where the **drift matrix** and noise vector are, respectively,

$$\Upsilon = \begin{pmatrix} 0 & -1 \\ 0 & \gamma \end{pmatrix} \quad \text{and} \quad \boldsymbol{\eta}(t) = \begin{pmatrix} 0 \\ \eta(t) \end{pmatrix}. \tag{12.9}$$

Here $\eta(t)$ is a stationary, Gaussian, δ-correlated Markov process (a Gaussian white noise) with zero mean: recall that

$$\langle \eta(t) \rangle_{\text{eq}} = 0, \quad \langle \eta(t)\,\eta(t') \rangle_{\text{eq}} = \Gamma\,\delta(t-t') = 2m\gamma k_B T\,\delta(t-t'). \tag{12.10}$$

The *pair* of variables $(x\,,\,v) = \mathbf{x}$ constitutes a two-dimensional Markovian diffusion process—in fact, a two-dimensional OU process.

In preceding chapters, we have stated and used the correspondence between the stochastic differential equation for a diffusion process and the Fokker-Planck equation for its conditional PDF—see Eqs. (6.6)-(6.7), or Eqs. (9.29)-(9.30). This SDE$\leftrightarrow$FPE correspondence can be extended to multi-component processes as well. We shall present its general form in Ch. 13, Sec. 13.2. In the present instance, however, we only need a simpler special case of the relationship. This simpler version obtains when the drift term in the Langevin equation is linear in $\mathbf{x}$ and the diffusion coefficient is a constant (i. e., the noise is additive), as in Eq. (12.8). This Langevin equation implies that the corresponding joint PDF $\rho(x\,,\,v\,,\,t)$ satisfies the two-dimensional Fokker-Planck equation[2]

$$\frac{\partial \rho}{\partial t} = \sum_{i,j=1}^{2} \left\{ \Upsilon_{ij}\frac{\partial}{\partial x_i}(x_j\,\rho) + \mathsf{D}_{ij}\frac{\partial^2 \rho}{\partial x_i\,\partial x_j} \right\}, \tag{12.11}$$

[1] Remember that we are considering a single Cartesian component of the motion.

[2] In a multi-component FPE such as Eq. (12.11), x_i and x_j are independent dynamical variables for $i \neq j$, so that $\partial x_i/\partial x_j = \delta_{ij}$.

where $x_1 \equiv x\,,\; x_2 \equiv v$. The drift matrix Υ has already been defined in Eq. (12.9). The **diffusion matrix** D is given by

$$\mathsf{D} = \begin{pmatrix} 0 & 0 \\ 0 & \gamma k_B T/m \end{pmatrix} = \begin{pmatrix} 0 & 0 \\ 0 & D\gamma^2 \end{pmatrix}. \tag{12.12}$$

Putting in these expressions for Υ and D, and reverting to the variables x and v, Eq. (12.11) can be written out as

$$\frac{\partial \rho}{\partial t} = -v\,\frac{\partial \rho}{\partial x} + \gamma\,\frac{\partial}{\partial v}(v\rho) + \frac{\gamma k_B T}{m}\,\frac{\partial^2 \rho}{\partial v^2}\,. \tag{12.13}$$

The form of the first term on the right-hand side suggests its origin: recall that the total time derivative operator acting on a scalar function is given by[3] $d/dt = \partial/\partial t + v\,\partial/\partial x$. We seek the solution of the FPE (12.13), for the sharp initial condition[4]

$$\rho(x\,,\, v\,,\, 0) = \delta(v - v_0)\,\delta(x - x_0) \tag{12.14}$$

and natural boundary conditions $\rho(x\,,\, v\,,\, t) \to 0$ as $|x| \to \infty$ or $|v| \to \infty$. This solution is a Gaussian in the two variables x and v, or a *bivariate* Gaussian[5]. It is useful to outline some of the intermediate steps in arriving at this solution.

12.3 Solution of the Langevin and Fokker-Planck equations

The Langevin equation (12.8) is solved with the help of the **matrix Green function**[6]

$$\mathsf{G}(t) = \exp\,(-\Upsilon t) \equiv \sum_{n=0}^{\infty} \frac{(-\Upsilon t)^n}{n!}\,. \tag{12.15}$$

Observe that $\Upsilon^2 = \gamma \Upsilon$ in the present instance. Hence $\Upsilon^n = \gamma^{n-1}\,\Upsilon$ for $n \geq 1$. This relation enables us to sum the infinite series easily. After simplification, we

[3]This is the one-dimensional version of the familiar expression $d/dt = \partial/\partial t + \mathbf{v}\cdot\nabla_{\mathbf{r}}$. The total time derivative is the partial time derivative plus the 'convective' derivative.

[4]Hence the solution is actually the conditional PDF $\rho(x\,,\, v\,,\, t\,|\, x_0\,,\, v_0\,)$. We omit the explicit dependence on the initial values, and denote the solution simply by $\rho(x, v, t)$ in order to keep the notation manageable.

[5]The definition of a Gaussian distribution in several variables, and its salient properties, are discussed in Sec. D.6, Appendix D. See, in particular, Sec. D.7 for the case of two variables.

[6]The simple exponential form in Eq. (12.15) is a consequence of the fact that the drift matrix Υ in Eq. (12.9) is a constant matrix, i. e., its elements are not functions of t. Otherwise $\mathsf{G}(t)$ would be a **time-ordered exponential**.

get

$$\mathsf{G}(t) = I + \frac{\Upsilon}{\gamma}\,(e^{-\gamma t} - 1) = \begin{pmatrix} 1 & (1 - e^{-\gamma t})/\gamma \\ 0 & e^{-\gamma t} \end{pmatrix}. \tag{12.16}$$

We are concerned with sharp initial conditions $\mathbf{x}_0 \equiv (x_0\,,\, v_0)$. The average value of $\mathbf{x}$, conditioned upon these initial values, is then given by

$$\overline{\mathbf{x}(t)} = \mathsf{G}(t)\,\mathbf{x}_0. \tag{12.17}$$

Therefore the mean values $\overline{x(t)}$ and $\overline{v(t)}$ are linear combinations of x_0 and v_0 with time-dependent coefficients. We find

$$\overline{x(t)} = x_0 + \frac{v_0}{\gamma}\,(1 - e^{-\gamma t}), \quad \overline{v(t)} = v_0\, e^{-\gamma t}. \tag{12.18}$$

Note, in particular, that the mean value of the position depends not only on the initial position x_0, but also on the initial velocity v_0. The implication of this fact will be discussed subsequently, in Sec. 12.5. The next step is to define the **covariance matrix**

$$\boldsymbol{\sigma}(t) = 2\int_0^t \mathsf{G}(t')\,\mathsf{D}\,\mathsf{G}^{\mathrm{T}}(t')\,dt'\,, \tag{12.19}$$

where G^{T} denotes the transpose of the matrix G. Using Eq. (12.12) for D and Eq. (12.16) for $\mathsf{G}(t)$ and simplifying, we get

$$\boldsymbol{\sigma}(t) = D\begin{pmatrix} C(t) & B(t) \\ B(t) & A(t) \end{pmatrix}, \tag{12.20}$$

where

$$\begin{aligned} A(t) &= \gamma\left(1 - e^{-2\gamma t}\right),\; B(t) = \left(1 - e^{-\gamma t}\right)^2, \\ C(t) &= \gamma^{-1}\left(2\gamma t - 3 + 4e^{-\gamma t} - e^{-2\gamma t}\right). \end{aligned} \tag{12.21}$$

The significance of these particular functions of t should be clear from earlier chapters, as we have already encountered them. But we shall comment further on this aspect at the end of this section.

As usual, let

$$\delta x = x - \overline{x(t)}, \quad \text{and} \quad \delta v = v - \overline{v(t)} \tag{12.22}$$

denote the deviations of x and v from their instantaneous mean values. The latter are given by Eqs. (12.18). Further, let

$$\Delta(t) = \det \boldsymbol{\sigma}(t), \quad \mathsf{S}(t) = \boldsymbol{\sigma}^{-1}(t). \tag{12.23}$$

Hence

$$\mathsf{S}(t) = \frac{D}{\Delta}\begin{pmatrix} A(t) & -B(t) \\ -B(t) & C(t) \end{pmatrix}, \tag{12.24}$$

where

$$\begin{aligned} \Delta(t) &= D^2\,(AC - B^2) \\ &= 2D^2\,\left(\gamma t - 2 + 4e^{-\gamma t} - 2e^{-2\gamma t} - \gamma t\, e^{-2\gamma t}\right). \end{aligned} \tag{12.25}$$

The formal solution of the Fokker-Planck equation (12.13) is then a Gaussian in $\delta\mathbf{x}$, given by[7]

$$\rho(x, v, t) = \frac{1}{2\pi\sqrt{\Delta}}\,\exp\left[-\tfrac{1}{2}\left\{\mathsf{S}_{11}\,(\delta x)^2 + 2\mathsf{S}_{12}\,(\delta x)(\delta v) + \mathsf{S}_{22}\,(\delta v)^2\right\}\right]. \tag{12.26}$$

In explicit terms,

$$\begin{aligned} \rho(x, v, t) &= \frac{1}{2\pi D\sqrt{AC - B^2}} \times \\ &\times \exp\left\{-\frac{A(t)\,(\delta x)^2 - 2B(t)\,(\delta x)(\delta v) + C(t)\,(\delta v)^2}{4D\,(\gamma t - 2 + 4e^{-\gamma t} - 2e^{-2\gamma t} - \gamma t\, e^{-2\gamma t})}\right\}. \end{aligned} \tag{12.27}$$

- Equation (12.27) is the joint probability density in the position and the velocity of a tagged particle obeying the Langevin equation in the absence of an external force.

We reiterate that the solution above corresponds to the sharp initial condition $\rho(x, v, 0) = \delta(v - v_0)\,\delta(x - x_0)$.

Let us turn now to the significance of the functions $A(t)$, $B(t)$ and $C(t)$ that appear in the solution above. From Eq. (12.20) we have, for the individual matrix elements of the covariance matrix,

$$\sigma_{11} = D\,C(t), \;\; \sigma_{22} = D\,A(t), \;\; \sigma_{12} = \sigma_{21} = D\,B(t). \tag{12.28}$$

Referring to Appendix D, Sec. D.7, we know that σ_{11} and σ_{22} are the variances of the components x and v of $\mathbf{x}$, while $\sigma_{12} = \sigma_{21}$ is the cross-correlation represented by the mean value of the product $\delta x(t)\,\delta v(t)$. We must also remember that the averages concerned are conditional ones. Recalling Eq. (6.13) and using the

[7] See Eq. (D.16), Appendix D, for the general expression for the PDF of a bivariate Gaussian distribution.

fact that $D = k_B T/(m\gamma)$, we see at once that $D\,A(t)$ is indeed the (conditional) variance of v, as required:

$$\overline{v^2(t)} - \big(\,\overline{v(t)}\,\big)^2 = \overline{\big(\delta v(t)\big)^2} = \sigma_{22}(t) = D\,A(t) = \frac{k_B T}{m}(1 - e^{-2\gamma t}). \tag{12.29}$$

Similarly, recalling Eq. (11.27), it follows that

$$\overline{\big(\delta x(t)\big)^2} = \sigma_{11}(t) = D\,C(t) = \frac{D}{\gamma}\left(2\gamma t - 3 + 4e^{-\gamma t} - e^{-2\gamma t}\right). \tag{12.30}$$

Finally, recalling Eq. (11.30), we see that

$$\overline{\delta x(t)\,\delta v(t)} = \sigma_{12}(t) = \sigma_{21}(t) = D\,B(t) = D\left(1 - e^{-\gamma t}\right)^2. \tag{12.31}$$

Each of the functions $A(t)$, $B(t)$ and $C(t)$ starts at the value 0 when $t = 0$. This is to be expected, as these conditional variances must vanish for the sharp initial conditions we have imposed. Hence $\Delta(t)$ also vanishes at $t = 0$: in fact, its leading small-t behavior is $\sim t^4$. All the three functions $A(t)$, $B(t)$ and $C(t)$ increase monotonically with t. For positive values of t, the determinant $\Delta(t)$ is also a positive definite, monotonically increasing function of t. As $t \to \infty$, $D\,A(t) \to k_B T/m = \langle v^2\rangle_{\text{eq}}$, as expected. Similarly, $D\,C(t) \to 2Dt$, which is the mean squared displacement in the diffusion limit.

12.4 PDFs of the velocity and position individually

What about the individual components of the process $\mathbf{x}$, namely, the velocity v and the position x? We already know, of course, that the velocity v by itself is a stationary, Gaussian, exponentially correlated, Markov process: namely, an OU process. This result can be recovered from Eq. (12.27) by integrating $\rho(x, v, t)$ over all x from $-\infty$ to ∞, and using the standard formulas for Gaussian integrals. Carrying out the algebra, the result is

$$\begin{aligned}\int_{-\infty}^{\infty} dx\,\rho(x, v, t) \equiv p(v, t) \;&=\; \left[\frac{m}{2\pi k_B\,T(1 - e^{-2\gamma t})}\right]^{\frac{1}{2}} \times \\ &\times\; \exp\left\{\frac{-m(v - v_0\,e^{-\gamma t})^2}{2k_B T(1 - e^{-2\gamma t})}\right\},\end{aligned} \tag{12.32}$$

which is precisely the OU density of Eq. (6.14) or (12.2). This PDF satisfies the Fokker-Planck equation (in this case, the Rayleigh equation) (6.9) or (12.1). In the limit $t \to \infty$, the OU density tends to the equilibrium Maxwellian density

$p^{\rm eq}(v)$ given in Eq. (2.1). The autocorrelation function of the velocity is a decaying exponential function of time.

In marked contrast, the position x (or the displacement) of the tagged particle is not a stationary random process, as we have already seen in some detail—in particular, in Ch. 11.

- On its own, the position $x(t)$ of a particle whose velocity satisfies the Langevin equation is not even a Markov process! It is simply the integral of an Ornstein-Uhlenbeck process.

It is instructive to see explicitly why $x(t)$ is non-Markovian. Integrating the joint density $\rho(x, v, t)$ over all values of v yields the result

$$\int_{-\infty}^{\infty} dv\, \rho(x,v,t) = \frac{1}{\sqrt{2\pi D C(t)}} \exp\left\{-\frac{(\delta x)^2}{2DC(t)}\right\}. \tag{12.33}$$

This is the exact expression for the PDF of the position of a tagged particle satisfying the Langevin equation in the absence of an external force. Writing out the x and t dependences explicitly, we have

$$\begin{aligned} \int_{-\infty}^{\infty} dv\, \rho(x,v,t) &= \left[\frac{\gamma}{2\pi D\big(2\gamma t - 3 + 4e^{-\gamma t} - e^{-2\gamma t}\big)}\right]^{1/2} \times \\ &\times \exp\left\{-\frac{\gamma\big(x - x_0 - v_0\gamma^{-1}(1-e^{-\gamma t})\big)^2}{2\pi D\left(2\gamma t - 3 + 4e^{-\gamma t} - e^{-2\gamma t}\right)}\right\}. \end{aligned} \tag{12.34}$$

We find that the PDF of the position is a Gaussian, as expected. But the solution also involves the initial value v_0 of the *velocity*, over and above the expected dependence on the initial position x_0. This fact suffices to show that the position $x(t)$ by itself cannot be a Markov process. Note, too, that we have *not* used the notation $\mathsf{p}(x,t)$ for this PDF. This is intentional. Purely as an aid to easy recognition, we reserve the notation $\mathsf{p}(x,t)$ for those cases in which the position $x(t)$ is a Markov process on its own. As we know, this holds good in the diffusion regime[8]. It is important to note that

- the PDF in Eq. (12.34), namely, the PDF of the position of a freely diffusing particle whose velocity satisfies the Langevin equation, does not satisfy any simple master equation by itself.

Finally, we have seen earlier that the autocorrelation of the displacement is given by Eq. (11.11).

[8]It will also be valid under certain conditions when an external force is present, as we shall see in Ch. 13. In those situations, we shall again denote the positional PDF by $\mathsf{p}(x,t)$.

12.5 The long-time or diffusion regime

We are now in a position to understand quantitatively what happens to the joint PDF $\rho(x, v, t)$ in the long-time limit, i. e., the diffusion regime given by $\gamma t \gg 1$. Terms such as $e^{-\gamma t}$ and $e^{-2\gamma t}$ can be neglected in this regime. The leading behavior of the various functions that appear in the solution for ρ is given by

$$A(t) \sim \gamma,\ B(t) \sim 1,\ C(t) \sim 2t,\ AC - B^2 \sim 2\gamma t. \tag{12.35}$$

The solution (12.27) then *factorizes* into the equilibrium Maxwellian density in v and the Gaussian solution to the diffusion equation in x, according to

$$\begin{aligned} \rho(x\,,\,v\,,\,t)\Big|_{\gamma t \gg 1} \quad \longrightarrow \quad & \left\{ \left(\frac{m}{2\pi k_B T}\right)^{\frac{1}{2}} \exp\left(-\frac{mv^2}{2k_B T}\right)\right\} \times \\ & \times \left\{ \frac{1}{\sqrt{4\pi D t}} \exp\left(-\frac{(x-x_0)^2}{4Dt}\right)\right\}. \end{aligned} \tag{12.36}$$

As we know already, in this long-time regime the displacement becomes a special kind of nonstationary Markovian diffusion process—namely, Brownian motion. Then, reiterating what has already been stated:

- Its PDF $\mathsf{p}(x\,,\,t)$ satisfies the diffusion equation (7.1) (repeated in Eqs. (9.1) and (12.4)).
- Its autocorrelation function is given by Eq. (9.7) (or Eq. (11.12)).
- The random variable $x(t)$ satisfies the stochastic differential equation (9.2).
- The strength of the white noise that drives x in this diffusion limit is related to that of the white noise η that drives the velocity in the Langevin equation by the FD relation. To repeat (for emphasis) the relationships involved,

$$2D = \frac{2k_B T}{m\gamma} = \frac{\Gamma}{m^2\gamma^2}\,. \tag{12.37}$$

As x is not a stationary process, the PDF $\mathsf{p}(x\,,\,t)$ does not tend to any nontrivial limit as $t \to \infty$; rather, it tends to zero for all x in this limit.

- In general, a density satisfying the Fokker-Planck equation (12.11) tends to a nontrivial stationary density if *all* the eigenvalues of the drift matrix Υ have positive real parts.
- In the present instance, the eigenvalues of Υ are 0 and γ, so that this condition is obviously not satisfied.

12.6 Exercises

12.6.1 Velocity and position PDFs from the joint density

(a) Starting from Eq. (12.27) for the joint PDF in x and v, integrate over all values of x to obtain the PDF $p(v,t)$ in the velocity, and show that this is precisely the OU density, Eq. (12.32).

(b) Similarly, integrate Eq. (12.27) over all values of v to obtain the probability density of the position, Eq. (12.33).

12.6.2 Phase space density for three-dimensional motion

For completeness, let us consider the motion of the tagged particle in its full six-dimensional phase space. We retain the notation $\mathbf{x}$ for a point in the phase space of the particle. The Langevin equation continues to be of the form given by Eq. (12.8), where the *six*-component vector $\mathbf{x} = (\mathbf{r}\,,\, \mathbf{v})$. The white noise that drives the velocity of the particle now comprises three independent Cartesian components. Each of these is a stationary, Gaussian, δ-correlated Markov process with zero mean:

$$\overline{\eta_k(t)} = 0\ ,\ \overline{\eta_k(t)\,\eta_l(t')} = \Gamma\,\delta_{kl}\,\delta(t-t') = 2m\gamma k_B T\,\delta_{kl}\,\delta(t-t'), \tag{12.38}$$

where the indices k, l run over the three Cartesian components[9]. The Fokker-Planck equation for the full six-dimensional phase space density $\rho(\mathbf{r}\,,\,\mathbf{v}\,,\,t)$ is again of the form given by Eq. (12.11), but the indices i and j now run over the values $1, \ldots, 6$. The drift and diffusion matrices are (6×6) matrices that can be written in terms of (3×3) block matrices as

$$\Upsilon = \begin{pmatrix} 0 & -I \\ 0 & \gamma\, I \end{pmatrix} \quad \text{and} \quad \mathsf{D} = \begin{pmatrix} 0 & 0 \\ 0 & D\gamma^2\, I \end{pmatrix}. \tag{12.39}$$

Here 0 and I stand for the (3×3) null matrix and unit matrix, respectively.

(a) Using these expressions for the drift and diffusion matrices, verify that the Fokker-Planck equation (12.11) becomes, in this case, the following generalization of Eq. (12.13):

$$\frac{\partial \rho}{\partial t} = -\mathbf{v}\cdot\nabla_{\mathbf{r}}\,\rho + \gamma\nabla_{\mathbf{v}}\cdot(\mathbf{v}\rho) + \frac{\gamma k_B T}{m}\,\nabla^2_{\mathbf{v}}\rho, \tag{12.40}$$

[9] The fluid is isotropic. Hence, owing to rotational invariance, the strengths of the three white noises are identical.

where $\nabla_{\mathbf{r}}$ and $\nabla_{\mathbf{v}}$ denote the gradient operators with respect to the position and velocity, respectively.

- Equation (12.40) is the complete master equation for the phase space density of a tagged particle satisfying the Langevin equation in the absence of an applied force. It can be solved exactly, as we shall see below.

(b) Verify that $\Upsilon^2 = \gamma\,\Upsilon$ in this case too. Hence show that the matrix Green function $\mathsf{G}(t) = \exp(-\Upsilon t)$ is given by

$$\mathsf{G}(t) = \begin{pmatrix} I & I\,(1-e^{-\gamma t})/\gamma \\ 0 & I\,e^{-\gamma t} \end{pmatrix}. \tag{12.41}$$

(c) For sharp initial conditions $\mathbf{x}_0 \equiv (\mathbf{r}_0\,,\,\mathbf{v}_0)$, we again have $\overline{\mathbf{x}(t)} = \mathsf{G}(t)\,\mathbf{x}_0$. Show that the mean values of $\mathbf{r}$ and $\mathbf{v}$ are given by

$$\begin{aligned} \overline{\mathbf{r}(t)} &= \mathbf{r}_0 + \frac{\mathbf{v}_0}{\gamma}\,(1-e^{-\gamma t}), \\ \overline{\mathbf{v}(t)} &= \mathbf{v}_0\,e^{-\gamma t}. \end{aligned} \tag{12.42}$$

(d) The covariance matrix is again defined as

$$\boldsymbol{\sigma}(t) = 2\int_0^t \mathsf{G}(t')\,\mathsf{D}\,\mathsf{G}^{\mathrm{T}}(t')\,dt'\,. \tag{12.43}$$

Similarly, the deviations of x and v from their instantaneous mean values are

$$\delta\mathbf{r} = \mathbf{r} - \overline{\mathbf{r}(t)}, \quad \delta\mathbf{v} = \mathbf{v} - \overline{\mathbf{v}(t)}. \tag{12.44}$$

The Gaussian solution of the Fokker-Planck equation (12.40) is given by

$$\begin{aligned} \rho(\mathbf{r}\,,\,\mathbf{v}\,,\,t) &= \frac{1}{\left(2\pi\sqrt{\Delta}\right)^3}\,\times \\ &\times\ \exp\left[-\tfrac{1}{2}\left\{\mathsf{S}_{11}\,(\delta\mathbf{r})^2 + 2\mathsf{S}_{12}\,\delta\mathbf{r}\cdot\delta\mathbf{v} + \mathsf{S}_{22}\,(\delta\mathbf{v})^2\right\}\right], \end{aligned} \tag{12.45}$$

where $\Delta(t) = \det\boldsymbol{\sigma}(t)$, $\mathsf{S}(t) = \boldsymbol{\sigma}^{-1}(t)$. Work through the algebra and show that the explicit solution for $\rho(\mathbf{r}\,,\,\mathbf{v}\,,\,t)$ is

$$\begin{aligned} \rho(\mathbf{r}\,,\,\mathbf{v}\,,\,t) &= \frac{1}{\left(2\pi D\sqrt{AC-B^2}\right)^3}\,\times \\ &\times\ \exp\left\{-\frac{A(t)\,(\delta\mathbf{r})^2 - 2B(t)\,\delta\mathbf{r}\cdot\delta\mathbf{v} + C(t)\,(\delta\mathbf{v})^2}{4D\left(\gamma t - 2 + 4e^{-\gamma t} - 2e^{-2\gamma t} - \gamma t\,e^{-2\gamma t}\right)}\right\}, \end{aligned} \tag{12.46}$$

where $A(t)$, $B(t)$ and $C(t)$ are the functions defined earlier in the single-component case, in Eqs. (12.21).

Chapter 13

Diffusion in an external potential

We write down the Langevin equation for a tagged particle in the presence of an external or applied force derived from a time-independent potential. The general form of the SDE↔FPE correspondence is introduced. Using this result, we obtain the Fokker-Planck equation for the phase space density (the Kramers equation). The case of a harmonically bound particle (the Brownian oscillator) is discussed in detail. In the long-time or high-friction regime, the Kramers equation leads to a modified diffusion equation, the Smoluchowski equation, for the PDF of the position of the tagged particle. Kramers' escape rate formula for thermally-activated diffusion over a potential barrier is derived. We also consider the case of a constant external field, applicable to the problem of sedimentation, in one of the exercises at the end of the chapter.

13.1 Langevin equation in an external potential

In this chapter, we extend the considerations of the preceding chapter to the case in which a time-*independent* external force is also present. This is quite distinct from the case of a time-*dependent* applied force, which we shall discuss in the next chapter in connection with dynamic response functions. The essential point is that

- the fluid remains in thermal equilibrium in the presence of the external force.

We are concerned with the manner in which the probability distributions of the velocity and position of the tagged particle are modified in the presence of such a force. We are interested, in particular, in the modification of the diffusion equation for the positional PDF in this case. For simplicity of notation, we revert to the case of a single Cartesian component.

We assume that the applied force is a conservative one, derived from a potential $V(x)$. Hence $F_{\text{ext}}(x) = -dV(x)/dx$. The Langevin equation for $\mathbf{x} = (x\,,\,v)$ is represented by the pair of equations

$$\dot{x} = v \quad \text{and} \quad m\dot{v} + m\gamma v = -V'(x) + \eta(t), \tag{13.1}$$

where $\eta(t) = \sqrt{\Gamma}\,\zeta(t)$, and $\zeta(t)$ is a Gaussian white noise with zero mean and a δ-function correlation. As discussed in Ch. 6, we assume that the heat bath is unaffected by the presence or otherwise of the external force. Hence the FD relation $\Gamma = 2m\gamma k_B T$ remains valid.

As we have seen earlier, it is more appropriate to write the stochastic differential equation for $\mathbf{x}$ in terms of a driving Wiener process rather than a white noise. Accordingly, we write the two-component SDE in the form

$$\begin{aligned} dx(t) &= v\,dt, \\ dv(t) &= = -\gamma v\,dt - \frac{1}{m}V'(x)\,dt + \left(\frac{2\gamma k_B T}{m}\right)^{1/2} dw(t), \end{aligned} \tag{13.2}$$

in terms of the standard Wiener process $w(t)$. Note that the presence of the term $V'(x)$ makes these equations of motion *nonlinear* (in $\mathbf{x}$), in general. The only exceptions are the cases of a constant applied force and a linear applied force. Both these cases are of physical importance, and will be considered individually in the sequel. Our immediate task, however, is to consider the general case of an arbitrary prescribed force $-V'(x)$. For this purpose, it is convenient to refer to the general form of the SDE↔FPE correspondence at this stage.

13.2 General SDE↔FPE correspondence

The formal correspondence between the stochastic differential equation and the Fokker-Planck equation for a *multi-component* diffusion process is as follows. This is the generalization of the relationship between Eqs. (6.6) and (6.7), or, more precisely, the SDE↔FPE correspondence indicated in Eqs. (9.29) and (9.30).

Let $\boldsymbol{\xi}(t) = \big(\xi_1(t), \ldots, \xi_n(t)\big)$ denote an n-component general diffusion process: that is, $\boldsymbol{\xi}(t)$ is specified by the stochastic differential equation

$$d\boldsymbol{\xi}(t) = \mathsf{f}(\boldsymbol{\xi}, t)\,dt + \mathsf{g}(\boldsymbol{\xi}, \mathsf{t})\,\mathsf{dw}(\mathsf{t}). \tag{13.3}$$

Here $\mathbf{f} = (f_1, \ldots, f_n)$ is an n-dimensional drift vector, while

$$\mathbf{w}(t) = \big(w_1(t), \ldots, w_\nu(t)\big) \tag{13.4}$$

is a ν-dimensional Wiener process, and g is an $(n \times \nu)$ matrix function of its arguments. The components of $\mathbf{f}$ and g are deterministic functions of their arguments. Note that a distinction has been made between n and ν. This generalization allows for the possibility that the number of independent driving Wiener processes (or Gaussian white noises) may be different from the number n of driven (or system) variables[1]. Now define the $(n \times n)$ diffusion matrix D with elements

$$\mathsf{D}_{ij}(\boldsymbol{\xi}, t) = \frac{1}{2} \sum_{\alpha=1}^{\nu} \mathsf{g}_{i\alpha}(\boldsymbol{\xi}, t)\, \mathsf{g}_{j\alpha}(\boldsymbol{\xi}, t) = \frac{1}{2}\big(\mathsf{g}\,\mathsf{g}^{\mathrm{T}}\big)_{ij}. \tag{13.5}$$

That is, the matrix $\mathsf{D} = \frac{1}{2}\mathsf{g}\,\mathsf{g}^{\mathrm{T}}$. Then the Fokker-Planck equation (or forward Kolmogorov equation) satisfied by the PDF $p(\boldsymbol{\xi}, t)$ of $\boldsymbol{\xi}$ is

$$\frac{\partial}{\partial t} p(\boldsymbol{\xi}, t) = -\sum_{i=1}^{n} \frac{\partial}{\partial \xi_i} \left[f_i(\boldsymbol{\xi}, t)\, p(\boldsymbol{\xi}, t)\right] + \sum_{i,j=1}^{n} \frac{\partial^2}{\partial \xi_i\, \partial \xi_j} \left[\mathsf{D}_{ij}(\boldsymbol{\xi}, t)\, p(\boldsymbol{\xi}, t)\right]. \tag{13.6}$$

Equations (13.3)-(13.6) represent the general SDE$\leftrightarrow$FPE correspondence we require for our purposes.

13.3 The Kramers equation

The application of the correspondence between Eqs. (13.3) and (13.6) to the case at hand is straightforward. It follows from Eqs. (13.2) that we again have $n = 2$ and $\nu = 1$ in this instance. Further,

$$\mathbf{f} = \begin{pmatrix} v \\ -\gamma v - V'(x)/m \end{pmatrix}, \quad \mathsf{g} = \begin{pmatrix} 0 \\ (2\gamma k_B T/m)^{1/2} \end{pmatrix}. \tag{13.7}$$

Hence the diffusion matrix is

$$\mathsf{D} = \frac{1}{2}\mathsf{g}\,\mathsf{g}^{\mathrm{T}} = \begin{pmatrix} 0 & 0 \\ 0 & \gamma k_B T/m \end{pmatrix} = \begin{pmatrix} 0 & 0 \\ 0 & D\gamma^2 \end{pmatrix}. \tag{13.8}$$

This D is exactly the same as the diffusion matrix that we had in the absence of $V(x)$ (see Eq. (12.12)). This is of course to be expected: we are again concerned

[1] We have already encountered a simple instance of this possibility in Ch. 12. In that case, $n = 2$ while $\nu = 1$, because the two-dimensional process (x, v) is driven by the single-component noise $\eta(t)$ appearing in the equation of motion for v.

with the situation in which the system is in thermal equilibrium, albeit in the presence of the potential $V(x)$. Using Eq. (13.6), the Fokker-Planck equation for the joint PDF $\rho(x, v, t)$ that corresponds to Eqs. (13.2) is then

$$\frac{\partial \rho}{\partial t} = -v\,\frac{\partial \rho}{\partial x} + \frac{1}{m}V'(x)\,\frac{\partial \rho}{\partial v} + \gamma\,\frac{\partial}{\partial v}(v\rho) + \frac{\gamma k_B T}{m}\,\frac{\partial^2 \rho}{\partial v^2}\,. \tag{13.9}$$

Equation (13.9) is also called the **Kramers equation**. In the absence of the external potential $V(x)$, it reduces to Eq. (12.13), as expected. It is not possible to write down an explicit closed-form solution to the Kramers equation for an arbitrary potential $V(x)$. Moreover, in general the solution is not a Gaussian in $\mathbf{x}$. However, a closed-form solution can be obtained in a small number of special cases. Prominent among these is the case of a harmonic potential $V(x) = \frac{1}{2}m\omega_0^2 x^2$. We shall work out this case in the next section. In this instance, the joint PDF ρ will again turn out to be a Gaussian in $\mathbf{x}$.

13.4 The Brownian oscillator

The potential $V(x) = \frac{1}{2}m\omega_0^2 x^2$ corresponds to the case of a harmonically bound tagged particle in a heat bath. The particle is thus under the influence of a restoring force $-m\omega_0^2 x$ that varies linearly with the distance from the origin. As this example is both important and instructive, we shall write out the results in some detail.

What do we expect to find in the case of the oscillator? First of all, translational invariance (i. e., the fact that all points on the x-axis are on the same footing) is lost owing to the presence of the applied force. The center of oscillation, located at $x = 0$ in this case, is obviously a point that is singled out. Moreover, the confining, parabolic-bowl shape of the potential suggests that the mean square displacement of the particle may actually turn out to be finite in the $t \to \infty$ limit. This would imply that the asymptotic behavior of the displacement of the particle is not diffusive, i. e., that there is no long-range diffusion of the tagged particle. In physical terms: we know that the random force exerted on the particle by the white noise term in the Langevin equation causes fluctuations in its motion. For a freely diffusing particle, these fluctuations are such that the variance of the position of the particle increases like t at long times (even though the mean displacement remains zero). A harmonically bound particle is subjected to a restoring force $-m\,\omega_0^2\,x$ that always tends to push the particle back toward the origin. This force may be strong enough to overcome the tendency of the noise to cause the variance of the position to diverge in the $t \to \infty$ limit. If that happens, we might also expect to find an equilibrium distribution for the *position* of the tagged particle, over and above that of its velocity. As both the kinetic energy and the potential energy of the particle are *quadratic* functions of the respective dynamical variables v and x, we have reason to expect that the various PDFs concerned will be Gaussian in form. Let us now see to what extent, and under what conditions, these expectations are borne out.

The Langevin equation for $\mathbf{x}(t)$ is given by the pair of equations

$$\begin{aligned} dx(t) &= v\,dt, \\ dv(t) &= = -\gamma v\,dt - \omega_0^2\,x\,dt + \left(\frac{2\gamma k_B T}{m}\right)^{1/2} dw(t). \end{aligned} \tag{13.10}$$

Thus, the SDE is linear in $\mathbf{x}$ in this case. The Fokker-Planck equation (13.9) for the phase space density is now given by

$$\frac{\partial \rho}{\partial t} = -v\,\frac{\partial \rho}{\partial x} + \omega_0^2 x\,\frac{\partial \rho}{\partial v} + \gamma\,\frac{\partial}{\partial v}(v\rho) + \frac{\gamma k_B T}{m}\,\frac{\partial^2 \rho}{\partial v^2}\,. \tag{13.11}$$

It is evident that the drift is again linear in $\mathbf{x}$ in this case. Hence the Fokker-Planck equation (13.11) is of the form given in Eq. (12.11), namely,

$$\frac{\partial \rho}{\partial t} = \sum_{i,j=1}^{2} \left\{ \Upsilon_{ij}\frac{\partial}{\partial x_i}\,(x_j\,\rho) + \mathsf{D}_{ij}\frac{\partial^2 \rho}{\partial x_i\,\partial x_j} \right\}, \tag{13.12}$$

where $x_1 \equiv x$, $x_2 \equiv v$ as before[2]. The diffusion matrix D is of course the same as that in Eq. (12.12) or (13.8). The drift matrix in this case is

$$\Upsilon = \begin{pmatrix} 0 & -1 \\ \omega_0^2 & \gamma \end{pmatrix}. \tag{13.13}$$

As in the case of the free Brownian particle, we shall use the matrix formalism to solve the FPE.

Since the elements of the drift matrix are constants, the matrix Green function is $\mathsf{G}(t) = \exp(-\Upsilon t)$. The exponential is not as simple to compute as it was in the absence of the potential. However, we can find it quite easily by using the general result for the exponential of an arbitrary (2×2) matrix given in Appendix F. Alternatively, we can find the eigenvalues and eigenvectors of Υ, and diagonalize this matrix—and hence $\exp(-\Upsilon t)$ as well. For definiteness, let us write down the

[2]Recall that $\partial x_i/\partial x_j = \delta_{ij}$.

results in the case $\omega_0 > \frac{1}{2}\gamma$, which corresponds to an *underdamped* oscillator[3].

It is convenient to define the shifted frequency

$$\omega_s = \left(\omega_0^2 - \tfrac{1}{4}\gamma^2\right)^{1/2}. \tag{13.14}$$

The eigenvalues of Υ are then given by

$$\lambda_\pm = \frac{1}{2}\gamma \pm i\omega_s\,. \tag{13.15}$$

Carrying out the algebra, we obtain the following expression for the matrix Green function $\mathsf{G}(t)$:

$$\mathsf{G}(t) = \frac{e^{-\gamma t/2}}{\omega_s}\begin{pmatrix} \omega_s \cos\omega_s t + \frac{1}{2}\gamma\sin\omega_s t & \sin\omega_s t \\ -\omega_0^2\sin\omega_s t & \omega_s\cos\omega_s t - \frac{1}{2}\gamma\sin\omega_s t \end{pmatrix}. \tag{13.16}$$

Once again, for sharp initial conditions $(x_0\,,\,v_0) = \mathbf{x}_0\,$, the conditional averages of x and v are given by Eq. (12.17), namely, $\overline{\mathbf{x}(t)} = \mathsf{G}(t)\,\mathbf{x}_0$. Hence

$$\begin{aligned}\overline{x(t)} &= \frac{e^{-\gamma t/2}}{\omega_s}\left[\left(\omega_s\cos\omega_s t + \tfrac{1}{2}\gamma\sin\omega_s t\right)x_0 + (\sin\omega_s t)\,v_0\right],\\ \overline{v(t)} &= \frac{e^{-\gamma t/2}}{\omega_s}\left[-\left(\omega_0^2\sin\omega_s t\right)x_0 + \left(\omega_s\cos\omega_s t - \tfrac{1}{2}\gamma\sin\omega_s t\right)v_0\right].\end{aligned} \tag{13.17}$$

Observe that both $\overline{x(t)}$ and $\overline{v(t)}$ tend to zero as $t \to \infty$, whatever be the initial values x_0 and v_0, in contrast to the case of a free particle (Eqs. (12.18)). This already suggests that the Brownian oscillator may perhaps have an equilibrium PDF in phase space. This will indeed turn out to be so.

Next, we define (as usual) the deviations from the mean values, $\delta x = x - \overline{x(t)}$ and $\delta v = v - \overline{v(t)}$. We know already that the equal-time averages of the squares of these deviations and their cross-correlation are represented by the covariance matrix $\boldsymbol{\sigma}(t)$, according to

$$\begin{pmatrix}\sigma_{11}(t) & \sigma_{12}(t)\\ \sigma_{21}(t) & \sigma_{22}(t)\end{pmatrix} = \begin{pmatrix}\overline{\left(\delta x(t)\right)^2} & \overline{\delta x(t)\,\delta v(t)}\\ \overline{\delta v(t)\,\delta x(t)} & \overline{\left(\delta v(t)\right)^2}\end{pmatrix}. \tag{13.18}$$

[3]It is straightforward to extend the results to the overdamped case, in which $\omega_0 < \frac{1}{2}\gamma$. See the exercises at the end of this chapter.

Recall that $\boldsymbol{\sigma}(t)$ can be computed from $\mathsf{G}(t)$ and the diffusion matrix D using the formula of Eq. (12.19), namely,

$$\boldsymbol{\sigma}(t) = 2\int_0^t \mathsf{G}(t')\,\mathsf{D}\,\mathsf{G}^{\mathrm{T}}(t')\,dt'. \tag{13.19}$$

Carrying out the integrations involved and simplifying, we find the following expressions for the matrix elements of the covariance matrix $\boldsymbol{\sigma}(t)$:

$$\begin{aligned}
\sigma_{11}(t) &= \frac{k_BT}{m\omega_0^2}\left[1-\frac{e^{-\gamma t}}{\omega_s^2}\left(\omega_0^2+\tfrac{1}{2}\gamma\,\omega_s\,\sin 2\omega_s t-\tfrac{1}{4}\gamma^2\cos 2\omega_s t\right)\right],\\
\sigma_{12}(t) &= \sigma_{21}(t)=\frac{\gamma k_BT}{m\omega_s^2}\,e^{-\gamma t}\,\sin^2\omega_s t\,,\\
\sigma_{22}(t) &= \frac{k_BT}{m}\left[1-\frac{e^{-\gamma t}}{\omega_s^2}\left(\omega_0^2-\tfrac{1}{2}\gamma\,\omega_s\,\sin 2\omega_s t-\tfrac{1}{4}\gamma^2\cos 2\omega_s t\right)\right].
\end{aligned} \tag{13.20}$$

Each of these averages starts with the value zero at $t=0$, consistent with the sharp initial conditions we have imposed. Further, the corresponding *equilibrium* averages are readily deduced from the foregoing expressions by passing to the limit $t\to\infty$. Since both $\overline{x(t)}$ and $\overline{v(t)}$ vanish as $t\to\infty$, we thus obtain

$$\lim_{t\to\infty}\sigma_{11}(t)=\left\langle(\delta x)^2\right\rangle_{\text{eq}}=\left\langle x^2\right\rangle_{\text{eq}}=\frac{k_BT}{m\omega_0^2} \tag{13.21}$$

and

$$\lim_{t\to\infty}\sigma_{22}(t)=\left\langle(\delta v)^2\right\rangle_{\text{eq}}=\left\langle v^2\right\rangle_{\text{eq}}=\frac{k_BT}{m}. \tag{13.22}$$

These expressions are precisely what we would expect from the equipartition theorem for the energy. The average kinetic energy $\frac{1}{2}m\left\langle v^2\right\rangle_{\text{eq}}$ and the average potential energy $\frac{1}{2}m\omega_0^2\left\langle x^2\right\rangle_{\text{eq}}$ are each equal to $\frac{1}{2}k_BT$. We also find

$$\lim_{t\to\infty}\sigma_{12}(t)=\langle\delta x\,\delta v\rangle_{\text{eq}}=\langle x\,v\rangle_{\text{eq}}=0. \tag{13.23}$$

Now, $\langle x\,v\rangle_{\text{eq}}=0$ is an expected result. We have shown in Ch. 11, Eq. (11.38), that $\langle\xi(t)\,\dot{\xi}(t)\rangle=0$ for a stationary random process $\xi(t)$; and, in a state of thermal equilibrium, the position $x(t)$ of the oscillator *is* a stationary random process. (We shall derive expressions for the various equilibrium PDFs shortly.) Observe that the result in Eq. (13.23) for the Brownian oscillator is in marked contrast to that of Eq. (11.36) for a freely diffusing particle, namely, $\langle x\,v\rangle_{\text{eq}}=D\left(1-e^{-\gamma t}\right)\not\equiv 0$. As we have pointed out more than once, $x(t)$ is *not* a stationary random process for a freely diffusing particle in thermal equilibrium.

From Eqs. (13.20), we find that the determinant of the covariance matrix $\boldsymbol{\sigma}(t)$ simplifies to

$$\Delta(t) = \left(\frac{k_B T}{m\omega_0}\right)^2 \left[1 - \left(2 + \frac{\gamma^2}{\omega_s^2}\sin^2\omega_s t\right) e^{-\gamma t} + e^{-2\gamma t}\right]. \tag{13.24}$$

The elements of the inverse $\mathsf{S}(t)$ of the covariance matrix are, of course, given by

$$S_{11}(t) = \frac{\sigma_{22}(t)}{\Delta(t)}, \quad S_{12}(t) = S_{21}(t) = -\frac{\sigma_{12}(t)}{\Delta(t)}, \quad S_{22}(t) = \frac{\sigma_{11}(t)}{\Delta(t)}. \tag{13.25}$$

Substitution of the expressions for the various quantities concerned from Eqs. (13.20), (13.24) and (13.25) in the formal solution

$$\rho(x, v, t) = \frac{1}{2\pi\sqrt{\Delta(t)}} \exp\left[-\left\{\tfrac{1}{2}\mathsf{S}_{11}(t)\,(\delta x)^2 + \mathsf{S}_{12}(t)\,(\delta x)(\delta v) + \tfrac{1}{2}\mathsf{S}_{22}(t)\,(\delta v)^2\right\}\right] \tag{13.26}$$

yields the explicit solution for the phase space PDF corresponding to the Brownian oscillator.

The PDF $\rho(x, v, t)$ in Eq. (13.26) is again a two-dimensional Gaussian, as in the case of free diffusion (see Eq. (12.27)). However, there is a significant difference between the two cases. The PDF obtained here does have an equilibrium limit. It is easy to see from the $t \to \infty$ behavior of the covariance matrix (recall Eqs. (13.21)-(13.23) above) that

$$\lim_{t\to\infty} \rho(x, v, t) \equiv \rho^{\text{eq}}(x, v) = p^{\text{eq}}(v)\,\mathsf{p}^{\text{eq}}(x) \tag{13.27}$$

where

$$p^{\text{eq}}(v) = \left(\frac{m}{2\pi k_B T}\right)^{1/2} \exp\left(-\frac{mv^2}{2k_B T}\right), \tag{13.28}$$

as already defined, and

$$\mathsf{p}^{\text{eq}}(x) = \left(\frac{m\omega_0^2}{2\pi k_B T}\right)^{1/2} \exp\left(-\frac{m\omega_0^2 x^2}{2k_B T}\right). \tag{13.29}$$

Observe that the equilibrium PDF can be written as

$$\rho^{\text{eq}}(x, v) = \left(\frac{m\omega_0}{2\pi k_B T}\right) \exp\left(-\frac{\varepsilon}{k_B T}\right), \tag{13.30}$$

where $\varepsilon = \frac{1}{2}mv^2 + \frac{1}{2}m\omega_0^2x^2$ is the total (kinetic plus potential) energy of the tagged particle. In other words, the equilibrium phase space density—as we might expect—is nothing but the Boltzmann factor $\exp(-\varepsilon/k_BT)$, multiplied by the corresponding normalization factor.

We have emphasized repeatedly that, in the absence of an applied force, the position (or displacement) of the tagged particle is *not* a stationary random process, although its velocity is such a process. In marked contrast, we now find that:

- Both the displacement and the velocity of a harmonically bound tagged particle are stationary random processes.
- The behavior of the tagged particle is no longer asymptotically diffusive (i. e., there is no long-range diffusion)! The mean squared displacement of the particle does not diverge like t as $t \to \infty$.

In fact, as a consequence of the confining potential $V(x) = \frac{1}{2}m\omega_0^2x^2$, the mean squared displacement does not diverge at all in this limit. Rather, it tends to a constant, as is evident from Eq. (13.21). In Ch. 15, Sec. 15.4.1, we shall show by an alternative method that the diffusion coefficient of the Brownian oscillator vanishes formally.

13.5 The Smoluchowski equation

Let us now return to the case of a general potential $V(x)$, and consider the PDF of the tagged particle in the case when $\gamma t \gg 1$. It is not immediately apparent whether we should call this the 'diffusion regime', because the tagged particle does not necessarily exhibit long-range diffusion in all cases. As we have just seen, a confining potential may ensure that the mean squared displacement does not diverge like t (or even diverge at all) as $t \to \infty$. Let us therefore use the term 'long-time limit' for the present. In this temporal regime, in which t is much larger than the Smoluchowski time γ^{-1}, the positional PDF of the tagged particle turns out to satisfy a generalization of the diffusion equation called the **Smoluchowski equation**.

The Smoluchowski equation may be derived rigorously from the Fokker-Planck equation for the full phase space PDF as the leading approximation in a systematic expansion in powers of γ^{-1}, the reciprocal of the friction coefficient[4]. However,

[4]It is therefore appropriate to use the term 'high-friction approximation' for the case $\gamma t \gg 1$.

we shall not present this derivation here. Instead, we state the physical conditions under which such a reduction is possible, and write down the result, taking recourse (once again!) to the SDE↔FPE correspondence.

In the presence of an applied force, a further assumption is necessary to properly define the long-time (or high-friction) regime. We have already discussed some aspects of the time scales involved in the Langevin model, in Ch. 4, Sec. 4.1 and in Ch. 13, Sec. 11.2. We found in Eq. (11.8) that $\langle v^2 \rangle_{\text{eq}}^{1/2} /\gamma = (k_B T/m\gamma^2)^{1/2}$ represents the characteristic length scale in the motion of the tagged particle.

- In the presence of a potential $V(x)$, we must assume further that the applied force does not vary significantly over distances of the order of the characteristic length[5] $\langle v^2 \rangle_{\text{eq}}^{1/2} /\gamma$.

Then, in the long-time limit $\gamma t \gg 1$, whatever be the initial phase space density $\rho_{\text{init}}(x_0\,,\, v_0\,,\, 0)$, the velocity at any point will be distributed according to the equilibrium PDF $p^{\text{eq}}(v)$. In other words, for t much larger than the Smoluchowski time, the velocity of the tagged particle may be assumed to have essentially 'thermalized'. The joint PDF then factorizes in this regime according to

$$\rho(x\,,\, v\,,\, t)\Big|_{\gamma t \gg 1} \longrightarrow p^{\text{eq}}(v) \times \mathsf{p}(x, t), \tag{13.31}$$

where $\mathsf{p}(x\,,\, t)$ again denotes the PDF of the position by itself. The master equation satisfied by $\mathsf{p}(x\,,\, t)$ can be written down by the following argument. Going back to the Langevin equations (13.1), let us eliminate v and write the stochastic differential equation for x in the form

$$m\ddot{x} + m\gamma\dot{x} = -V'(x) + \eta(t). \tag{13.32}$$

The leading high-friction approximation corresponds to neglecting the inertia term in this equation, namely, the $\ddot{x}$ term. The result is the SDE

$$\dot{x} \simeq -\frac{1}{m\gamma}V'(x) + \frac{1}{m\gamma}\,\eta(t). \tag{13.33}$$

In the form analogous to Eq. (13.2), this is

$$dx(t) = -\frac{1}{m\gamma}V'(x)\,dt + \left(\frac{2k_B T}{m\gamma}\right)^{1/2} dw(t). \tag{13.34}$$

[5]In Ch. 11, Sec. 11.2, we have seen that this length scale is essentially of atomic dimensions ($\sim 10^{-10}$ m) for a micron-sized tagged particle. Hence the assumption above is really a very weak one, and is certainly justified in practice.

Hence the SDE↔FPE correspondence immediately yields, for the corresponding PDF $\mathsf{p}(x,t)$, the Fokker-Planck equation

$$\frac{\partial \mathsf{p}(x,t)}{\partial t} = \frac{1}{m\gamma}\frac{\partial}{\partial x}\big[V'(x)\,\mathsf{p}(x,t)\big] + \frac{k_B T}{m\gamma}\,\frac{\partial^2 \mathsf{p}(x,t)}{\partial x^2}. \tag{13.35}$$

This generalization of the diffusion equation is the Smoluchowski equation.

The presence of the first term (the drift term) on the right-hand side of the Smoluchowski equation opens up the possibility of a long-time behavior that is very different from the diffusive behavior of a free particle in a fluid. This may be seen as follows. Equation (13.35) can be written in the form of an equation of continuity,

$$\frac{\partial \mathsf{p}(x,t)}{\partial t} + \frac{\partial j(x,t)}{\partial x} = 0, \tag{13.36}$$

where the 'diffusion current density' $j(x,t)$ is given by

$$j(x,t) = -\frac{1}{m\gamma}\,\big[V'(x)\,\mathsf{p}(x,t)\big] - \frac{k_B T}{m\gamma}\,\frac{\partial \mathsf{p}(x,t)}{\partial x}. \tag{13.37}$$

Therefore, if a nontrivial *equilibrium* PDF $\mathsf{p}^{\mathrm{eq}}(x)$ exists in any given case, it must correspond to a constant or stationary current j^{st}. Further, these two quantities must be related according to

$$\frac{d\mathsf{p}^{\mathrm{eq}}(x)}{dx} + \frac{V'(x)}{k_B T}\,\mathsf{p}^{\mathrm{eq}}(x) = -\frac{m\gamma}{k_B T}\,j^{\mathrm{st}}. \tag{13.38}$$

In addition to satisfying the ordinary differential equation (13.38), $\mathsf{p}^{\mathrm{eq}}(x)$ must of course satisfy the physical requirements of non-negativity and normalizability. Whether such a solution exists or not depends on the particular potential concerned. If Eq. (13.38) does have such a solution, this is given formally by

$$\mathsf{p}^{\mathrm{eq}}(x) = \left(C - \frac{m\gamma}{k_B T}\,j^{\mathrm{st}}\int dx\; e^{V(x)/(k_B T)}\right) e^{-V(x)/(k_B T)}, \tag{13.39}$$

where C is a constant of integration. The values of the two constants j^{st} and C in the solution above are determined, of course, by the boundary conditions and the normalization of p^{eq}. In the special case when the constant $j^{\mathrm{st}} = 0$, we have

$$\mathsf{p}^{\mathrm{eq}}(x) = C\,\exp\left(-\frac{V(x)}{k_B T}\right). \tag{13.40}$$

As we might expect, this is just the Boltzmann factor corresponding to the potential energy of the particle.

It is obvious that, whenever an equilibrium PDF $\mathsf{p}^{\mathrm{eq}}(x)$ exists, the variance of the position cannot diverge as $t \to \infty$, but must tend to a constant value instead. There is then no long-range diffusion. In particular, consider the solution in Eq. (13.40). If the range of x is $(-\infty, \infty)$, and $V(x)$ is integrable, and moreover increases like $|x|^{\alpha}$ (where α is any positive number, however small) as $|x| \to \infty$, a normalizable $\mathsf{p}^{\mathrm{eq}}(x)$ exists. The particle remains confined by the potential without undergoing long-range diffusion.

13.6 Smoluchowski equation for the oscillator

Once again, let us revert to the case of the harmonic oscillator potential $V(x) = \frac{1}{2}m\omega_0^2 x^2$, and consider the leading high-friction approximation. The Langevin equation (13.33) is, in this instance,

$$\dot{x} \simeq -\frac{\omega_0^2}{\gamma}\,x + \frac{1}{m\gamma}\,\eta(t). \tag{13.41}$$

The Smoluchowski equation (13.35), i. e., the FPE corresponding to the SDE (13.41), is then given by

$$\frac{\partial \mathsf{p}(x,t)}{\partial t} = \frac{\omega_0^2}{\gamma}\frac{\partial}{\partial x}\big[x\,\mathsf{p}(x,t)\big] + \frac{k_B T}{m\gamma}\,\frac{\partial^2 \mathsf{p}(x,t)}{\partial x^2}. \tag{13.42}$$

Before we discuss this equation and its solution, we reiterate that Eq. (13.42) can be derived rigorously from the FPE for the phase space density as the leading approximation in a systematic expansion in powers of γ^{-1}, the reciprocal of the friction coefficient.

- In the present case, it can be shown that this expansion converges only if $\gamma^{-1} < (2\omega_0)^{-1}$, i. e., $\omega_0 < \frac{1}{2}\gamma$. Hence the results of this section pertain to the case of an *overdamped* oscillator.

A very interesting fact that is evident from an inspection of Eq. (13.42) is the following:

- The Langevin equation (13.41) for the *position* $x(t)$ of the Brownian oscillator in the high-friction regime is exactly the same in form as the original Langevin equation (6.1) for the *velocity* $v(t)$ of a free tagged particle.

- Correspondingly, Eq. (13.42) for the PDF of the position of the Brownian oscillator (in the high-friction regime) is exactly the same in form as the original Fokker-Planck equation (or Rayleigh equation) for the PDF of the velocity of a free Langevin particle, Eq. (6.9).

Repeating the latter equation for ready comparison, we have

$$\frac{\partial p(v,t)}{\partial t} = \gamma\,\frac{\partial}{\partial v}\,[v\,p(v,t)] + \frac{\gamma k_B T}{m}\,\frac{\partial^2 p(v,t)}{\partial v^2}. \tag{13.43}$$

It is now obvious that, apart from the constant factors, Eqs. (13.42) and (13.43) are identical in form.

- Hence, for a sharp initial condition $\mathsf{p}(x,0) = \delta(x - x_0)$ and free boundary conditions, the solution for the positional PDF $\mathsf{p}(x,t)$ of the oscillator is precisely the Ornstein-Uhlenbeck density.

- For this reason, the OU process is sometimes also called the **oscillator process**.

The normalized solution to Eq. (13.42) can thus be written down by analogy with the known solution to Eq. (6.9) or (13.43), given by Eq. (6.14). However, to make sure that we get all the constant factors right, let us deduce the solution to Eq. (13.42) afresh, using the fact that the solution is a Gaussian for all t.

We need, then, to find the (conditional) mean and variance of x in order to write down the Gaussian explicitly. This is easily done by a method that should be quite familiar by now[6]. Multiplying both sides of Eq. (13.42) by x and integrating over all x, we find that the mean value $\overline{x(t)}$ satisfies the differential equation

$$\frac{d}{dt}\,\overline{x(t)} = -\,\frac{\omega_0^2}{\gamma}\,\overline{x(t)}. \tag{13.44}$$

Hence, with the sharp initial condition $\mathsf{p}(x,0) = \delta(x - x_0)$ which implies that $\overline{x(0)} = x_0$, the conditional mean value of the position is

$$\overline{x(t)} = e^{-\omega_0^2 t/\gamma}\,x_0\,. \tag{13.45}$$

Similarly, multiplying both sides of Eq. (13.42) by x^2 and integrating over all x, we obtain the equation

$$\frac{d}{dt}\,\overline{x^2(t)} + \frac{2\omega_0^2}{\gamma}\,\overline{x^2(t)} = \frac{2k_BT}{m\gamma}\,. \tag{13.46}$$

The solution corresponding to the initial condition $\overline{x^2(0)} = x_0^2$ is

$$\overline{x^2(t)} = e^{-2\omega_0^2 t/\gamma}\,x_0^2 + \frac{k_BT}{m\omega_0^2}\left(1 - e^{-2\omega_0^2 t/\gamma}\right). \tag{13.47}$$

[6]You need to use integration by parts, and the fact that the normalized PDF $\mathsf{p}(x,t)$ vanishes as $x \to \pm\infty$ faster than any inverse power of x.

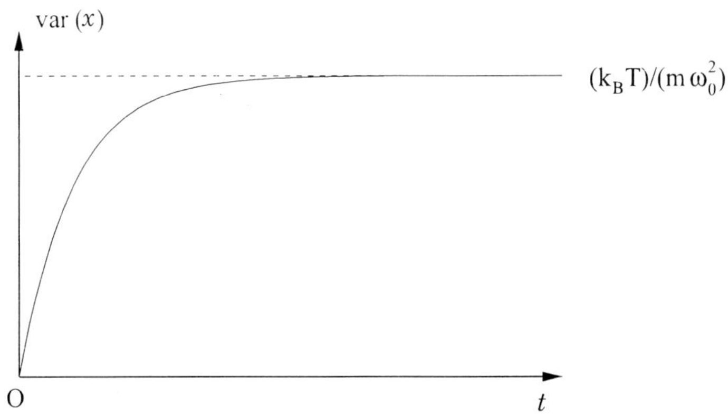

Figure 13.1: The variance of the position of an overdamped Brownian oscillator as a function of time, as given by Eq. (13.48). The behavior shown is valid in the high-friction regime, in which the Smoluchowski equation (13.42) for the positional PDF $\mathsf{p}(x,t)$ is valid. The characteristic relaxation time of the position is then given by γ/ω_0^2.

Hence the variance of the position is given by

$$\overline{x^2(t)} - \left(\overline{x(t)}\right)^2 = \frac{k_B T}{m\omega_0^2}\left(1 - e^{-2\omega_0^2 t/\gamma}\right). \tag{13.48}$$

Figure 13.1 shows how this variance saturates to its equilibrium value as $t \to \infty$. Once again, observe that the dependence on the initial position x_0 cancels out in the variance. Note, too, that although exponentially decaying terms of the form $\exp(-\gamma t)$ and $\exp(-2\gamma t)$ are neglected in the approximation under discussion, damped exponentials of the form $\exp(-\omega_0^2 t/\gamma)$ and $\exp(-2\omega_0^2 t/\gamma)$ are retained. This is completely consistent with the fact that $\omega_0 \ll \gamma$ in the high-friction regime.

Given the conditional mean and variance of x, the PDF of the position can now be written down. Indicating the initial condition explicitly in the argument of p, the conditional PDF is given by

$$\mathsf{p}(x,t\,|x_0) = \left[\frac{m\omega_0^2}{2\pi k_B T(1 - e^{-2\omega_0^2 t/\gamma})}\right]^{\frac{1}{2}} \exp\left[\frac{-m\omega_0^2\left(x - x_0 e^{-\omega_0^2 t/\gamma}\right)^2}{2k_B T(1 - e^{-2\omega_0^2 t/\gamma})}\right]. \tag{13.49}$$

- We reiterate that this is *not* the exact positional PDF of a harmonically bound tagged particle: it is an approximation to this PDF that is valid at times $t \gg \gamma^{-1}$, in the highly overdamped case.

Finally, note that the limit $t \to \infty$ in Eq. (13.49) correctly yields the expression in Eq. (13.29) for the equilibrium PDF of the position of the oscillator.

- The relaxation time of the velocity of the oscillator is $1/\gamma$, as in the case of a freely diffusing tagged particle. The relaxation time of the position of the oscillator is γ/ω_0^2, as is evident from the expression for $\overline{x(t)}$ in Eq. (13.45), as well as the form of the conditional PDF in Eq. (13.49).

- The high friction approximation and the corresponding Smoluchowski equation discussed in this section pertain to the case when *the velocity of the tagged particle relaxes much more rapidly than the position*, which is equivalent to saying that we are in the highly overdamped regime, $\omega_0 \ll \gamma$.

13.7 Escape over a potential barrier: Kramers' escape rate formula

An important application of the Smoluchowski equation pertains to the rate of escape, by diffusion, of (tagged) particles trapped in a potential well. This problem arises in numerous situations involving thermally activated processes. Chemical reactions offer a prime example. There is an extensive literature on the subject, covering different regimes of physical parameters such as the shape of the barrier, the ratio of its height to the thermal energy k_BT, and so on. An interesting and commonly-occurring situation is one in which the height of the potential barrier is much larger than k_BT. Even so, thermal fluctuations can lead to a small but nonzero flux of particles across the barrier. A simple approximate formula for the escape rate in this case is obtained as follows.

The potential barrier we consider is shown in Fig. 13.2. As stated above, we assume that the height of the barrier $\Delta V \equiv V(x_1) - V(x_0) \gg k_BT$. We wish to calculate the rate at which particles diffuse from the neighborhood of the minimum at x_0 to a point like x_2 on the other side of the maximum at x_1. A quasi-static situation is assumed: imagine that there is a steady input of particles into the neighborhood $(a\,,\,b)$ of x_0, together with a steady removal of particles at x_2. The escape rate over the barrier, $\lambda_{\rm esc}$, is calculated by noting that the small (but nonzero) stationary flux at x_2 is given by the product of the 'probability mass' around x_0 and the rate of escape over the barrier. Hence

$$\lambda_{\rm esc} = \frac{\text{flux at } x_2}{\text{probability mass around } x_0}. \tag{13.50}$$

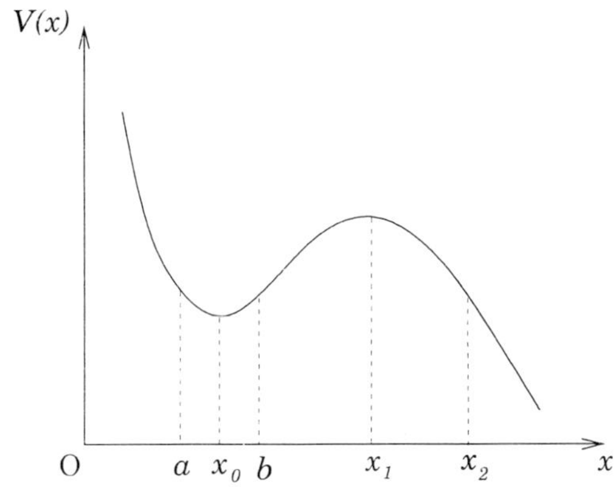

Figure 13.2: Schematic of a potential barrier. Owing to thermal fluctuations, tagged particles can diffuse from the vicinity of the potential well around x_0, escaping over the maximum at x_1 to a point such as x_2. This happens even if the barrier height $\Delta V = V(x_1) - V(x_0)$ is much larger than the thermal energy $k_B T$. The Kramers formula, Eq. (13.58), gives an estimate of the rate of escape over the barrier. The rate turns out to be proportional to an Arrhenius-type factor $\exp(-\Delta V/k_B T)$. The prefactor involves the curvature of the potential at x_0 and at x_1, as well as the friction constant γ.

The probability mass around x_0 is given by $\int_a^b dx\, p(x,t)$, but we can approximate it by using the stationary distribution $\mathsf{p}^{\text{eq}}(x)$ for the PDF. The latter is given by Eq. (13.39). But this solution again involves the stationary current j^{st}—and this latter quantity is precisely the flux we require as an *independent* input for the calculation of λ_{esc}. However, since j^{st} is small, we may drop the term proportional to it (i. e., the particular integral) in Eq. (13.39). Hence $\mathsf{p}^{\text{eq}}(x)$ in the region $(a\,,\,b)$ is effectively given by Eq. (13.40). The proportionality constant C may be eliminated by observing that it cancels out in the ratio of the PDFs at two different points. Thus

$$\mathsf{p}^{\text{eq}}(x) = \mathsf{p}^{\text{eq}}(x_0)\,\exp\left(\frac{V(x_0)-V(x)}{k_B T}\right). \tag{13.51}$$

Therefore the probability mass is

$$\int_a^b dx\,\mathsf{p}^{\text{eq}}(x) = \mathsf{p}^{\text{eq}}(x_0)\int_a^b dx\, e^{[V(x_0)-V(x)]/(k_B T)}. \tag{13.52}$$

But the integral on the right-hand side is precisely of the form that is amenable to a Gaussian approximation, as explained in Appendix A, Sec. A.4. Using the result in Eq. (A.14), we get

$$\int_a^b dx\,\mathsf{p}^{\text{eq}}(x) \simeq \mathsf{p}^{\text{eq}}(x_0)\left(\frac{2\pi k_B T}{V''(x_0)}\right)^{1/2}. \tag{13.53}$$

Next, we turn to the estimation of the stationary flux or current at x_2. We note that Eq. (13.38) for the stationary current can be re-written as

$$j^{\text{st}} = -\frac{k_B T}{m\gamma}\, e^{-V(x)/(k_B T)}\,\frac{d}{dx}\left(e^{V(x)/(k_B T)}\,\mathsf{p}^{\text{eq}}(x)\right). \tag{13.54}$$

Moving the factor $\exp[-V(x)/(k_B T)]$ to the left-hand side and integrating both sides from $x = x_0$ to $x = x_2$, we get

$$j^{\text{st}}\int_{x_0}^{x_2} dx\, e^{V(x)/(k_B T)} = \frac{k_B T}{m\gamma}\left[e^{V(x_0)/(k_B T)}\,\mathsf{p}^{\text{eq}}(x_0) - e^{V(x_2)/(k_B T)}\,\mathsf{p}^{\text{eq}}(x_2)\right]. \tag{13.55}$$

The stationary density $\mathsf{p}^{\text{eq}}(x_2)$ at the endpoint x_2 is negligible, and may be dropped. Further, the integral on the left-hand side is dominated by the contribution from the neighborhood of the *maximum* of the potential at x_1. Once again, a Gaussian approximation to the integral may be made. The result is

$$\int_{x_0}^{x_2} dx\, e^{V(x)/(k_B T)} \simeq \left(\frac{2\pi k_B T}{|V''(x_1)|}\right)^{1/2} e^{V(x_1)/(k_B T)}. \tag{13.56}$$

(Note that $V''(x_1) < 0$ because $V(x)$ has a maximum at x_1.) Therefore

$$j^{\text{st}} \simeq \frac{k_B T}{m\gamma} \left(\frac{2\pi k_B T}{|V''(x_1)|} \right)^{-1/2} \mathsf{p}^{\text{eq}}(x_0)\, e^{-\Delta V/(k_B T)}. \tag{13.57}$$

Using Eqs. (13.53) and (13.57) in Eq. (13.50), we get the formula sought for the escape rate. The quantity $\mathsf{p}^{\text{eq}}(x_0)$ cancels out, and we obtain

$$\lambda_{\text{esc}} = \frac{\left[V''(x_0)\, |V''(x_1)|\right]^{1/2}}{2\pi m\gamma}\, e^{-\Delta V/(k_B T)}. \tag{13.58}$$

This is the Kramers formula for the rate of thermally activated escape over a potential barrier. It involves the curvature of the potential at the minimum and at the maximum of the potential. These quantities are of course directly related to the frequency of harmonic oscillations (or small oscillations) in the corresponding potential wells: writing $V''(x_0)/m = \omega_0^2$ and $|V''(x_1)|/m = \omega_1^2$, we have the simple formula

$$\lambda_{\text{esc}} = \left(\frac{\omega_0\, \omega_1}{2\pi\gamma} \right) e^{-\Delta V/(k_B T)}. \tag{13.59}$$

13.8 Exercises

13.8.1 Phase space PDF for the overdamped oscillator

Extend the results obtained in Sec. 13.4 to the case of the overdamped oscillator, $\omega_0 < \frac{1}{2}\gamma$. Instead of the shifted frequency $\omega_s = \left(\omega_0^2 - \frac{1}{4}\gamma^2\right)^{1/2}$, it is now convenient to define the shifted friction coefficient

$$\gamma_s = \left(\gamma^2 - 4\omega_0^2\right)^{1/2}, \tag{13.60}$$

so that $\omega_s \equiv \frac{1}{2} i\gamma_s$. You have now merely to use the relations $\sin iz = i \sinh z$ and $\cos iz = \cosh z$ in all the expressions derived earlier. Show that the elements of the covariance matrix are given by

$$\begin{aligned}
\sigma_{11}(t) &= \frac{k_B T}{m\omega_0^2} \left[1 + \frac{4e^{-\gamma t}}{\gamma_s^2} \left(\omega_0^2 - \tfrac{1}{4}\gamma\gamma_s \sinh \gamma_s t - \tfrac{1}{4}\gamma^2 \cosh \gamma_s t \right) \right], \\
\sigma_{12}(t) &= \sigma_{21}(t) = \frac{2\gamma k_B T}{m\gamma_s^2}\, e^{-\gamma t} \left(\cosh \gamma_s t - 1 \right), \\
\sigma_{22}(t) &= \frac{k_B T}{m} \left[1 + \frac{4e^{-\gamma t}}{\gamma_s^2} \left(\omega_0^2 + \tfrac{1}{4}\gamma\gamma_s \sinh \gamma_s t - \tfrac{1}{4}\gamma^2 \cosh \gamma_s t \right) \right].
\end{aligned} \tag{13.61}$$

Similarly, obtain the counterpart of Eq. (13.24) for the determinant $\Delta(t)$, and write down the result for the phase space PDF $\rho(x, v, t)$ in this case.

13.8.2 Critically damped oscillator

Considerable simplification of all the expressions involved in the foregoing occurs in the critically damped case $\omega_0 = \frac{1}{2}\gamma$. Show that one now obtains

$$\begin{aligned}
\sigma_{11}(t) &= \frac{4k_BT}{m\gamma^2}\left[1-\left(1+\gamma t+\tfrac{1}{2}\gamma^2t^2\right)e^{-\gamma t}\right], \\
\sigma_{12}(t) &= \sigma_{21}(t) = \frac{\gamma k_BT}{m}\,t^2e^{-\gamma t}, \\
\sigma_{22}(t) &= \frac{k_BT}{m}\left[1-\left(1-\gamma t+\tfrac{1}{2}\gamma^2t^2\right)e^{-\gamma t}\right],
\end{aligned} \tag{13.62}$$

and hence

$$\Delta(t) = \left(\frac{2k_BT}{m\gamma}\right)^2\left[1-\left(2+\gamma^2t^2\right)e^{-\gamma t}+e^{-2\gamma t}\right]. \tag{13.63}$$

13.8.3 Diffusion in a constant force field: Sedimentation

The problem of the diffusion of a tagged particle in a constant force field, i. e., under a linear potential, is of considerable practical importance in many applications. A common example is the problem of the sedimentation of particles in a fluid held in a container, under the influence of gravity. If x stands for the vertical coordinate of a tagged particle, and the 'floor' of the container is taken to be at $x = 0$, the potential in this case is $V(x) = mgx$ with $x \geq 0$. A reflecting boundary condition must be imposed at $x = 0$, because there is no 'leakage' through the floor.

Let us therefore consider the linear potential $V(x) = Kx$ (where K is a positive constant) in the region $x \geq 0$. We shall assume that $x = 0$ is a perfectly reflecting boundary[7]. The Smoluchowski equation in this case is

$$\frac{\partial \mathsf{p}(x,t)}{\partial t} = \frac{K}{m\gamma}\frac{\partial \mathsf{p}(x,t)}{\partial x} + \frac{k_BT}{m\gamma}\frac{\partial^2 \mathsf{p}(x,t)}{\partial x^2}. \tag{13.64}$$

Note that the combination

$$\frac{K}{m\gamma} \equiv c \tag{13.65}$$

has the physical dimensions of a velocity. It represents the *drift velocity* arising from the external force. For $K > 0$ (or $c > 0$), this drift is directed toward the origin, i. e., in the direction of decreasing x. From what has been said at the end of Sec. 13.5, it follows that the particle will remain confined in this potential, in

[7]Equivalently, we may set $V(x) = +\infty$ for $x < 0$.

the sense that the long-time behavior of the variance of its position will not be diffusive. However, it is notationally convenient to write $k_B T/m\gamma = D$, with the understanding that D denotes (as always) the diffusion constant in the *absence* of the potential. This enables us to re-write the Smoluchowski equation in the suggestive form

$$\frac{\partial \mathsf{p}(x,t)}{\partial t} = c\,\frac{\partial \mathsf{p}(x,t)}{\partial x} + D\,\frac{\partial^2 \mathsf{p}(x,t)}{\partial x^2}. \qquad (13.66)$$

It is quite evident that the two terms on the right-hand side of Eq. (13.66) represent the effects of drift and diffusion, respectively. The ratio c/D provides (the reciprocal of) a natural or characteristic length scale in this problem. Loosely speaking, it is a measure of the relative importance of the drift and diffusion contributions to the motion of the particle[8]. In the sedimentation problem, $c/D = mg/k_B T$.

A perfectly reflecting boundary at $x = 0$ implies that the current through the boundary vanishes identically. Hence the boundary condition imposed at $x = 0$ is given by $j(x,t)\big|_{x=0} = 0$. Referring to Eq. (13.37) for the current density, we find that the boundary condition at $x = 0$ is given by

$$\left[\frac{\partial \mathsf{p}(x,t)}{\partial x} + \frac{c}{D}\,\mathsf{p}(x,t)\right]_{x=0} = 0. \qquad (13.67)$$

At the other boundary, we impose the natural boundary condition

$$\lim_{x\to\infty} \mathsf{p}(x,t) = 0. \qquad (13.68)$$

The task now is to find the solution to Eq. (13.66) with the boundary conditions (13.67) and (13.68), together with the initial condition

$$\mathsf{p}(x,0) = \delta(x - x_0) \qquad (13.69)$$

where x_0 is any given positive number.

(a) Before undertaking this task, find the equilibrium PDF $\mathsf{p}^{\rm eq}(x)$ in this case. Since $\mathsf{p}^{\rm eq}(x)$ is independent of t, it follows from Eq. (13.36) that $j^{\rm st}$ must be a constant independent of x. But the boundary condition at $x = 0$ requires the current to be zero at that point. Hence $j^{\rm st}$ must vanish identically. Use this fact

[8] If there is already a length scale in the problem, as in the case of a column of finite height L, this measure is provided by the dimensionless ratio Lc/D. This is (one of the definitions of) the **Péclet number**.

to show that the normalized equilibrium PDF in the region $0 \leq x < \infty$ is given by

$$\mathsf{p}^{\text{eq}}(x) = \frac{c}{D}\,\exp\left(-\frac{cx}{D}\right). \tag{13.70}$$

This expression is just a version of the so-called **barometric distribution** for the density of the atmosphere, assuming the latter to be isothermal: the ratio $c/D = mg/k_B T$ in that case, where m is the mass of a molecule of air.

(b) Show that

$$\langle x \rangle_{\text{eq}} = D/c\,, \quad \langle x^2 \rangle_{\text{eq}} = 2(D/c)^2\,, \quad \text{so that} \quad \langle x^2 \rangle_{\text{eq}} - \langle x \rangle_{\text{eq}}^2 = (D/c)^2\,. \tag{13.71}$$

Hence the variance of x does not diverge as $t \to \infty$. There is no long-range diffusion in this case, corroborating our earlier assertion.

(c) Let us now turn to the time-dependent Smoluchowski equation (13.66). A convenient way to solve it is to use Laplace transforms. Denoting the transform of $\mathsf{p}(x,t)$ by $\widetilde{\mathsf{p}}(x,s)$, show that this quantity satisfies the equation

$$\frac{d^2\widetilde{\mathsf{p}}}{dx^2} + \frac{c}{D}\frac{d\widetilde{\mathsf{p}}}{dx} - \frac{s}{D}\widetilde{\mathsf{p}} = -\frac{1}{D}\delta(x - x_0). \tag{13.72}$$

Equation (13.72) has the form of a Green function equation. Its solution using a standard procedure is straightforward[9]. Show that

$$\widetilde{\mathsf{p}}(x,s) = \frac{2}{r(r-c)}\,\exp\left(-\frac{c(x-x_0) + rx_>}{2D}\right)\left(r\,\cosh\frac{rx_<}{2D} - c\,\sinh\frac{rx_<}{2D}\right), \tag{13.73}$$

where $x_< = \min(x_0\,,\,x)$, $x_> = \max(x_0\,,\,x)$, and $r(s) = (c^2 + 4sD)^{1/2}$.

(d) Check that $\int_0^\infty dx\widetilde{\mathsf{p}}(x,s) = 1/s$, thus verifying that the PDF $\mathsf{p}(x,t)$ is correctly normalized according to $\int_0^\infty dx\,\mathsf{p}(x,t) = 1$ for all $t \geq 0$.

[9]Here is a quick recapitulation of this procedure. The δ-function vanishes for $x > x_0$ and also for $x < x_0$. In each of these two regions, Eq. (13.72) then becomes a homogeneous differential equation with constant coefficients, whose solution is a linear combination of two exponentials. Hence there are four constants of integration to be determined. The solution for $x > x_0$ cannot involve an increasing exponential in x because p must vanish as $x \to \infty$. To find the remaining three constants of integration, match the two solutions at $x = x_0$, and use the fact that the discontinuity in the first derivative $d\widetilde{\mathsf{p}}/dx$ at $x = x_0$ is equal to $-1/D$. Finally, impose the boundary condition at $x = 0$ to determine the solution completely.

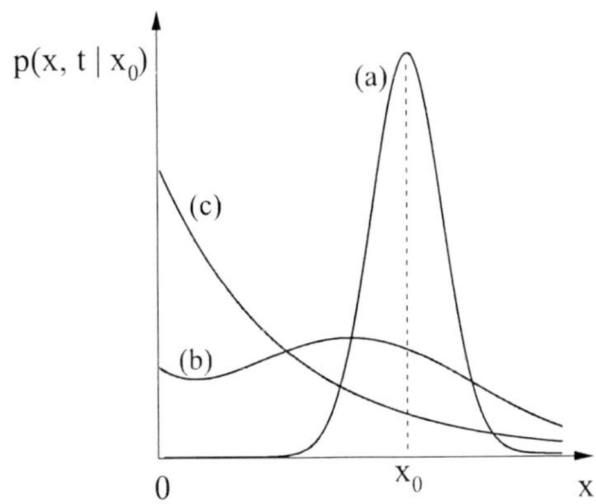

Figure 13.3: The probability density of the position of the tagged particle (which starts at an arbitrary point x_0) in the presence of a reflecting boundary at $x = 0$, under the combined influence of diffusion as well as a constant force directed toward the origin. Sedimentation under gravity is a physical example of this situation. (The PDF $\mathsf{p}(x,t)$ essentially represents the concentration of particles in that case.) (a) The initial δ-function peak at x_0 spreads out at very short times as in free diffusion. (b) The effect of reflection at the boundary at the origin is clearly seen in the profile at intermediate times. (c) The limiting density as $t \to \infty$ is an exponential function that decays with increasing x, with a characteristic length D/c that is just the ratio of the free diffusion constant D to the drift velocity c. The variance of x does not diverge as $t \to \infty$. Instead, it tends to the finite value $(D/c)^2$ in this limit.

(e) Remarkably enough, the complicated expression in Eq. (13.73) for the Laplace transform of the PDF can be inverted, and $\mathsf{p}(x,t)$ can be obtained explicitly. The result is

$$\begin{aligned} \mathsf{p}(x,t|x_0) &= \frac{e^{-[2c(x-x_0)+c^2t]/(4D)}}{\sqrt{4\pi Dt}} \left[e^{-(x-x_0)^2/(4Dt)} + e^{-(x+x_0)^2/(4Dt)} \right] \\ &+ \frac{c}{2D}\, e^{-cx/D}\, \mathrm{erfc}\left(\frac{x+x_0-ct}{\sqrt{4Dt}} \right), \end{aligned} \tag{13.74}$$

where $\mathrm{erfc}\, x$ is the complementary error function[10]. Verify that this solution satisfies the Smoluchowski equation (13.66) and the boundary conditions in Eqs. (13.67) and (13.68). Figure 13.3 depicts the solution in Eq. (13.74) at three different values of t, showing how the initial δ-function at $x = x_0$ evolves into the

[10]This function is defined as $\mathrm{erfc}\, x = 1 - \mathrm{erf}\, x = (2/\sqrt{\pi}) \int_x^{\infty} du\, e^{-u^2}$. See Appendix A.

equilibrium PDF $\mathsf{p}^{\mathrm{eq}}(x)$.

Chapter 14

Diffusion in a magnetic field

Returning to the problem of a charged tagged particle in a magnetic field, we write down the corresponding Fokker-Planck equation in velocity space. This is solved to obtain a modified form of Ornstein-Uhlenbeck distribution for the conditional PDF of the velocity. The conditional mean velocity is found, and the result is analyzed in physical terms. Passing to the high-friction approximation, the diffusion equation satisfied by the PDF of the position of the particle in the long-time limit is derived. The diffusion tensor is calculated. Its longitudinal and transverse parts are identified and shown to behave as expected on physical grounds.

In Ch. 4, Sec. 4.3.2, we studied the equilibrium velocity correlation of a charged tagged particle in a uniform, constant magnetic field. In this chapter, we return to this very instructive physical problem, and study it in greater depth. We now have the machinery to analyze this multi-component situation completely.

14.1 The PDF of the velocity

14.1.1 The Fokker-Planck equation

In this section, we shall write down and solve the Fokker-Planck equation for the conditional PDF of the velocity of the particle. The solution will turn out to be a modified Ornstein-Uhlenbeck distribution.

We start with the Langevin equation (4.18) for the particle in a magnetic field $\mathbf{B}$, written in component form:

$$\dot{v}_j(t) = -\gamma\, v_j(t) + \frac{q}{m}\,\epsilon_{jkl}\, v_k(t)\, B_l + \frac{1}{m}\,\eta_j(t). \tag{14.1}$$

Observe that:

- This equation is still linear in $\mathbf{v}$, although the different components of the velocity are 'mixed up' by the velocity-dependent Lorentz force that acts on the particle.

- The noise is additive.

We may therefore use the version of the SDE$\leftrightarrow$FPE correspondence that pertains to such a case, and write down at once the Fokker-Planck equation for the conditional PDF $p(\mathbf{v}, t|\mathbf{v}_0)$ of the velocity: This FPE is of the form given in Eq. (12.11) or Eq. (13.12), with the indices now running from 1 to 3. Retaining the summation convention for repeated indices, we have

$$\frac{\partial p}{\partial t} = -\frac{\partial}{\partial v_j}\left[\left(-\gamma\, v_j + \frac{q}{m}\,\epsilon_{jkl}\, v_k\, B_l\right)\, p\right] + \frac{\Gamma}{2m^2}\,\delta_{jk}\,\frac{\partial^2 p}{\partial v_j\,\partial v_k}\,. \tag{14.2}$$

Before we turn to the solution of Eq. (14.2), some remarks are in order on the nature of the drift terms in the present instance.

14.1.2 Detailed balance

The drift terms in the Langevin equation (14.1) are $-\gamma\, v_j$ and $(q/m)\,\epsilon_{jkl}\, v_k\, B_l$, respectively. Repeating what has been said in Sec. 4.3.2, the first of these represents the dissipative effect of the medium, and is the *irreversible* part of the drift. The second arises from the Lorentz force on the particle, and is the *reversible* part of the drift. The former changes sign under time reversal, while the latter does not do so[1]. Now, in Ch. 5, Sec. 5.3.3, we have introduced the concept of detailed balance, and shown how this condition enables us to find the equilibrium probability distribution of a (discrete) Markov process in terms of the transition probabilities governing the process. In the present situation, too, the detailed balance condition is applicable. In Ch. 6, Sec. 6.2 we saw that, in the absence of an external force, the Fokker-Planck equation for the velocity of the tagged particle could be written in the form of a continuity equation, Eq. (6.10). The probability current in that case was given by Eq. (6.11). We then found, by the argument given in Sec. 6.4, that the equilibrium distribution could be deduced by setting the stationary probability current J^{st} to zero. In the case at hand, we can again write the FPE in the form of a continuity equation, identify the stationary current, and impose the detailed balance condition. The FPE (14.2) becomes

$$\frac{\partial p}{\partial t} = -\frac{\partial J_j(\mathbf{v}, t)}{\partial v_j} \equiv -\nabla_{\mathbf{v}}\cdot\mathbf{J}(\mathbf{v}, t)\,, \tag{14.3}$$

[1]These 'time parities' apply to the particular problem at hand. They are not general statements.

where $\nabla_{\mathbf{v}}$ is the gradient operator in velocity space[2]. The probability current is given by

$$\mathbf{J}(\mathbf{v},t) = \Big[-\gamma\,\mathbf{v} + \frac{q}{m}\,(\mathbf{v}\times\mathbf{B})\Big]\,p(\mathbf{v},t) + \frac{\Gamma}{2m^2}\,\nabla_{\mathbf{v}}\,p(\mathbf{v},t). \tag{14.4}$$

The stationary current is therefore given by

$$\begin{aligned}\mathbf{J}^{\text{st}}(\mathbf{v}) &= \Big[-\gamma\,\mathbf{v}\,p^{\text{st}}(\mathbf{v}) + \frac{\Gamma}{2m^2}\,\nabla_{\mathbf{v}}p^{\text{st}}(\mathbf{v})\Big] + \frac{q}{m}\,(\mathbf{v}\times\mathbf{B})\,p^{\text{st}}(\mathbf{v})\\ &\equiv \mathbf{J}^{\text{st}}_{\text{irrev}} + \mathbf{J}^{\text{st}}_{\text{rev}}\,,\end{aligned} \tag{14.5}$$

where we have separated the irreversible and reversible contributions. As a consequence of the fact that the current is made up of these two parts, the detailed balance condition in this case is a generalization of the condition $J^{\text{st}}(v) = 0$ that was used in Sec. 6.4. The necessary and sufficient conditions for detailed balance to hold good (for a continuous process) are as follows, in general:

- The *irreversible* part of the stationary probability current must vanish, i. e., $\mathbf{J}^{\text{st}}_{\text{irrev}} = 0$.
- The *reversible* part of the stationary probability current must have a vanishing divergence, i. e., $\nabla_{\mathbf{v}}\cdot\mathbf{J}^{\text{st}}_{\text{rev}} = 0$.

The former condition leads to the *equilibrium* probability density function $p^{\text{eq}}(\mathbf{v}) \propto \exp\left(-m^2\gamma\,\mathbf{v}^2/\Gamma\right)$, on following the steps leading to Eq. (6.18). Incorporating the consistency condition (or FD relation) $\Gamma = 2m\gamma k_BT$ then yields the normalized equilibrium Maxwellian PDF

$$p^{\text{eq}}(\mathbf{v}) = \left(\frac{m}{2\pi k_BT}\right)^{3/2}\,\exp\left(-\frac{m\mathbf{v}^2}{2k_BT}\right), \tag{14.6}$$

as expected. It remains to check the second part of the detailed balance condition, namely, the vanishing of the divergence of the reversible part of the stationary current. This is left as an exercise for the reader[3].

14.1.3 The modified OU distribution

We now turn to the solution of the Fokker-Planck equation (14.2), with the initial condition $p(\mathbf{v},0|\mathbf{v}_0) = \delta^{(3)}(\mathbf{v}-\mathbf{v}_0)$, and natural boundary conditions $p \to 0$ as

[2]This notation has been used earlier, in Eq. (12.40), Sec. 12.6.2 .

[3]See the exercises at the end of this chapter.

$|\mathbf{v}| \to \infty$. In order to keep the focus on the physical features of the expression obtained, and its implications, we shall merely write down the solution here. A rather simple way to derive it is described in Appendix G.

The drift is linear in the velocity, and the noise is additive. Hence we may expect the solution to the FPE to be related to a multi-dimensional OU process (in the three components of the velocity). As in Ch. 4, Sec. 4.3.2, let the magnetic field be directed along an arbitrary unit vector $\mathbf{n}$. Recall also that we have defined the antisymmetric matrix M with elements $M_{kj} = \epsilon_{kjl}\, n_l$ (Eq. (4.19)), and the cyclotron frequency $\omega_c = qB/m$ (Eq. (4.15)). The significance of M should be clear. Imagine, for a moment, that the tagged particle moves under the influence of the magnetic field alone, in the *absence* of the medium (and hence the frictional drag as well as the noise). Then the equation of motion of the particle, Eq. (4.20), reduces to $\dot{v}_j(t) = \omega_c\, M_{jk}\, v_k(t)$, or

$$\dot{\mathbf{v}} = \omega_c\, \mathsf{M}\, \mathbf{v}. \tag{14.7}$$

Hence the initial velocity $\mathbf{v}_0$ of the tagged particle evolves deterministically in time. It is given at any instant t by the vector[4]

$$\mathbf{u}(t\,;\, \mathbf{v}_0) = e^{\mathsf{M}\omega_c t}\, \mathbf{v}_0 = \left[I + \mathsf{M}\, \sin\, \omega_c t + \mathsf{M}^2\, (1 - \cos\, \omega_c t)\right]\, \mathbf{v}_0\,. \tag{14.8}$$

We have used Eq. (4.27) for the exponential of the matrix M to write down the second equation. Equation (14.8) is a succinct way of expressing the following well-known fact: the component of the initial velocity along the magnetic field remains unchanged, while the two transverse components undergo simple harmonic motion, with a mutual phase difference of $\frac{1}{2}\pi$ at all times[5]. Even more briefly: $\mathbf{u}(t; \mathbf{v}_0)$ is simply the initial velocity vector $\mathbf{v}_0$ rotated about the axis represented by the magnetic field vector $\mathbf{B}$, through an angle $\omega_c t$.

Returning to the Fokker-Planck equation, the solution obtained for the conditional PDF of the velocity involves the vector $\mathbf{u}$, modulated by the usual damping factor $e^{-\gamma t}$. It reads:

$$p(\mathbf{v}, t\,|\, \mathbf{v}_0) = \left[\frac{m}{2\pi k_B T(1 - e^{-2\gamma t})}\right]^{3/2} \exp\left\{-\,\frac{m\{\mathbf{v} - \mathbf{u}(t\,;\, \mathbf{v}_0)\, e^{-\gamma t}\}^2}{2k_B T\,(1 - e^{-2\gamma t})}\right\}. \tag{14.9}$$

It is immediately evident that

$$\lim_{t\to\infty} p(\mathbf{v}, t\,|\, \mathbf{v}_0) = \left(\frac{m}{2\pi k_B T}\right)^{3/2} \exp\left(-\,\frac{m\mathbf{v}^2}{2k_B T}\right) = p^{\rm eq}(\mathbf{v}), \tag{14.10}$$

[4]The notation implied in Eq. (14.8) should be obvious: the vector $\mathbf{v}_0$ is to be written as a column vector, and the matrices act on it from the left to produce the column vector $\mathbf{u}$.

[5]See the exercises at the end of this chapter.

the Maxwellian PDF, as expected. The (conditional) mean velocity at any time $t > 0$ is given by

$$\overline{\mathbf{v}(t)} = \mathbf{u}(t\,;\,\mathbf{v}_0)\,e^{-\gamma t}. \tag{14.11}$$

This is nothing but the initial velocity $\mathbf{v}_0$ rotated about the direction of the magnetic field through an angle $\omega_c t$, and diminished in magnitude by the damped exponential factor $\exp(-\gamma t)$. The variance of the velocity behaves exactly as it does in the absence of the magnetic field: it starts with the value zero, and increases monotonically to the saturation value $3k_BT/m$ that is characteristic of the equilibrium Maxwellian distribution.

14.2 Diffusion in position space

Let us turn, now, to the positional probability density function of the tagged particle in the long-time or diffusion regime $t \gg \gamma^{-1}$. Without loss of generality, we assume that the particle starts from the origin of coordinates at $t = 0$. We use the abbreviated notation $\mathsf{p}(\mathbf{r}, t)$ for the conditional PDF $\mathsf{p}(\mathbf{r}, t\,|\,\mathbf{0}, 0)$. The questions that arise are:

- What sort of Fokker-Planck equation, if any, does $\mathsf{p}(\mathbf{r}, t)$ satisfy in this regime? Is it a simple diffusion equation?
- If so, how do the diffusion coefficients in different directions get modified owing to the presence of the magnetic field?

It turns out that $\mathsf{p}(\mathbf{r}, t)$ does indeed satisfy a three-dimensional diffusion equation. In the next section, we shall derive this equation, once again by using a simple physical argument and exploiting the familiar SDE↔FPE correspondence. We continue the derivation in Sec. 14.2.2 and determine the set of coefficients comprising the **diffusion tensor** D_{ij}. There are several other ways of identifying this tensor. In Ch. 15, we shall point out how these alternatives also yield the same result. This agreement is yet another check on the consistency of the entire formalism that forms the theme of this book.

14.2.1 The diffusion equation

In Ch. 4, Sec. 4.3.2, we deduced the velocity correlation of the tagged particle in the presence of the magnetic field. The behavior of this quantity, and the behavior of the conditional mean velocity as found in Sec. 14.1.3, show clearly that

- the magnetic field does not affect the relaxation time of the velocity, which remains equal to the Smoluchowski time γ^{-1}.

We may therefore pass to the high-friction approximation that is valid in the diffusion regime $t \gg \gamma^{-1}$ exactly as we did in the absence of the magnetic field[6].

We go back to the Langevin equation satisfied by the tagged particle. It is convenient to use the form of the LE in Eq. (4.20). Repeating this equation for ready reference,

$$\dot{v}_j(t) = -\gamma v_j(t) + \omega_c\, M_{jk}\, v_k(t) + \frac{1}{m}\, \eta_j(t). \tag{14.12}$$

In the high-friction approximation, the inertia term $\dot{\mathbf{v}}$ in the LE may be neglected. We thus have

$$\gamma v_j(t) - \omega_c\, M_{jk}\, v_k \simeq \frac{1}{m}\, \eta_j(t). \tag{14.13}$$

In vector form, this may be written as $(\gamma\, I - \omega_c\, \mathsf{M})\, \mathbf{v} = (1/m)\, \boldsymbol{\eta}$. Hence

$$\dot{\mathbf{r}}(t) = \frac{1}{m}\, (\gamma I - \omega_c\, \mathsf{M})^{-1}\, \boldsymbol{\eta}(t) \equiv \mathbf{h}(t), \tag{14.14}$$

say[7]. $\mathbf{h}(t)$ is also a (three-component) stationary white noise with zero mean. The stationarity of $\mathbf{h}(t)$ follows from that of $\boldsymbol{\eta}(t)$ because the factor to the left of $\boldsymbol{\eta}$ in the definition of $\mathbf{h}$ is time-independent. In component form, the SDE (14.14) is $\dot{x}_i = h_i(t)$, where x_i stands for a Cartesian component of the position of the tagged particle, and $h_i(t)$ is a white noise, i. e., it is δ-correlated. The SDE↔FPE correspondence then implies at once that the corresponding PDF satisfies a Fokker-Planck equation of the form

$$\frac{\partial \mathsf{p}}{\partial t} = D_{ij}\, \frac{\partial^2 \mathsf{p}}{\partial x_i\, \partial x_j}, \tag{14.15}$$

where D_{ij} denotes the ij^{th} component of the diffusion tensor[8]. It is of course related to the autocorrelation function of the driving white noise by the equation

$$\overline{h_i(t_1)\, h_j(t_2)} = 2D_{ij}\, \delta(t_1 - t_2)\,. \tag{14.16}$$

[6]This point should be fairly clear by now. But it is not necessarily obvious *à priori*. After all, the magnetic field does introduce *another* time scale into the problem, namely, ω_c^{-1}, in addition to the already-present time scale γ^{-1}. In principle, this could have changed the relaxation time of the velocity: the latter could then have turned out to be γ^{-1} multiplied by some function of the dimensionless ratio γ/ω_c. But this does not happen.

[7]M itself does not have an inverse, because one of its eigenvalues is equal to zero. But we can assert that $(\gamma I - \omega_c\, \mathsf{M})$ does have an inverse, for all positive values of the parameters γ and ω_c. (Why? Hint: recall what the eigenvalues of M are.)

[8]A summation over the repeated indices in Eq. (14.15) is implied, as usual.

- The implication of Eq. (14.15) is that the motion of the tagged particle remains diffusive even in the presence of a magnetic field. The mean squared displacement of the particle in a time interval t is guaranteed to be proportional to t, this being the primary characteristic of normal diffusion.

14.2.2 The diffusion tensor

Having shown that the positional PDF $\mathsf{p}(\mathbf{r}, t)$ satisfies a three-dimensional diffusion equation in the presence of a magnetic field, the next task is to determine the set of diffusion coefficients $\{D_{ij}\}$. There is no reason to expect D_{ij} to be of the form $D\,\delta_{ij}$, because all directions in space are no longer equivalent in the presence of the magnetic field.

Here is what we should expect on simple physical grounds: Since the field does not affect the component of the motion of the particle along $\mathbf{B}$, we expect the 'longitudinal diffusion coefficient' to take on its original value, $D = k_B T/m\gamma$. On the other hand, the tendency of the field to always bend the path of the particle in the transverse plane into a circular orbit may be expected to diminish the distance it diffuses in a given time. We may therefore expect the 'transverse diffusion coefficient' to be smaller than the 'free' value $k_B T/m\gamma$. Further, we expect this reduction to be higher, the higher the strength of the field is.

Let us now see how these expectations are borne out. The autocorrelation function of the noise $\mathbf{h}(t)$ is given by

$$\overline{h_i(t_1)\,h_j(t_2)} = \frac{1}{m^2}\,[(\gamma I - \omega_c\,\mathsf{M})^{-1}]_{ik}\,[(\gamma I - \omega_c\,\mathsf{M})^{-1}]_{jl}\;\overline{\eta_k(t_1)\,\eta_l(t_2)}, \qquad (14.17)$$

from the definition of $\mathbf{h}(t)$. But we know (see Eq. (14.18)) that

$$\overline{\eta_k(t_1)\,\eta_l(t_2)} = \Gamma\,\delta_{kl}\,\delta(t_1 - t_2). \qquad (14.18)$$

Hence

$$\overline{h_i(t_1)\,h_j(t_2)} = \frac{\Gamma}{m^2}\,[(\gamma I - \omega_c\,\mathsf{M})^{-1}]_{ik}\,[(\gamma I - \omega_c\,\mathsf{M})^{-1}]_{jk}\,\delta(t_1 - t_2). \qquad (14.19)$$

Comparing this expression with that in Eq. (14.16), and further using the FD relation $\Gamma = 2m\gamma k_B T$, we get

$$D_{ij} = \frac{\gamma k_B T}{m}\,[(\gamma I - \omega_c\,\mathsf{M})^{-1}]_{ik}\,[(\gamma I - \omega_c\,\mathsf{M})^{-1}]_{jk}\,. \qquad (14.20)$$

The matrix inverse that appears above can be evaluated in more than one way[9]. The result we obtain finally is

$$D_{ij} = \frac{k_B T}{m\gamma} \left[n_i\, n_j + \frac{\gamma^2}{\gamma^2 + \omega_c^2} \left(\delta_{ij} - n_i\, n_j \right) \right]. \tag{14.21}$$

There is a specific reason why we have combined terms so as to write D_{ij} in the form above. It corresponds to separating the longitudinal and transverse components of the diffusion tensor[10]. Hence the **longitudinal diffusion coefficient** is the coefficient of $n_i\, n_j$ in D_{ij} , and is given by

$$D_{\text{long}} = \frac{k_B T}{m\gamma} = D. \tag{14.22}$$

Thus, the longitudinal diffusion coefficient remains equal to the free-particle value, as expected. The **transverse diffusion coefficient** is the coefficient of $(\delta_{ij} - n_i\, n_j)$ in D_{ij} , and is given by

$$D_{\text{trans}} = \frac{k_B T}{m\gamma} \frac{\gamma^2}{\gamma^2 + \omega_c^2} = D\, \frac{\gamma^2}{\gamma^2 + \omega_c^2}\,. \tag{14.23}$$

This result shows precisely how the cyclotron motion of the tagged particle in the plane transverse to the magnetic field reduces the effective diffusion coefficient in that plane. The attenuation factor starts at unity for zero field, and decreases to zero monotonically as $B \to \infty$. Figure 14.1 illustrates this diminution of D_{trans} with increasing field strength, for a fixed value of γ.

14.3 Phase space distribution

In the preceding sections, we have discussed the velocity distribution of the tagged particle in the presence of a magnetic field, as well as its positional PDF in the

[9]See the exercises at the end of this chapter.

[10]A simple but basic point is involved here. An explanation may be in order, in case this is unfamiliar to you. Given a unit vector $\mathbf{n}$, the tensor product (or *dyadic*, in somewhat older terminology) $\mathbf{n}\,\mathbf{n}$ is the projection operator that projects out the component (of any other vector) along $\mathbf{n}$. The unit operator I is the sum of all the projection operators corresponding to the basis set of vectors in the space concerned. Hence $(I - \mathbf{n}\,\mathbf{n})$ projects out the complete transverse part of any vector upon which it operates. When written in component form, these operators are represented by tensors (whose components are) given by $n_i\, n_j$ and $(\delta_{ij} - n_i\, n_j)$, respectively. To see the projection explicitly, 'apply' these to the unit vector $\mathbf{n}$. In component form, this amounts to multiplying n_j by these tensors. We thus have $(n_i\, n_j)n_j = n_i$, while $(\delta_{ij} - n_i\, n_j)n_j = 0$, on summing over the repeated index j and using the fact that $n_j\, n_j = 1$.

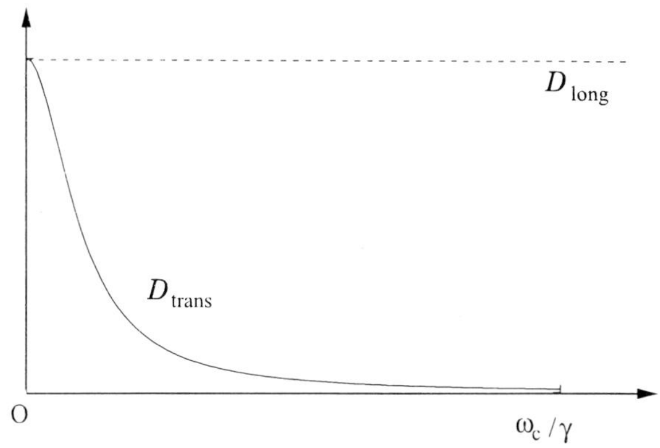

Figure 14.1: The variation of the transverse diffusion coefficient D_{trans} as a function of the ratio of the cyclotron frequency to the friction constant, $\omega_c/\gamma = qB/(m\gamma)$. The curve is (the right half of) a Lorentzian. The longitudinal diffusion coefficient D_{long} remains constant at the value $k_BT/(m\gamma)$, which is the value of the diffusion coefficient in the absence of a magnetic field.

diffusion regime. These are special cases derived from the full phase space probability density function of the particle. In Ch. 12, we have found the phase space distribution of the tagged particle in the absence of any external force, both for a single Cartesian component of its motion (Eq. (12.27)), as well as in the full three-dimensional case (Eq. (12.46)). Can we find the phase space PDF in the presence of a magnetic field?

It turns out that an exact expression can indeed be obtained for the full phase space density $\rho(\mathbf{r}, \mathbf{v}, t \,|\, \mathbf{0}, \mathbf{v}_0, 0)$. We will not go through this solution in detail. Instead, we shall merely indicate the noteworthy aspects that are involved.

The Langevin equations for the pair $(\mathbf{r}, \mathbf{v})$ are of course

$$\begin{aligned} \dot{\mathbf{r}} &= \mathbf{v}, \\ \dot{\mathbf{v}} &= -\gamma\,\mathbf{v}(t) + \frac{q}{m}\,(\mathbf{v}\times\mathbf{B}) + \frac{1}{m}\,\boldsymbol{\eta}(t). \end{aligned} \tag{14.24}$$

The SDE↔FPE correspondence then yields the following Fokker-Planck equation for the PDF ρ:

$$\frac{\partial\rho}{\partial t} = -\mathbf{v}\cdot\nabla_{\mathbf{r}}\,\rho + \gamma\nabla_{\mathbf{v}}\cdot(\mathbf{v}\rho) - \frac{q}{m}\,(\mathbf{v}\times\mathbf{B})\cdot\nabla_{\mathbf{v}}\rho + \frac{\gamma k_BT}{m}\,\nabla_{\mathbf{v}}^2\rho, \tag{14.25}$$

where (as before) $\nabla_{\mathbf{r}}$ and $\nabla_{\mathbf{v}}$ denote the gradient operators with respect to the position and velocity, respectively. As in the field-free case (Eq. (12.40)), we have already put in the FD relation to eliminate Γ. In comparison with that case, the only extra term here is of course the one involving $\mathbf{B}$. But this suffices to make the solution considerably more complicated algebraically, essentially because of the way the magnetic field mixes up the transverse components of the motion. However, two crucial features do continue to be valid: the linearity in $\mathbf{v}$ of the drift terms, and the additivity of the noise. Hence $\rho(\mathbf{r}, \mathbf{v}, t \,|\, \mathbf{0}, \mathbf{v}_0, 0)$ is again a multivariate Gaussian in the deviations of $\mathbf{v}$ and $\mathbf{r}$ from their conditional mean values. These deviations are given by[11]

$$\delta\mathbf{v} \equiv \mathbf{v} - \overline{\mathbf{v}(t)} = \mathbf{v} - e^{-(\gamma I - \mathsf{M}\omega_c)t}\,\mathbf{v}_0 \tag{14.26}$$

and

$$\delta\mathbf{r} \equiv \mathbf{r} - \overline{\mathbf{r}(t)} = \mathbf{r} - (\gamma I - \mathsf{M}\,\omega_c)^{-1}\left(I - e^{-(\gamma I - \mathsf{M}\omega_c)t}\right)\mathbf{v}_0\,. \tag{14.27}$$

Again, as we did in the field-free case, we may go on to calculate the (6×6) covariance matrix, its determinant and inverse. The multivariate Gaussian expression for ρ can then be written down. The algebra is somewhat tedious and the resulting expression quite lengthy. We shall not present it here.

14.4 Exercises

14.4.1 Velocity space FPE in vector form

Show that the Fokker-Planck equation (14.2) for the PDF of the velocity can be written in vector form as

$$\frac{\partial p}{\partial t} = \gamma\,\nabla_{\mathbf{v}} \cdot (\mathbf{v}p) - \frac{q}{m}\,(\mathbf{v} \times \mathbf{B}) \cdot (\nabla_{\mathbf{v}} p) + \frac{\Gamma}{2m^2}\,\nabla_{\mathbf{v}}^2 p. \tag{14.28}$$

14.4.2 Divergence of the reversible stationary current

Show that the divergence of the reversible part of the stationary current vanishes identically. Essentially, you must show that

$$\nabla_{\mathbf{v}} \cdot \left[(\mathbf{v} \times \mathbf{B})\; e^{-(\text{const.})\,\mathbf{v}^2}\right] = 0. \tag{14.29}$$

[11]See Sec. 14.4.3 below.

14.4.3 Conditional mean velocity and displacement

(a) Let the tagged particle start with velocity $\mathbf{v}_0$ at $t = 0$. Show that Eqs. (14.8) and (14.11) then imply that the conditional mean velocity at any time t is given by the explicit expression

$$\overline{\mathbf{v}(t)} = \big[\mathbf{v}_0 \cos \omega_c t + (\mathbf{n}\cdot\mathbf{v}_0)\,\mathbf{n}\,(1 - \cos \omega_c t) + (\mathbf{v}_0 \times \mathbf{n}) \sin \omega_c t\big]\, e^{-\gamma t}. \quad (14.30)$$

Re-arrange this expression so as to obtain

$$\overline{\mathbf{v}(t)} = \big[(\mathbf{n}\cdot\mathbf{v}_0)\,\mathbf{n} + \mathbf{n}\times(\mathbf{v}_0 \times \mathbf{n}) \cos \omega_c t + (\mathbf{v}_0 \times \mathbf{n}) \sin \omega_c t\big]\, e^{-\gamma t}. \quad (14.31)$$

The final formula shows, *in a manner independent of the orientation of the coordinate axes and the choice of the direction of the magnetic field*, how the average value of the longitudinal component of the initial velocity along the magnetic field decays monotonically with time; and how the two transverse components damp out in an oscillatory fashion, while remaining out of phase with each other by $\frac{1}{2}\pi$.

(b) Let the tagged particle start from an arbitrary point $\mathbf{r}_0$ with a velocity $\mathbf{v}_0$ at $t = 0$. Show that the conditional mean displacement of the tagged particle at time t is given formally by

$$\overline{\mathbf{r}(t)} - \mathbf{r}_0 = (\gamma I - \mathsf{M}\,\omega_c)^{-1}\left(I - e^{-(\gamma I - \mathsf{M}\omega_c)t}\right)\mathbf{v}_0\,. \quad (14.32)$$

Simplify this expression to obtain the mean displacement as an explicit function of t, $\mathbf{v}_0$ and the field direction $\mathbf{n}$. Verify that the result matches what you get by directly integrating the right-hand side of Eq. (14.31) with respect to t, as it ought to.

14.4.4 Calculation of D_{ij}

Starting from Eq. (14.20) for D_{ij}, carry out the calculation necessary to arrive at the formula in Eq. (14.21). You can do this in either of two slightly different ways.

(a) Find the matrix inverse $(\gamma I - \omega_c\,\mathsf{M})^{-1}$. An easy way to do this is to observe that the Laplace transform of $\exp(\mathsf{M}\,\omega_c t)$ is formally given by

$$\int_0^\infty dt\, e^{-st+\mathsf{M}\,\omega_c t} = (sI - \omega_c\,\mathsf{M})^{-1}. \quad (14.33)$$

But $\exp(\mathsf{M}\,\omega_c t)$ has already been found, in Eq. (4.27). Hence its Laplace transform is easily written down. The matrix inverse required is just this transform, evaluated at the value $s = \gamma$. Now put in the expressions for the matrix elements of M and M^2 from Eqs. (4.19) and (4.28), namely, $\mathsf{M}_{ij} = \epsilon_{ijk}\, n_k$ and $(\mathsf{M}^2)_{ij} = n_i\, n_j - \delta_{ij}$, and simplify.

(b) Alternatively, first use the antisymmetry property $\mathsf{M}^{\mathrm{T}} = -\mathsf{M}$ to show that

$$[(\gamma I - \omega_c\, \mathsf{M})^{-1}]_{ik}\, [(\gamma I - \omega_c\, \mathsf{M})^{-1}]_{jk} = [(\gamma^2 I - \omega_c^2\, \mathsf{M}^2)^{-1}]_{ij}\,. \tag{14.34}$$

The matrix inverse in the above can now be computed easily if you use the following property: since $\mathsf{M}^3 = -\mathsf{M}$, we have $\mathsf{M}^4 = -\mathsf{M}^2$, and hence

$$\mathsf{M}^{2n} = (-1)^{n-1}\, \mathsf{M}^2, \; n \geq 1. \tag{14.35}$$

Using this fact, show that

$$(\gamma^2 I - \omega_c^2\, \mathsf{M}^2)^{-1} = \frac{1}{\gamma^2}\, I + \frac{\omega_c^2}{\gamma^2 + \omega_c^2}\, \mathsf{M}^2\,. \tag{14.36}$$

Now put in the expression for the matrix element of M^2 and simplify.

(c) Based on the comments made regarding the decomposition of the diffusion tensor into its longitudinal and transverse parts, write down the eigenvalues of the matrix with elements D_{ij}.

14.4.5 Phase space FPE in vector form

Write out the pair of Langevin equations (14.24) in component form. Use the SDE↔FPE correspondence in the multidimensional case to obtain the corresponding Fokker-Planck equation for the phase space density. Express this FPE in vector notation, to arrive at Eq. (14.25).

Chapter 15

Kubo-Green formulas

A general formula is derived, expressing the diffusion constant as the time integral of the equilibrium velocity autocorrelation In three dimensions, the diffusion tensor is the time integral of the symmetric part of the velocity correlation tensor. The static and dynamic mobilities are defined and related to certain integrals of the velocity autocorrelation. This leads to the fundamental Kubo-Green formula for the dynamic mobility. The formalism is applied to the case of the Brownian oscillator, and also to that of a charged particle in a magnetic field. Additional remarks are made on causality and stationarity in the Langevin model.

15.1 Relation between D and the velocity autocorrelation

We return to the Langevin model for a single Cartesian component of the motion of the tagged particle in the absence of an applied force. As we have seen, there is a direct relationship (Eq. (11.6)) between the diffusion coefficient D and the dissipation coefficient γ, namely, $D = k_B T/(m\gamma)$. It is clear that this relationship is specific to the model concerned, i. e., to the Langevin equation that we have used to describe the random motion of the tagged particle. However, if we go back a step in the derivation of this relation, a very general formula can be obtained for D. This general formula is not restricted to any particular model for the random motion concerned. It follows, in fact, from linear response theory (LRT). However, the derivation of this formula is quite simple, and can be given without getting into the general formalism of LRT.

We know that the mean squared displacement tends to $2Dt$ in the diffusion regime, whenever there is long-range diffusion[1]. D may therefore be defined as

$$D = \lim_{t \to \infty} \frac{1}{2t} \left\langle X^2(t) \right\rangle_{\text{eq}} . \tag{15.1}$$

But the mean squared displacement can be written as in Eq. (11.3). This equation is actually an identity that follows from the very definition of the displacement. Repeating it for ready reference, we have

$$\left\langle X^2(t) \right\rangle_{\text{eq}} = \int_0^t dt_1 \int_0^t dt_2 \, \langle v(t_1)\, v(t_2) \rangle_{\text{eq}} . \tag{15.2}$$

Substituting this expression in Eq. (15.1), we have the formal relation

$$D = \lim_{t \to \infty} \frac{1}{2t} \int_0^t dt_1 \int_0^t dt_2 \, \langle v(t_1)\, v(t_2) \rangle_{\text{eq}} . \tag{15.3}$$

Equation (15.3) is the starting point for the derivation of the general formula for D. The steps that follow are typical of the kind of algebraic simplification that is frequently necessary in this subject. It is therefore instructive to write them out in detail, at least this once.

For brevity, let us set[2]

$$\langle v(t_1)\, v(t_2) \rangle_{\text{eq}} = \phi(t_2 - t_1), \quad t_2 \geq t_1 . \tag{15.4}$$

The basic property we exploit is the following: the velocity of the tagged particle is a stationary random process when the system is in thermal equilibrium. We have shown in Ch. 4 (see Eq. (4.8)) that $\langle v(t_1)\, v(t_2) \rangle_{\text{eq}}$ is then a function of $|t_2 - t_1|$. It must be emphasized that this is a general property, *and is not restricted to the Langevin equation model for the velocity.* In order to keep this property manifest, let us write $\phi(|t_2 - t_1|)$ for the correlation function. Then

$$\begin{aligned} \int_0^t dt_1 \int_0^t dt_2 \, \langle v(t_1)\, v(t_2) \rangle_{\text{eq}} &= \left(\int_0^t dt_1 \int_0^{t_1} dt_2 + \int_0^t dt_1 \int_{t_1}^t dt_2 \right) \phi(|t_2 - t_1|) \\ &= \left(\int_0^t dt_1 \int_0^{t_1} dt_2 + \int_0^t dt_2 \int_0^{t_2} dt_1 \right) \phi(|t_2 - t_1|), \end{aligned} \tag{15.5}$$

[1]In this book we do not consider cases involving **anomalous diffusion**. An example of the latter is subdiffusive behavior: the variance of the displacement then diverges like t^{α}, where $0 < \alpha < 1$. However, see the remarks following Eq. (15.10) below. See also the final remarks in Appendix E, Sec. E.4.

[2]Recall that we have reserved the notation $C(t)$ for the *normalized* autocorrelation function $\langle v(0)\, v(t) \rangle_{\text{eq}} / \left\langle v^2 \right\rangle_{\text{eq}}$.

on reversing the order of integration in the second term on the right-hand side[3]. Next, in this term we may re-label t_1 as t_2, and t_2 as t_1, as these are merely dummy variables of integration. This gives

$$\begin{aligned}\int_0^t dt_1 \int_0^t dt_2 \, \langle v(t_1)\, v(t_2)\rangle_{\text{eq}} &= 2\int_0^t dt_1 \int_0^{t_1} dt_2 \, \phi(|t_2 - t_1|) \\ &= 2\int_0^t dt_1 \int_0^{t_1} dt' \, \phi(t'), \end{aligned} \tag{15.6}$$

on changing the variable of integration in the second integral from t_2 to $t' = t_1 - t_2$. Reversing the order of integration once again, we get

$$\int_0^t dt_1 \int_0^t dt_2 \, \langle v(t_1)\, v(t_2)\rangle_{\text{eq}} = 2\int_0^t dt' \, \phi(t') \int_{t'}^t dt_1 \,. \tag{15.7}$$

The second integral on the right-hand side is a trivial one, and so we get

$$\int_0^t dt_1 \int_0^t dt_2 \, \langle v(t_1)\, v(t_2)\rangle_{\text{eq}} = 2\int_0^t dt' \, (t - t')\, \phi(t'). \tag{15.8}$$

Equation (15.3) then becomes

$$D = \lim_{t\to\infty} \frac{1}{t} \int_0^t dt'(t - t') \, \langle v(0)\, v(t')\rangle_{\text{eq}} \,. \tag{15.9}$$

But the autocorrelation $\langle v(0)\, v(t')\rangle_{\text{eq}}$ is expected to decay to zero as $t' \to \infty$, in any physically realistic model of the random motion—the velocity of the tagged particle will "lose the memory of its initial value" at long times, as a consequence of a very large number of random collisions. This means that the contribution to the integral from large values of t' is negligible. Hence the factor $(t - t')$ in the integrand in Eq. (15.9) may be replaced[4] simply by t in the limit $t \to \infty$. Thus, we arrive at the important formula

$$D = \int_0^\infty dt \, \langle v(0)\, v(t)\rangle_{\text{eq}} \,. \tag{15.10}$$

- This is the basic example of a **Kubo-Green formula**, connecting a transport coefficient (here, D) to an *equilibrium* correlation function.

Some remarks are in order at this stage. They are technical in nature, but relevant in physical applications.

[3]Note how the limits of integration change as a consequence.

[4]Clearly, this step is valid provided the autocorrelation function $\phi(t')$ falls off faster than $1/t'$ for very large t'. Once again we refer to the remarks following Eq. (15.10).

- The formula in Eq. (15.10) is valid, *provided* the integral on the right-hand side converges to a finite value. This depends, of course, on how rapidly the velocity autocorrelation function decays to zero as its argument tends to infinity. There do occur instances when the integral diverges, because of a slow power-law decay of $\langle v(0)\, v(t)\rangle_{\text{eq}}$ as $t \to \infty$. In such cases, *provided the motion is diffusive* (see below), the rigorous expression for D is found as follows. We first obtain the Laplace transform of the velocity autocorrelation in equilibrium, i. e.,

$$\widetilde{\phi}(s) = \int_0^\infty dt\, e^{-st}\, \phi(t) = \int_0^\infty dt\, e^{-st}\, \langle v(0)\, v(t)\rangle_{\text{eq}}\,. \tag{15.11}$$

This quantity certainly exists for all $\text{Re}\, s > 0$, no matter how slowly the autocorrelation function $\langle v(0)\, v(t)\rangle_{\text{eq}}$ decays with increasing t. The diffusion constant D is then given by the *analytic continuation* of $\widetilde{\phi}(s)$ to the point $s = 0$ in the complex s-plane: symbolically, we may write this as

$$D = \lim_{s=0} \widetilde{\phi}(s). \tag{15.12}$$

- There also occur situations in which the mean squared displacement of the tagged particle grows asymptotically *faster* than linearly in t. In such cases, the integral in Eq. (15.9) will diverge more strongly than t as $t \to \infty$, and the limit indicated in that equation does not exist (it becomes formally infinite). This serves to signal the fact that the motion is, in fact, *superdiffusive.*

- Finally, it may turn out that D as evaluated from the formula of Eq. (15.10) vanishes. This is an indication that the motion is *subdiffusive*: the mean squared displacement grows asymptotically *slower* than linearly in t. This includes the case in which the mean squared displacement actually does not diverge at all as $t \to \infty$, but tends to a constant value, instead. We have already come across such a situation in the case of the harmonically bound tagged particle[5].

15.2 Generalization to three dimensions

The Kubo-Green formula $D = \int_0^\infty dt\, \langle v(0) v(t)\rangle_{\text{eq}}$ appears to suggest a straightforward generalization to three dimensions. One may conjecture that the diffusion

[5]See the exercises at the end of this chapter, where the Kubo-Green formula for the diffusion coefficient is applied to the Brownian oscillator.

tensor is perhaps given by the following expression:

$$D_{ij} \stackrel{?}{=} \int_0^\infty dt\, \langle v_i(0) v_j(t) \rangle_{\text{eq}}\,. \tag{15.13}$$

However, this is not quite correct. To obtain the correct formula, we must start with the generalization of Eq. (15.1), which is based on the asymptotic behavior of the variance of the displacement of the tagged particle. Thus

$$D_{ij} \equiv \lim_{t\to\infty} \frac{1}{2t}\, \langle X_i(t)\, X_j(t) \rangle_{\text{eq}}\,. \tag{15.14}$$

The definition of the displacement yields

$$\langle X_i(t)\, X_j(t) \rangle_{\text{eq}} = \int_0^t dt_1 \int_0^t dt_2\, \langle v_i(t_1)\, v_j(t_2) \rangle_{\text{eq}}\,. \tag{15.15}$$

Generalizing Eq. (15.4), let us define the unnormalized velocity correlation tensor

$$\langle v_i(t_1)\, v_j(t_2) \rangle_{\text{eq}} = \phi_{ij}(t_2 - t_1), \quad t_2 \geq t_1\,. \tag{15.16}$$

Using the stationarity of the velocity in equilibrium, we then have

$$\langle X_i(t)\, X_j(t) \rangle_{\text{eq}} = \int_0^t dt_1 \int_0^{t_1} dt_2\, \phi_{ji}(t_1 - t_2) + \int_0^t dt_2 \int_0^{t_2} dt_1\, \phi_{ij}(t_2 - t_1). \tag{15.17}$$

This expression may be simplified exactly as was done in the preceding section, in going from Eq. (15.5) to Eq. (15.8). The result is

$$\langle X_i(t)\, X_j(t) \rangle_{\text{eq}} = \int_0^t dt'\, (t - t')\, \left[\phi_{ij}(t') + \phi_{ji}(t')\right]. \tag{15.18}$$

We substitute this expression in Eq. (15.14), and repeat the argument leading from Eq. (15.9) to Eq. (15.10). The outcome is the Kubo-Green formula for the diffusion tensor,

$$D_{ij} = \frac{1}{2} \int_0^\infty dt\, \left[\phi_{ij}(t) + \phi_{ji}(t)\right]. \tag{15.19}$$

We have also established in Eq. (4.36), Ch. 4, that the equilibrium velocity correlation matrix satisfies the property $\phi_{ij}(t) = \phi_{ji}(-t)$. Therefore the formula above may be written in the alternative form

$$D_{ij} = \frac{1}{2} \int_0^\infty dt\, \left[\phi_{ij}(t) + \phi_{ij}(-t)\right]. \tag{15.20}$$

- Thus, the diffusion tensor is the integral over time, from 0 to ∞, of the *symmetric* part of the equilibrium velocity correlation tensor. Equivalently, it is the time integral of the *even* part of this tensor (as a function of t).

15.3 The mobility

The formula derived above relating the diffusion coefficient to the integral of the equilibrium velocity autocorrelation function is a special case of a more general class of relationships (or Kubo-Green formulas) that are derived from LRT. We turn our attention now to this aspect, using once again the vehicle of the Langevin model to arrive at the formulas sought.

We started our discussion of the Langevin equation in Ch. 2 by considering the motion of the tagged particle in the presence of an external force $F_{\text{ext}}(t)$. However, shortly after writing down the formal solution to this equation, we focused our attention on the approach (starting from arbitrary initial conditions) to thermal equilibrium, in the absence of a time-dependent external force. This 'digression' has taken us through several intervening chapters, but it has served a purpose. We now have the machinery, and are ready, at last, to consider the LE in the presence of an externally applied time-dependent force.

In general, if a system in thermal equilibrium is perturbed by a (possibly time-dependent) force switched on at some instant of time[6], the system will move out of the equilibrium state. In some cases, and under suitable conditions, it may then tend to a new nonequilibrium stationary state. Way back in Ch. 1, we made the following general remark: for a sufficiently weak external stimulus or force, this nonequilibrium steady state may be 'deduced' from the properties of the system in equilibrium. The class of Kubo-Green relationships mentioned above are precise versions of this general remark.

The quantity of primary interest in the presence of a time-dependent external force is the mean velocity of the tagged particle. This is evidently a nonequilibrium quantity. Its dependence on time can be quite arbitrary, depending on the manner in which the applied force varies with time. The problem is made more precise by the following observation. A general time-dependent force can be decomposed into individual harmonics (i. e., sinusoidally varying components) by writing it as a Fourier integral. The idea is that, if we can find the solution to the problem for a single frequency ω, then the general solution can be obtained by superposing the individual solutions. As the superposition principle is invoked, it is immediately evident that we have restricted ourselves to the *linear* response regime. The quantity we seek is the **mobility** of the particle. It is essentially the average value

[6] Recall the convention agreed upon in Ch. 3: any external force is assumed to be switched on at $t = 0$, unless otherwise specified.

of the velocity of the particle per unit applied force, *in the limit when all transient effects have died out.* The mobility is an example of a **transport coefficient**. When F_{ext} is constant in time (for all $t > 0$, of course), we are concerned with the **static mobility**, also called the DC mobility, or zero-frequency mobility. When $F_{\text{ext}}(t)$ varies periodically in time with an angular frequency ω, the corresponding mobility is the **dynamic mobility** (or AC mobility, or frequency-dependent mobility), denoted by $\mu(\omega)$.

15.3.1 Relation between D and the static mobility

We begin by computing the static mobility $\mu(0)$, which we denote by μ_0 for brevity. We have, *by definition,*

$$\mu_0 = \lim_{t\to\infty} \frac{\overline{v(t)}}{F_{\text{ext}}}, \quad \text{with} \quad F_{\text{ext}} = \text{constant}. \tag{15.21}$$

The limit $t \to \infty$ ensures that all transient effects (caused by switching on the force at $t = 0$) have died out, and only the steady state effects remain. But we have already found the (conditional) mean velocity of the tagged particle in the presence of an applied force, in Eq. (3.3):

$$\overline{v(t)} = v_0\, e^{-\gamma t} + \frac{1}{m}\int_0^t dt_1\, e^{-\gamma(t-t_1)}\, F_{\text{ext}}(t_1)\,. \tag{15.22}$$

For a constant force F_{ext}, the integration over time is trivially carried out, to give

$$\overline{v(t)} = v_0\, e^{-\gamma t} + \frac{1}{m\,\gamma}\left(1 - e^{-\gamma t}\right) F_{\text{ext}}\,. \tag{15.23}$$

Therefore, passing to the limit $t \to \infty$ as required by Eq. (15.21), we find

$$\mu_0 = \frac{1}{m\,\gamma}\,. \tag{15.24}$$

Thus, the static mobility is proportional to the reciprocal of the friction coefficient. Using Eq. (15.24) in Eq. (11.6) for the diffusion coefficient, namely, $D = k_B T/(m\gamma)$, we get the important relation

$$D = \mu_0\, k_B\, T. \tag{15.25}$$

Equation (15.25) connects the diffusion coefficient with the static mobility. Combining it with the Kubo-Green formula of Eq. (15.10) for D, we arrive at the

Kubo-Green formula for the static mobility in terms of the equilibrium velocity autocorrelation,

$$\mu_0 = \frac{1}{k_B T} \int_0^\infty dt \, \langle v(0)\, v(t) \rangle_{\text{eq}} \,. \tag{15.26}$$

Note that the friction constant γ, a parameter that is specific to the Langevin model, has been eliminated in Eq. (15.26).

- Once again, we assert that the relationships exhibited in Eqs. (15.25) and (15.26) are not specific to the Langevin model, but are valid in general[7].

Equation (15.26) is a special case of a more general formula connecting the dynamic mobility to the equilibrium velocity autocorrelation function. This is considered next.

15.3.2 The dynamic mobility

We have already mentioned that a general time-dependent applied force may be written as a Fourier integral over all frequencies. Recall that we have specified our Fourier transform and inverse transform conventions in Eqs. (8.18). In the case of a time↔frequency transform pair, we shall uniformly use the conventions

$$f(t) = \int_{-\infty}^{\infty} d\omega \, e^{-i\omega t}\, \widetilde{f}(\omega) \quad \text{and} \quad \widetilde{f}(\omega) = \frac{1}{2\pi} \int_{-\infty}^{\infty} dt \, e^{i\omega t}\, f(t). \tag{15.27}$$

A real-valued applied force $F_{\text{ext}}(t)$ can therefore be written as

$$F_{\text{ext}}(t) = \text{Re} \left\{ \int_{-\infty}^{\infty} d\omega \, e^{-i\omega t} \widetilde{F}_{\text{ext}}(\omega) \right\}. \tag{15.28}$$

Note the following (elementary) points with regard to Eq. (15.28): We need a Fourier integral rather than a Fourier series, because the applied force need not be periodic, in general. We use the complex form of the Fourier integral, involving the exponential function $e^{-i\omega t}$ rather than the trigonometric functions $\cos \omega t$ and $\sin \omega t$, because the former is more convenient for our purposes. In consequence, the amplitude $\widetilde{F}_{\text{ext}}(\omega)$ corresponding to the frequency ω is complex, in general. As the applied force $F_{\text{ext}}(t)$ is a real quantity, we take the real part of the Fourier integral.

To find the dynamic mobility, we need to compute the mean velocity when the external force varies with just a single frequency ω, say. Let us therefore put aside

[7]Within the framework of LRT, of course. See Appendix I, Sec. I.4.

the general form in Eq. (15.28) for the moment, and consider the case

$$F_{\text{ext}}(t) = \text{Re}\left\{\widetilde{F}_{\text{ext}}(\omega)\,e^{-i\omega t}\right\}. \tag{15.29}$$

Inserting this expression in Eq. (15.22) for the mean velocity and carrying out the integration over t_1, we get

$$\overline{v(t)} = v_0\,e^{-\gamma t} + \text{Re}\left\{\frac{\widetilde{F}_{\text{ext}}(\omega)\left(e^{-i\omega t} - e^{-\gamma t}\right)}{m\,(\gamma - i\omega)}\right\}. \tag{15.30}$$

As we have already stated, the dynamic mobility $\mu(\omega)$ is a measure of the response of the system, as given by the average velocity, to an applied force of unit amplitude and a given frequency ω, after all the transient effects have died out. In this limit, we expect *the steady state response to oscillate in time with the same frequency* ω *as the applied force.* To extract this response from Eq. (15.30), we have merely to let the damped exponential terms die out[8]. Thus

$$\overline{v(t)}\Big|_{\substack{\text{steady}\\ \text{state}}} = \text{Re}\left\{\frac{\widetilde{F}_{\text{ext}}(\omega)\,e^{-i\omega t}}{m\,(\gamma - i\omega)}\right\} \equiv \text{Re}\left\{\mu(\omega)\,\widetilde{F}_{\text{ext}}(\omega)\,e^{-i\omega t}\right\}. \tag{15.31}$$

As expected, this response is indeed a pure sinusoidal term with frequency ω. The dynamic mobility in the Langevin model is thus given by

$$\mu(\omega) = \frac{1}{m\,(\gamma - i\omega)}. \tag{15.32}$$

The static mobility μ_0 is recovered, of course, on setting $\omega = 0$ in this expression.

We can now write down the average velocity corresponding to a *general* time-dependent external force $F_{\text{ext}}(t)$, as given by Eq. (15.28). This is, in fact, the primary purpose of determining the dynamic mobility $\mu(\omega)$. Since the Langevin equation is a *linear* equation for the velocity, the **superposition principle** holds good. The average velocity is then given by

$$\overline{v(t)} = v_0\,e^{-\gamma t} + \text{Re}\left\{\int_{-\infty}^{\infty} d\omega\,\frac{\widetilde{F}_{\text{ext}}(\omega)\left(e^{-i\omega t} - e^{-\gamma t}\right)}{m\,(\gamma - i\omega)}\right\}. \tag{15.33}$$

Hence the steady state response, after the transients have died out, is simply

$$\overline{v(t)}\Big|_{\substack{\text{steady}\\ \text{state}}} = \text{Re}\left\{\int_{-\infty}^{\infty} d\omega\,e^{-i\omega t}\,\mu(\omega)\,\widetilde{F}_{\text{ext}}(\omega)\right\}. \tag{15.34}$$

[8] Alternatively, we could have solved the Langevin equation with the initial condition imposed at $t = -\infty$ rather than $t = 0$, so that all transient effects would have died out at any finite value of t. But the present approach is simpler, and just as effective.

In other words, it is essentially the real part of the inverse Fourier transform of the *product* $\mu(\omega)\,\widetilde{F}_{\text{ext}}(\omega)$.

15.3.3 Kubo-Green formula for the dynamic mobility

Finally, we turn to the Kubo-Green formula connecting the dynamic mobility to the equilibrium velocity autocorrelation function. As we did in the case of the static mobility, we shall establish this relation using the Langevin equation—and then assert that the relation is true in general[9]. A simple way to derive the relationship sought is as follows.

We begin with the Langevin equation (6.1) for the tagged particle in the *absence* of an external force, and write it at an instant of time $t_0 + t$, where $t_0 > 0$ and $t > 0$. We have[10]

$$\dot{v}(t_0 + t) + \gamma\, v(t_0 + t) = \frac{1}{m}\,\eta(t_0 + t). \tag{15.35}$$

Multiply both sides of this equation by $v(t_0)$, and take the complete average in equilibrium—that is, average over all realizations of the random noise η as well as over all initial conditions $v(0) = v_0$, distributed according to the Maxwellian distribution $p^{\text{eq}}(v_0)$. Then

$$\langle v(t_0)\,\dot{v}(t_0 + t)\rangle_{\text{eq}} + \gamma\,\langle v(t_0)\,v(t_0 + t)\rangle_{\text{eq}} = \frac{1}{m}\,\langle v(t_0)\,\eta(t_0 + t)\rangle_{\text{eq}}\,. \tag{15.36}$$

We now invoke the principle of causality to argue that the right-hand side of Eq. (15.36) must vanish for all $t > 0$. We have already used this argument in Sec. 4.3.2 for precisely this purpose (see Eq. (4.23)). As causality is a fundamental principle, the argument bears repetition and some elaboration.

The force (or *cause*) η drives the velocity response (or *effect*) v in accordance with the Langevin equation. But the cause at a *later* instant of time $t_0 + t$ cannot

---

[9]The general proof is not difficult, but it requires the machinery of Linear Response Theory. We do not go into this here, but see the comments following Eq. (15.43) below. The topic is an important one. We therefore pursue it further in Appendix I.

[10]The reason for introducing an arbitrary instant of time $t_0 > 0$ is merely to distinguish this instant from the instant $t = 0$ at which the initial condition on the LE is imposed. Had we imposed the initial condition at $t = -\infty$, rather than $t = 0$, all transient effects would have died out at any finite instant of time. We could then have simply set $t_0 = 0$. But the initial condition has been imposed at $t = 0$ in all the preceding chapters. We retain this convention in order to avoid any unnecessary confusion.

be correlated to the effect at an *earlier* instant of time t_0. Hence we must have[11]

$$\langle v(t_0)\,\eta(t_0+t)\rangle_{\text{eq}} = \langle v(t_0)\rangle_{\text{eq}}\,\langle \eta(t_0+t)\rangle_{\text{eq}} = 0 \quad \text{for all} \quad t>0. \tag{15.37}$$

Equation (15.36) therefore simplifies to

$$\langle v(t_0)\,\dot{v}(t_0+t)\rangle_{\text{eq}} + \gamma\,\langle v(t_0)\,v(t_0+t)\rangle_{\text{eq}} = 0. \tag{15.38}$$

Recall that we have already established that the velocity is a stationary random variable in equilibrium, in the absence of an external force: the autocorrelation function $\langle v(t_0)\,v(t_0+t)\rangle_{\text{eq}} \equiv \phi(t)$ is a function of t alone. Writing $\dot{v}(t_0+t)$ as $(d/dt)\,v(t_0+t)$, we now find that it satisfies the equation[12]

$$\frac{d}{dt}\,\langle v(t_0)\,v(t_0+t)\rangle_{\text{eq}} + \gamma\,\langle v(t_0)\,v(t_0+t)\rangle_{\text{eq}} = 0 \quad \text{for all} \quad t>0. \tag{15.39}$$

Multiply both sides of this equation by $e^{i\omega t}$, and integrate over t from 0 to ∞. The first term can be simplified by integration by parts. The 'boundary' terms are

$$\lim_{t\to\infty}\langle v(t_0)\,v(t_0+t)\rangle_{\text{eq}} = \langle v(t_0)\rangle_{\text{eq}}\,\langle v(\infty)\rangle_{\text{eq}} = 0 \quad \text{and} \quad \left\langle v^2(t_0)\right\rangle_{\text{eq}} = k_BT/m. \tag{15.40}$$

Hence

$$-\frac{k_BT}{m} + (\gamma - i\omega)\int_0^\infty dt\, e^{i\omega t}\,\langle v(t_0)\,v(t_0+t)\rangle_{\text{eq}} = 0. \tag{15.41}$$

We may re-write this in the form

$$\frac{1}{m\,(\gamma - i\omega)} = \mu(\omega) = \frac{1}{k_BT}\int_0^\infty dt\, e^{i\omega t}\,\langle v(t_0)\,v(t_0+t)\rangle_{\text{eq}}\,. \tag{15.42}$$

But the fiducial instant t_0 in the above is arbitrary, and can take any value without altering the integrand, owing to the stationarity property of the velocity in thermal equilibrium. In particular, we may simply set it equal to zero. We thus arrive, finally, at the dynamical counterpart of the relation in Eq. (15.26) that connects the static mobility (and hence the diffusion constant) to the time-integral of the velocity autocorrelation function. This more general relation is

$$\mu(\omega) = \frac{1}{k_BT}\int_0^\infty dt\, e^{i\omega t}\,\langle v(0)\,v(t)\rangle_{\text{eq}} = \frac{1}{k_BT}\int_0^\infty dt\, e^{i\omega t}\,\phi(t). \tag{15.43}$$

---

[11]The principle of causality is a basic physical requirement. We will consider the condition in Eq. (15.37)—in particular, its limiting case as $t \to 0$ from above—more closely in the first exercise at the end of this chapter, and add further remarks in Sec. 15.5. We shall return to the matter once again in Ch. 17.

[12]A minor technicality: differentiation and statistical averaging are both linear operations. Hence they commute with each other, and can be performed in either order.

Once again, if the integral in Eq. (15.43) does not exist as it stands for real values of ω, we must take the Laplace transform $\widetilde{\phi}(s)$ of $\phi(t)$, and continue it analytically to $s = -i\omega$. That is,

$$\mu(\omega) = \frac{1}{k_B T} \lim_{s=-i\omega} \widetilde{\phi}(s)\,. \tag{15.44}$$

Thus:

- The Kubo-Green formula of Eq. (15.43) connects the dynamic mobility to the Fourier-Laplace transform of the equilibrium velocity autocorrelation function.

- Again, we claim that the foregoing relationship is valid in general, and is not restricted to the purview of the Langevin equation model. It is derivable from LRT. This is done in Appendix I, where we identify the generalized susceptibility represented by the mobility (see Sec. I.4).

15.4 Exercises

15.4.1 Application to the Brownian oscillator

In Ch. 13, Sec. 13.4, we have shown that the mean squared displacement of a harmonically bound tagged particle has the finite equilibrium value $k_BT/(m\omega_0^2)$ (Eq. (13.21)), as we would expect on the basis of the equipartition theorem for the energy. This means that a Brownian oscillator does not undergo long-range diffusion, as we have already pointed out. Let us now see what happens if we apply the Kubo-Green formula for the diffusion coefficient to this case.

(a) Find the equilibrium velocity autocorrelation function $\phi(t) = \langle v(0)\, v(t)\rangle_{\text{eq}}$ for the oscillator. The shortest way to do this is by the method[13] used in Ch. 4, Sec. 4.3.2, to find the velocity autocorrelation matrix for a charged tagged particle in a magnetic field.

Start with the Langevin equation for the oscillator,

$$\dot{v}(t) = -\gamma v(t) - \omega_0^2 x(t) + \frac{1}{m}\,\eta(t), \quad t > 0. \tag{15.45}$$

[13]But we have to watch our step in order to avoid a pitfall, as you will see! This example will serve to illustrate certain important matters of principle. We shall therefore discuss it at some length.

Pre-multiply both sides of the equation by $v(0)$ and take equilibrium (or complete) averages, to get

$$\langle v(0)\,\dot{v}(t)\rangle_{\text{eq}} = -\gamma\,\langle v(0)\,v(t)\rangle_{\text{eq}} - \omega_0^2\,\langle v(0)\,x(t)\rangle_{\text{eq}} + \frac{1}{m}\,\langle v(0)\,\eta(t)\rangle_{\text{eq}}\,. \tag{15.46}$$

The term $\langle v(0)\,\eta(t)\rangle_{\text{eq}}$ vanishes for all $t > 0$ by causality[14]. Hence Eq. (15.46) may be written as

$$\frac{d\phi(t)}{dt} = -\gamma\,\phi(t) - \omega_0^2\,\langle v(0)\,x(t)\rangle_{\text{eq}}\,. \tag{15.47}$$

To eliminate the cross-correlation in the last term, differentiate both sides with respect to t. Thus

$$\frac{d^2\phi(t)}{dt^2} + \gamma\,\frac{d\phi(t)}{dt} + \omega_0^2\,\phi(t) = 0. \tag{15.48}$$

Consider the case of an underdamped oscillator, for definiteness. The general solution to Eq. (15.48) is a linear combination

$$\phi(t) = c_1\,e^{-\lambda_+ t} + c_2\,e^{-\lambda_- t}, \tag{15.49}$$

where (repeating Eqs. (13.15) and (13.14) for ready reference)

$$\lambda_\pm = \frac{1}{2}\gamma \pm i\omega_s\,, \quad \omega_s = \left(\omega_0^2 - \tfrac{1}{4}\gamma^2\right)^{1/2}. \tag{15.50}$$

The constants c_1 and c_2 are determined by the initial values $\phi(0)$ and $\dot{\phi}(0)$. We have[15]

$$c_1 = \frac{\lambda_-\,\phi(0) + \dot{\phi}(0)}{\lambda_- - \lambda_+} \quad \text{and} \quad c_2 = \frac{\lambda_+\,\phi(0) + \dot{\phi}(0)}{\lambda_+ - \lambda_-}\,. \tag{15.51}$$

$\phi(0)$ can be written down at once, because it is simply the mean squared velocity in equilibrium:

$$\phi(0) = \left\langle v^2 \right\rangle_{\text{eq}} = \frac{k_B T}{m}\,. \tag{15.52}$$

Turning to $\dot{\phi}(0)$, we know that, in a state of thermal equilibrium, $v(t)$ is a stationary random process. Hence, by the general result for a stationary process that is expressed in Eq. (11.38), Sec. 11.7.3, we might expect that

$$\dot{\phi}(0) = \langle v(0)\dot{v}(0)\rangle_{\text{eq}} = \langle v(t)\dot{v}(t)\rangle_{\text{eq}} = 0. \tag{15.53}$$

---

[14] Recall Eqs. (4.23), (15.37) and the remarks made in connection with these equations.

[15] The reason for writing out these simple intermediate steps will become clear subsequently.

But this is inconsistent with the Langevin equation itself! For, pre-multiplying both sides of the Langevin equation (15.45) by $v(t)$ and taking complete averages, we find

$$\langle v(t)\,\dot{v}(t)\rangle_{\text{eq}} = -\gamma \left\langle v^2(t)\right\rangle_{\text{eq}} - \omega_0^2 \left\langle v(t)\, x(t)\right\rangle_{\text{eq}} + \frac{1}{m} \left\langle v(t)\,\eta(t)\right\rangle_{\text{eq}}. \tag{15.54}$$

But we have already shown (see Eq. (13.23), Ch. 13) that the equal-time correlator

$$\langle v(t)\, x(t)\rangle_{\text{eq}} = \langle \dot{x}(t)\, x(t)\rangle_{\text{eq}} = 0. \tag{15.55}$$

This is completely consistent with what we should expect on the basis of the stationarity of the random process $x(t)$ for a harmonically bound particle. Therefore

$$\langle v(t)\,\dot{v}(t)\rangle_{\text{eq}} = -\gamma \left\langle v^2(t)\right\rangle_{\text{eq}} + \frac{1}{m} \left\langle v(t)\,\eta(t)\right\rangle_{\text{eq}}. \tag{15.56}$$

The crucial question is: what are we to do with the second term on the right-hand side of Eq. (15.56), the *equal*-time correlation between the random force $\eta(t)$ and the velocity $v(t)$? By causality, we know that $\langle v(t)\,\eta(t+\epsilon)\rangle_{\text{eq}} = 0$ for *arbitrarily* small $\epsilon > 0$. We argue that this must remain true in the limit $\epsilon \to 0$ from above. On physical grounds, too, we know that the white noise $\eta(t)$ directly controls the acceleration $\dot{v}$, rather than the velocity $v(t)$ (as we have already stressed). We therefore set

$$\langle v(t)\,\eta(t)\rangle_{\text{eq}} = 0. \tag{15.57}$$

Using Eqs. (15.55) and (15.57) in Eq. (15.56), we get

$$\dot{\phi}(0) = \langle v(0)\dot{v}(0)\rangle_{\text{eq}} = \langle v(t)\,\dot{v}(t)\rangle_{\text{eq}} = -\gamma \left\langle v^2(t)\right\rangle_{\text{eq}} = -\frac{\gamma k_B T}{m} \neq 0. \tag{15.58}$$

Hence, although $\langle v(t)\,\dot{v}(t)\rangle_{\text{eq}}$ is certainly independent of t, it is not equal to zero, as we would expect by the stationarity of the random process $v(t)$. But $\dot{\phi}(0) = -\gamma k_B T/m$ is what the LE gives, and we must accept it here[16].

Put in the initial values found above for $\phi(0)$ and $\dot{\phi}(0)$ in Eq. (15.51), and substitute the result in Eq. (15.49), to arrive at the result

$$\phi(t) = \langle v(0)\, v(t)\rangle_{\text{eq}} = \frac{k_B T}{m}\, e^{-\gamma t/2} \left(\cos\,\omega_s t - \frac{\gamma}{2\omega_s}\,\sin\,\omega_s t\right), \quad t \geq 0. \tag{15.59}$$

---

[16]We defer further discussion of this point to Sec. 15.5. The consequences of using Eq. (15.53) rather than Eq. (15.58) will be worked out at length there. It will be shown that unacceptable inconsistencies arise as a result.

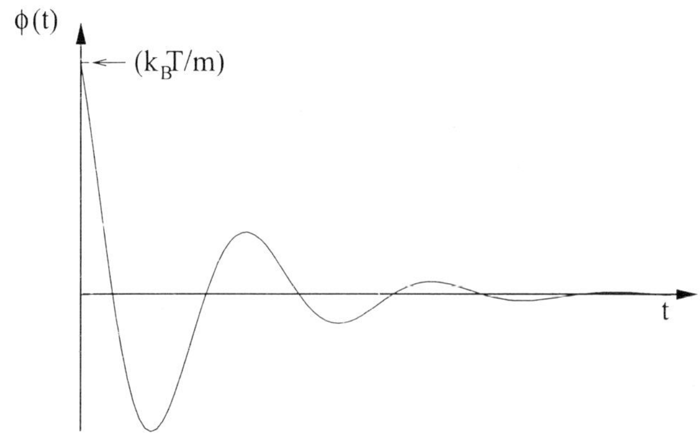

Figure 15.1: The velocity autocorrelation function of an underdamped Brownian oscillator of natural frequency ω_0, as a function of time for $t \geq 0$ (see Eq. (15.59)). The sinusoidal variation with a shifted frequency $\omega_s = \left(\omega_0^2 - \frac{1}{4}\gamma^2\right)^{1/2}$ is modulated by the damping factor $\exp\left(-\frac{1}{2}\gamma t\right)$. Note that (i) $\phi(t)$ is not a monotonically decreasing function of t, and (ii) the total (signed) area under the curve is equal to zero. (Both these statements remain valid in the critically damped and overdamped cases as well.) The second property signifies that a harmonically bound tagged particle does not undergo long-range diffusion.

This is the equilibrium velocity autocorrelation function of a harmonically bound tagged particle in the underdamped case. As a quick check, we may verify that the boundary values in Eqs. (15.52) and (15.58) are recovered. Figure 15.1 shows the variation with time of the velocity autocorrelation function $\phi(t)$ as given by Eq. (15.59).

(b) Substitute Eq. (15.59) in the Kubo-Green formula for the diffusion coefficient, Eq. (15.10), to show that

$$\int_0^\infty dt \; \langle v(0)\, v(t)\rangle_{\text{eq}} = 0 \tag{15.60}$$

for the Brownian oscillator. Hence a harmonically bound tagged particle does not undergo long-range diffusion. It is obvious that this conclusion will be all the more applicable in the critically damped and overdamped cases, but check this out anyway. Show that the velocity autocorrelation function of the overdamped oscillator is given by

$$\phi(t) = \frac{k_B T}{m}\, e^{-\gamma t/2} \left(\cosh \tfrac{1}{2}\gamma_s t - \frac{\gamma}{\gamma_s} \sinh \tfrac{1}{2}\gamma_s t\right), \quad (t \geq 0) \tag{15.61}$$

where $\gamma_s = -2i\omega_s = (\gamma^2 - 4\omega_0^2)^{1/2}$, as defined in Eq. (13.60); and by

$$\phi(t) = \frac{k_B T}{m}\, e^{-\gamma t/2} \left(1 - \tfrac{1}{2}\gamma t\right) \quad (t \geq 0) \tag{15.62}$$

in the critically damped case. Observe that $\phi(t)$ changes sign in these cases too. Verify that $\int_0^\infty dt\, \phi(t) = 0$ in both instances.

From the result in Eq. (15.61) for the overdamped oscillator, we may pass to the limit $\omega_0 = 0$, corresponding to a freely diffusing tagged particle. Then $\gamma_s = \gamma$. Verify that the familiar result $\langle v(0)\, v(t)\rangle_{\text{eq}} = (k_B T/m)\, e^{-\gamma t}$ $(t \geq 0)$ for free diffusion is correctly recovered in that case.

15.4.2 Application to a particle in a magnetic field

In Eq. (4.29) (Ch. 4, Sec. 4.3.2), we have obtained the matrix elements of the velocity correlation matrix for a tagged particle in a constant, uniform magnetic field. Re-writing this result for $\phi_{ij}(t)$ for ready reference, we have (for $t \geq 0$)

$$\phi_{ij}(t) = \frac{k_B T}{m}\, e^{-\gamma t}\, [n_i\, n_j + (\delta_{ij} - n_i\, n_j)\, \cos \omega_c t - \epsilon_{ijk}\, n_k\, \sin \omega_c t]. \tag{15.63}$$

(a) Substitute this expression in the Kubo-Green formula (15.19) for D_{ij}, and show that the result is precisely that already found in Eq. (14.21), namely,

$$D_{ij} = \frac{k_B T}{m\gamma}\left[n_i\, n_j + \frac{\gamma^2}{\gamma^2+\omega_c^2}\,(\delta_{ij} - n_i\, n_j)\right]. \qquad (15.64)$$

We have already explained, in Sec. 14.2.2, that the coefficients proportional to $n_i\, n_j$ and $(\delta_{ij} - n_i\, n_j)$ represent, respectively, the longitudinal and transverse parts of the diffusion tensor. Observe that the antisymmetric part of the velocity correlation, represented by the term proportional to $\epsilon_{ijk}\, n_k$, does *not* contribute to the diffusion tensor at all. As a consequence,

- the diffusion matrix with elements D_{ij} (a real symmetric matrix) becomes a diagonal matrix if the orientation of the coordinate system is chosen such that the magnetic field lies along one of the axes.

(b) The exact covariance matrix of the displacement of the tagged particle is also computed quite easily from Eq. (15.18), namely,

$$\sigma_{ij}(t) \equiv \langle X_i(t)\, X_j(t)\rangle_{\text{eq}} = \int_0^t dt'\,(t-t')\,\left[\phi_{ij}(t') + \phi_{ji}(t')\right]. \qquad (15.65)$$

Use the expression in Eq. (15.63) for the velocity correlation matrix elements to show that $\sigma_{ij}(t)$ can be written as

$$\sigma_{ij}(t) = n_i\, n_j\, \sigma_{\text{long}} + (\delta_{ij} - n_i\, n_j)\, \sigma_{\text{trans}}\,, \qquad (15.66)$$

where the longitudinal and transverse parts are as follows:

$$\sigma_{\text{long}} = \frac{2k_B T}{m\gamma^2}\left(\gamma t - 1 + e^{-\gamma t}\right). \qquad (15.67)$$

This result is only to be expected, because the motion in the direction of the magnetic field is essentially the same as that of a freely diffusing tagged particle[17]. The transverse part, on the other hand, is given by

$$\begin{aligned}\sigma_{\text{trans}} &= \frac{2k_B T}{m(\gamma^2+\omega_c^2)}\Big[\gamma t\,(\gamma^2+\omega_c^2) - (\gamma^2-\omega_c^2) + \\ &+ e^{-\gamma t}\left\{(\gamma^2-\omega_c^2)\,\cos\,\omega_c t - 2\gamma\omega_c\,\sin\,\omega_c t\right\}\Big]. \qquad (15.68)\end{aligned}$$

This part, too, comprises a term that increases linearly with t, a constant (or t^0) term, and a term that decays like $e^{-\gamma t}$. The coefficients of these terms depend, of course, on both γ and ω_c.

[17]Compare the expression for σ_{long} in Eq. (15.67) with that for the mean squared value of any component of the displacement of a freely diffusing tagged particle, Eq. (11.5).

15.5 Further remarks on causality and stationarity

Equations (15.53) and (15.58) are obviously inconsistent with each other. It is important to appreciate the fact that there is, indeed, an issue to be addressed here. Let us repeat the facts of the case somewhat elaborately, for a clear understanding of the situation.

- The position of a Brownian oscillator is a stationary random process in thermal equilibrium. Hence, by the general result we have established for such processes (Eq. (11.38)), the equal-time correlation $\langle x(t)\, \dot{x}(t)\rangle_{\rm eq} = 0$ for all t. This relation has also been derived explicitly for the oscillator, in Eq. (13.23).
- The velocity of a Brownian oscillator is also a stationary random process in thermal equilibrium. Hence we should expect the equal-time correlation $\langle v(t)\, \dot{v}(t)\rangle_{\rm eq}$ to be zero for all t. But the Langevin equation gives the result $-\gamma k_B T/m$ for this correlation function, if we invoke causality and continuity to equate the equal-time correlator $\langle v(t)\, \eta(t)\rangle_{\rm eq}$ to zero.

However, it is also possible to argue that there is really no problem, and that $\langle v(t)\, \dot{v}(t)\rangle_{\rm eq}$ is indeed equal to zero, as follows. Let us go back to Eq. (15.56), namely,

$$\langle v(t)\, \dot{v}(t)\rangle_{\rm eq} = -\gamma \left\langle v^2(t)\right\rangle_{\rm eq} + \frac{1}{m} \langle v(t)\, \eta(t)\rangle_{\rm eq}. \tag{15.69}$$

Instead of invoking causality to argue that $\langle v(t)\, \eta(t)\rangle_{\rm eq} = 0$, let us try to calculate this quantity from the Langevin equation itself! We note that the quantity ω_0 has dropped out in Eq. (15.69). Hence the issue to be addressed is common to both a harmonically bound tagged particle and a free tagged particle[18]. We may therefore start with the formal solution to the LE for a free tagged particle, Eq. (3.7). We have

$$v(t) = v(0)\, e^{-\gamma t} + \frac{1}{m}\int_0^t dt'\, e^{-\gamma(t-t')}\, \eta(t') \tag{15.70}$$

for all $t > 0$. Multiply each side of this equation by $\eta(t)$ and take complete averages. Hence

$$\langle v(t)\, \eta(t)\rangle_{\rm eq} = \langle v(0)\, \eta(t)\rangle_{\rm eq}\, e^{-\gamma t} + \frac{1}{m}\int_0^t dt'\, e^{-\gamma(t-t')} \left\langle \eta(t')\, \eta(t)\right\rangle_{\rm eq}. \tag{15.71}$$

We may of course set $\langle v(0)\, \eta(t)\rangle_{\rm eq} = 0$ for $t > 0$ by appealing to causality. We now use the fact that $\langle \eta(t')\, \eta(t)\rangle_{\rm eq} = \Gamma\, \delta(t-t') = 2m\gamma k_B T\, \delta(t-t')$ to carry out the

---

[18] Although the relation $\langle x(t)\, \dot{x}(t)\rangle_{\rm eq} = 0$ that is valid for the former is not valid for the latter.

integral over t'. However, since the δ-function 'fires' precisely *at* the upper limit of integration, we must be careful to include only *half* its contribution! Thus[19]

$$\langle v(t)\,\eta(t)\rangle_{\text{eq}} \overset{?}{=} \gamma k_B T, \tag{15.72}$$

rather than zero. Inserting this result in Eq. (15.69) and using the fact that $\langle v^2(t)\rangle_{\text{eq}} = k_B T/m$ (for all t), we get $\langle v(t)\,\dot{v}(t)\rangle_{\text{eq}} = 0$, as required by the stationarity of the velocity process.

On the face of it, we seem to have got around the difficulty. But the source of the inconsistency lies a little deeper. After all, we could equally well have started with the well-established result

$$\langle v(t)\,v(t+t')\rangle_{\text{eq}} = \frac{k_B T}{m}\, e^{-\gamma t'} \;\; (t' \geq 0) \tag{15.73}$$

for the freely diffusing tagged particle. Now differentiate each side with respect to t', and pass to the limit $t' \to 0$ from above. It follows at once that $\langle v(t)\,\dot{v}(t)\rangle_{\text{eq}} = -\gamma k_B T/m$, rather than zero.

The point appears to be a technical one. All possible loopholes in the argument must therefore be eliminated. You might again object to the conclusion of the preceding paragraph by pointing out that $\langle v(t)\,v(t+t')\rangle_{\text{eq}}$ is actually equal to $(k_B T/m)\, e^{-\gamma|t'|}$. This is a function of the *modulus* $|t'|$, rather than t' itself, and therefore does not have a unique derivative at $t' = 0$. But we are only concerned with the *right* derivative in our calculation, i. e., with the derivative as $t' \to 0$ from the positive side. This quantity is certainly well-defined, and nonzero. The contradiction therefore persists.

But what if we take the conclusion $\langle v(t)\,\dot{v}(t)\rangle_{\text{eq}} = 0$ to be valid for the oscillator, and go ahead with the calculation of its velocity correlation function? We must then put $\dot{\phi}(0) = 0$ in Eq. (15.51), and re-work the calculation. After simplification, we find

$$\phi(t) \overset{?}{=} \frac{k_B T}{m}\, e^{-\gamma t/2} \left(\cos\, \omega_s t + \frac{\gamma}{2\omega_s}\, \sin\, \omega_s t \right), \;\; t \geq 0. \tag{15.74}$$

The only difference between the expression above and the result derived in Eq. (15.59) is the sign of the second term in the brackets. But then the time integral $\int_0^\infty dt\, \phi(t)$ no longer vanishes—instead, it becomes equal to $k_B T/(m\gamma)$, the value of the diffusion constant for the free tagged particle. This would imply

[19]We use a question mark above the equality sign in Eq. (15.72) and in Eqs. (15.74) and (15.76) below, to highlight the dubious nature of the results concerned!

that the Brownian oscillator undergoes long-range diffusion, which is simply incorrect. Even more telling is the following unphysical result: Set $\omega_0 = 0$ in Eq. (15.74) for $\phi(t)$ (or in its counterpart for the overdamped case). One obtains $\langle v(0)\, v(t)\rangle_{\text{eq}} = k_B T/m$ (independent of t) in the case of a free tagged particle!

We conclude that Eq. (15.59) is the correct expression for the equilibrium velocity autocorrelation function of an underdamped, harmonically bound tagged particle. Equations (15.61) and (15.62) express this autocorrelation function in the overdamped and critically damped cases. The expression in Eq. (15.74) is incorrect. The lesson from the foregoing is that, when we extrapolate the Langevin model to very short times, the equations

$$\langle v(t)\, \eta(t)\rangle_{\text{eq}} = 0, \quad \langle v(t)\, \dot{v}(t)\rangle_{\text{eq}} = -\gamma k_B T/m \tag{15.75}$$

are tenable, while the equations

$$\langle v(t)\, \eta(t)\rangle_{\text{eq}} \stackrel{?}{=} \gamma k_B T, \quad \langle v(t)\, \dot{v}(t)\rangle_{\text{eq}} \stackrel{?}{=} 0 \tag{15.76}$$

are not. In effect, this means that

- *causality must be preserved*, even at the expense of stationarity.

As we have already mentioned more than once, it is the Langevin model itself that requires amendment. It involves an idealized white noise with *zero* correlation time, and this is unphysical. The acceleration $\dot{v}(t)$ is directly driven by the white noise, which is why the problem gets carried over to the velocity-acceleration correlation function $\langle v(t)\, \dot{v}(t)\rangle_{\text{eq}}$. What happens at the next stage, i. e., in the case of the position-velocity correlation function $\langle x(t)\, \dot{x}(t)\rangle_{\text{eq}}$? The question is meaningful, of course, only if $x(t)$ is a stationary process in equilibrium. This is so in the case of the Brownian oscillator. We find that the problem disappears in this instance: the equal-time correlation $\langle x(t)\, \dot{x}(t)\rangle_{\text{eq}}$ does vanish, as required. The reason is that, after the original LE is integrated once, the output variable $v(t)$ actually has a finite (*nonzero*) correlation time (given by γ^{-1}).

Thus, we are left with no choice but to go back and reconsider the Langevin equation for the velocity. The remedy lies in appropriately modifying this equation itself, such that the short-time behavior it predicts does not lead to a contradiction. This modification will be taken up in Ch. 17.

Chapter 16

Mobility as a generalized susceptibility

The concept of the power spectrum of a stationary random process is introduced. The Wiener-Khinchin Theorem relates this quantity to the Fourier transform of the autocorrelation of the process. The fluctuation-dissipation theorem is obtained, relating the dissipation in the system to the power spectrum of the response, in this case the velocity of the tagged particle. The analyticity properties of the dynamic mobility are deduced. These arise from the causal nature of the response to an external force. Dispersion relations are derived for the real and imaginary parts of the mobility. These considerations are extended to the three-dimensional multi-component case.

16.1 The power spectral density

16.1.1 Definition of the power spectrum

At this point, it is convenient to digress briefly to introduce the concept of the **power spectral density**, or **power spectrum** for short, of a stationary random process. This quantity (for which we shall use the abbreviation PSD whenever convenient) is an important quantitative characterizer of the random process.

Consider a general *stationary* random process[1], $\xi(t)$. In general, the plot of $\xi(t)$ versus t may be expected to be a highly irregular curve—the random process

[1] We assume the process is real-valued, as in all the preceding chapters. The discussion can be extended to the case when $\xi(t)$ is complex-valued, but we do not need this generalization for our purposes.

is, after all, a 'noise' in a sense. We may ask: perhaps a Fourier transform of the function $\xi(t)$ would give us some insight into its time variation, by decomposing it into individual harmonics? But $\xi(t)$, regarded as a function of t, need not be integrable, in general[2]. Strictly speaking, therefore, its Fourier transform may not exist. However, we have seen that a much more regular function of t can be associated with the process $\xi(t)$: namely, its autocorrelation function. The latter function may be Fourier-decomposed into frequency components. The component corresponding to any given frequency ω is a measure of how much of the intensity of the fluctuating signal $\xi(t)$ lies in an infinitesimal frequency window centered at the frequency ω. This is precisely what the power spectrum $S_\xi(\omega)$ of a stationary random process $\xi(t)$ represents. These general remarks are made more precise as follows.

Suppose we monitor the process (or 'signal') $\xi(t)$ over a very long interval of time, say from $t = 0$ up to $t = T$. Consider the integral of $\xi(t)$, weighted with the factor $e^{i\omega t}$, from $t = 0$ to $t = T$. The power spectrum $S_\xi(\omega)$ is then defined as

$$S_\xi(\omega) = \lim_{T\to\infty} \frac{1}{2\pi T} \left| \int_0^T dt\, e^{i\omega t}\, \xi(t) \right|^2 . \tag{16.1}$$

- It is evident from the definition above that $S_\xi(\omega)$ is real and positive.

16.1.2 The Wiener-Khinchin theorem

Observe, too, that the right-hand side of Eq. (16.1) does not involve any average over all realizations of the random process. An averaging of this sort is not necessary because the process is supposed to be ergodic: given a sufficient amount of time (ensured by passing to the limit $T \to \infty$), $\xi(t)$ will take on all values in its sample space[3]. Consequently, we may expect $S_\xi(\omega)$ to be expressible in terms of a statistical average. Indeed, this is the content of the Wiener-Khinchin theorem, which states that

- the power spectrum $S_\xi(\omega)$ of a stationary random process $\xi(t)$ is equal to the Fourier transform of its autocorrelation function.

[2]To be precise, $\int_{-\infty}^{\infty} dt\, |\xi(t)|$ may not be finite.

[3]It might be helpful, at this stage, to read once again the remarks in Sec. 5.5 on ergodicity and related aspects.

That is[4],

$$S_\xi(\omega) = \frac{1}{2\pi}\int_{-\infty}^{\infty} dt\, e^{i\omega t}\, \phi_\xi(t), \tag{16.2}$$

where

$$\phi_\xi(t) = \langle \xi(0)\, \xi(t) \rangle\,. \tag{16.3}$$

The steps leading from Eq. (16.1) to Eq. (16.2) are given in Appendix H. Recall that the autocorrelation function of a stationary process is a function of a *single* time argument. This is why we can define its Fourier transform at all—the transform involves an integration over all values of this argument[5]. Going back a step in the derivation of Eq. (16.2) given in Appendix H, we have (repeating Eq. (H.7) for ready reference)

$$S_\xi(\omega) = \frac{1}{\pi}\int_{0}^{\infty} dt\, \phi_\xi(t)\, \cos\omega t. \tag{16.4}$$

It is evident from this form of the relationship that

$$S_\xi(-\omega) = S_\xi(\omega), \tag{16.5}$$

i. e., the PSD is an even function of the frequency.

We may invert the Fourier transform in Eq. (16.2), to obtain the relation

$$\phi_\xi(t) = \langle \xi(0)\, \xi(t) \rangle = \int_{-\infty}^{\infty} d\omega\, e^{-i\omega t}\, S_\xi(\omega) = 2\int_{0}^{\infty} d\omega\, S_\xi(\omega)\, \cos\omega t. \tag{16.6}$$

The last equation follows from the fact that $S_\xi(\omega)$ is an even function of ω. Setting $t = 0$ in Eq. (16.6), we get a very useful formula, or a **sum rule**, for the mean squared value of the stationary process ξ as an integral of its power spectrum over all physical frequencies[6]:

$$\langle \xi^2 \rangle = 2\int_{0}^{\infty} d\omega\, S_\xi(\omega). \tag{16.7}$$

---

[4]We follow the Fourier transform convention given in Eqs. (15.27) for functions of t and ω, respectively. We also assume, for simplicity, that $\langle \xi \rangle = 0$. Hence the variance of ξ is just its mean squared value. The assumption that the mean value is zero is easily relaxed.

[5]If a random process $x(t)$ is *non*stationary (such as the displacement of a freely diffusing tagged particle), we cannot define its power spectrum in the rigorous sense. However, there often occur physical situations in which the autocorrelation function $\langle x(t_1)\, x(t_2)\rangle$ of a nonstationary process has a relatively rapidly varying dependence on the difference $(t_1 - t_2)$ of its time arguments, and a much slower, secular dependence on the individual arguments t_1 or t_2 or the sum $(t_1 + t_2)$. In such cases, it may still be meaningful to speak of its 'power spectrum', if only in an appropriate frequency window.

[6]Provided, of course, that the integral converges! It will not do so in the case of white noise—as we shall see shortly.

We now evaluate the power spectra of the stationary random processes that we have encountered in the Langevin model for the motion of a tagged particle in a fluid.

16.1.3 White noise; Debye spectrum

Consider first the random force $\eta(t)$. The statistical properties of this stationary, Gaussian, δ-correlated Markov process have already been specified in earlier chapters[7]. Inserting the expression $\langle\eta(0)\,\eta(t)\rangle_{\text{eq}} = \Gamma\,\delta(t)$ in Eq. (16.2) trivially yields

$$S_\eta(\omega) = \frac{\Gamma}{2\pi} = \frac{m\gamma k_B T}{\pi}, \tag{16.8}$$

on using the FD relation $\Gamma = 2m\gamma k_B T$. Thus S_η is actually a constant, *independent* of the frequency ω.

- The constancy of the power spectrum is the most basic characteristic of a *white* noise. Indeed, it is the origin of the very name itself.

It then follows from Eq. (16.7) that the mean squared value of $\eta(t)$ is formally infinite, as we know already from the fact that its autocorrelation function is a δ-function.

It is instructive to consider, at this point, a physical situation that is analogous to the Langevin model for the motion of a tagged particle in a fluid. A relation similar to Eq. (16.8) may perhaps be already familiar to you in the context of the thermal noise or Johnson noise from a resistor R at an absolute temperature T. The electrons in the resistor undergo diffusive motion. In the absence of an external voltage applied to the resistor, this motion sets up a random, rapidly fluctuating EMF $V(t)$ across it. As a consequence, a fluctuating current $I(t)$ flows in the resistor. If we assume that the voltage and the current are related to each other via the same laws (in terms of lumped circuit parameters) as the usual macroscopic voltage and current, then

$$L\frac{dI}{dt} + RI = V(t), \tag{16.9}$$

where L is the effective self-inductance of the resistor. Regarded as a stochastic differential equation, Eq. (16.9) is of precisely the same form as the Langevin equation $m\dot{v} + m\gamma\, v = \eta(t)$. If the random EMF is assumed to be a white noise like $\eta(t)$, it follows that its power spectrum must be

$$S_V(\omega) = \frac{R\,k_B T}{\pi}. \tag{16.10}$$

[7]Recall, in particular, Eqs. (6.4)-(6.5) and the accompanying discussion in Sec. 6.1.

(All we have to do is to replace the parameters in the problem according to $m \to L$, $\gamma \to R/L$.) Equation (16.10) is the **Nyquist Theorem** for thermal noise in a resistor[8]. The left-hand side of Eq. (16.10) measures the strength of the voltage fluctuations, while R is obviously a measure of the dissipation in the system.

What about the power spectrum of the *driven* or 'output' variable[9] in the Langevin equation, namely, the velocity (or the current, in Eq. (16.9) above)? There is no reason to expect these output spectra also to be flat (or independent of ω)—we have already seen that the equilibrium velocity autocorrelation is not a δ-function, but rather a decaying exponential, given by Eq. (4.5). Using this in Eq. (16.4), we find

$$S_v(\omega) = \frac{k_B T}{m\pi} \int_0^\infty dt\, e^{-\gamma t} \cos \omega t = \frac{\gamma k_B T}{m\pi\,(\gamma^2 + \omega^2)}\,. \tag{16.11}$$

As a check on Eq. (16.11), we may substitute this expression in Eq. (16.7) and integrate over ω, to verify that we recover the result[10] $\langle v^2 \rangle_{\text{eq}} = k_B T/m$. Similarly, the power spectrum of the current arising from thermal noise in a resistor is given by

$$S_I(\omega) = \frac{R k_B T}{\pi\,(R^2 + L^2\omega^2)}\,. \tag{16.12}$$

The functional form in Eq. (16.11) or (16.12) is a Lorentzian, also called a **Debye spectrum**, and is depicted in Figure 16.1. It is encountered very frequently in the context of diverse phenomena such as magnetic, dielectric and mechanical relaxation in condensed matter physics.

---

[8] A trivial but relevant remark: The numerical constant in this relation (here, π^{-1}) is dependent on the Fourier transform convention used, and on the numerical factor in the definition of the power spectrum. Our convention is that given in Eq. (15.27), and we have included a factor of $(2\pi)^{-1}$ in Eq. (16.1). Another convention used frequently in the literature is as follows: the factor $(2\pi)^{-1}$ in the definition of a Fourier transform pair is shifted to the first of the equations in Eqs. (15.27), and the power spectrum is defined such that it equals *twice* the Fourier transform of the autocorrelation function. Hence our formulas would have be multiplied by a factor 4π to obtain the corresponding formulas in this other convention. This is why the Nyquist Theorem frequently appears in the literature in the form

$$S_V(\omega) = 4R\, k_B T.$$

[9] Recall that, in general, the output is a *stationary* random process only in a state of thermal equilibrium, i. e., when there is no time-dependent external force or applied EMF term in the Langevin equation.

[10] This merely confirms that all our factors of 2π have been deployed consistently!

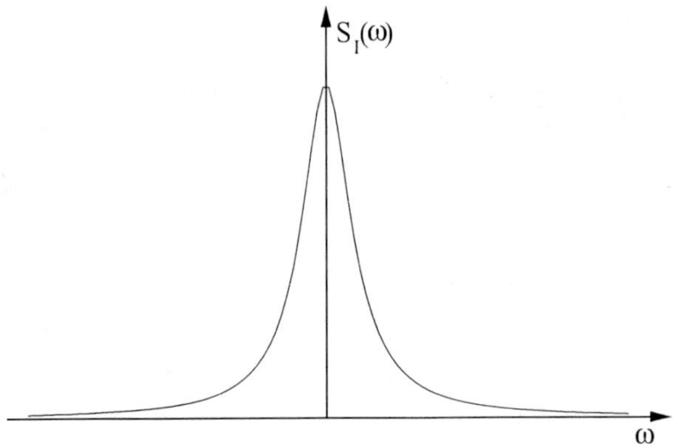

Figure 16.1: The Debye spectrum as a function of the frequency. Typically, such a power spectrum indicates that the underlying relaxation phenomenon is controlled by a single time scale rather than a set of time scales. The same curve also represents a probability density function, corresponding to a stable distribution with characteristic exponent 1 (see Appendix K).

A Debye spectrum results when the relaxation is controlled by a mechanism that is characterized by a single time scale or relaxation rate or frequency. It is clear that such a model is likely to be applicable to an actual physical system (i. e., to fit observed data with some accuracy) only in a restricted frequency interval about this characteristic frequency. The latter is given by $\omega_{\text{relax}} = \gamma$ and R/L, respectively, in the examples considered above.

More generally, when several distinct time scales are involved in a physical problem, a superposition of such spectra with a finite number of distinct relaxation rates (or resonance frequencies) would provide a better fit to the data. Even more generally, a continuous superposition of relaxation rates with a certain spectral weight factor is likely to be required in many cases. There is more than one way of modeling such situations[11]. We shall return to this possibility in Ch. 17, when we consider the generalized Langevin equation.

[11] Recall, for instance, the kinds of non-exponential behavior of the autocorrelation function that are possible in the case of the kangaroo process, discussed in Ch. 5. See, in particular, Eq. (5.58) and the comments following it.

16.2 Fluctuation-dissipation theorems

We turn now to a very important aspect of nonequilibrium statistical mechanics: the deep formal connection between the fluctuations in a system in thermal equilibrium, and the response of the system to an applied time-dependent force. Such a connection is basic to linear response theory. As always, our approach is via the Langevin model of a tagged particle in a fluid. Our system is the tagged particle, and its dynamic mobility is the response with which we are concerned.

In the presence of an externally applied time-dependent force, the Langevin equation (Eq. (2.18)) for the velocity of the tagged particle is

$$m\frac{dv}{dt} + m\,\gamma\,v = F_{\text{ext}}(t) + \eta(t). \tag{16.13}$$

In order to wash out the effect of the initial condition and to eliminate transients, let us (now that we have gained adequate experience in this regard!) take the initial condition to be specified at $t = -\infty$ rather than at $t = 0$. Taking complete (rather than conditional) averages[12] on both sides of Eq. (16.13), we get

$$m\frac{d\,\langle v\rangle}{dt} + m\gamma\,\langle v\rangle = F_{\text{ext}}(t). \tag{16.14}$$

Multiplying both sides by $\exp(i\omega t)$ and integrating over t from $-\infty$ to ∞ (i. e., equating the corresponding Fourier transforms), we get

$$\langle\widetilde{v}(\omega)\rangle = \frac{\widetilde{F}_{\text{ext}}(\omega)}{m\,(\gamma - i\omega)} = \mu(\omega)\,\widetilde{F}_{\text{ext}}(\omega), \tag{16.15}$$

in terms of the dynamic mobility $\mu(\omega)$ identified in Eq. (15.32). This is simply a re-confirmation of the fact that

- the average steady-state velocity response to a sinusoidal applied force of frequency ω is governed by the dynamic mobility: the latter may in fact be *defined* as the ratio

$$\mu(\omega) = \frac{\langle\widetilde{v}(\omega)\rangle}{\widetilde{F}_{\text{ext}}(\omega)}. \tag{16.16}$$

[12]The average is over all initial conditions and all realizations of the noise $\eta(t)$, of course. But now it is formally a 'nonequilibrium average', because of the presence of $F_{\text{ext}}(t)$. This is why we use the notation $\langle\cdots\rangle$ rather than $\langle\cdots\rangle_{\text{eq}}$, in keeping with the remarks made in Sec. 3.1.

Thus, we have essentially recovered the result $\mu(\omega) = 1/[m\,(\gamma - i\omega)]$ that obtains in the Langevin model. Similarly, in the case of the resistor subjected to an applied voltage $V_{\rm ext}(t)$, we find similarly

$$\langle \widetilde{I}(\omega) \rangle = \frac{\widetilde{V}_{\rm ext}(\omega)}{R - i\omega L} \tag{16.17}$$

for the average value[13] of the current in the resistor. We thus recover the well-known formula for the complex admittance of a series LR circuit, namely,

$$Y(\omega) = \frac{1}{R - i\omega L}\,. \tag{16.18}$$

Both the dynamic mobility and the complex admittance are examples of **generalized susceptibilities**. Such susceptibilities are quantities of central interest in LRT. They have many interesting and important general properties (some of which we will touch upon very shortly). This is why we devote Appendix I to a brief discussion of the salient features of generalized susceptibilities in LRT at the classical level, in spite of having stated more than once in the foregoing that we shall not enter into a discussion of LRT itself in this book.

In Ch. 15, we have obtained a relation connecting the dynamic mobility to an integral of the equilibrium velocity correlation function: recall Eq. (15.43) and the comments accompanying it. For ready reference, let us write down this Kubo-Green formula once again:

$$\mu(\omega) = \frac{1}{k_B T} \int_0^\infty dt\, e^{i\omega t}\, \langle v(0)\, v(t) \rangle_{\rm eq}\,. \tag{16.19}$$

The dynamic velocity response is thus directly related to the fluctuations of the velocity in the absence of an applied force, and in a state of thermal equilibrium. As we have already stated, this sort of relationship is by no means restricted to the Langevin model. It is derivable from LRT, and is an instance of what is sometimes called the **first fluctuation-dissipation theorem**.

We now equate the real parts of Eq. (16.19), and use the form given in Eq. (16.4) for the power spectrum of the velocity. This yields the relation

$$S_v(\omega) = \frac{k_B T}{\pi}\, \mathrm{Re}\; \mu(\omega). \tag{16.20}$$

[13] Recall that this average is over all realizations of the noise voltage $V(t)$ arising from the random motion of the charge carriers in the resistor at a temperature T. The total voltage across the resistor is $V_{\rm ext}(t) + V(t)$, and $\langle V(t) \rangle = 0$.

Equation (16.20) is the basic **response-relaxation relationship** in the problem at hand, namely, the motion of a tagged particle in a fluid in thermal equilibrium. The mobility measures the response to an applied force, while the power spectrum (in the absence of the force) characterizes the manner in which the dissipation causes the relaxation of the mean velocity from any initial value to its equilibrium value. It is important to understand that the relationship above is not restricted to the specific Langevin model that we have used throughout this book[14].

- More generally, we may regard *the relation between the power spectrum of the response and the dissipation in the system as the fluctuation-dissipation relation.*

The natural question that arises now is: just as we could relate the fluctuations of the *driven* variable (the velocity, in our case) to the response of the system, can we relate the fluctuations of the *driving* noise to a physical property of the system? The answer is already well known to us. In fact, it is precisely what we have called the FD relation in all the preceding chapters—namely, the relation $\Gamma = 2m\gamma k_B T$. This formula is trivially recast in a form analogous to Eq. (16.19). Remembering that $\langle \eta(0)\,\eta(t)\rangle_{\text{eq}} = \Gamma\,\delta(t)$, we have

$$m\gamma = \frac{1}{2k_B T}\int_{-\infty}^{\infty} dt\,\langle \eta(0)\,\eta(t)\rangle_{\text{eq}} = \frac{1}{k_B T}\int_{0}^{\infty} dt\,\langle \eta(0)\,\eta(t)\rangle_{\text{eq}}\,, \tag{16.21}$$

on using the evenness property of the autocorrelation function of the stationary process $\eta(t)$. Equation (16.21) is sometimes referred to as the **second fluctuation-dissipation theorem**. Evidently, it is (in its present form) specific to the Langevin model, because both the random force $\eta(t)$ and the friction coefficient γ pertain to this model.

The alert reader would have noted that Eq. (16.19) involves the one-sided Fourier transform (or the Fourier-Laplace transform) of the velocity autocorrelation, while Eq. (16.21) does not involve the corresponding transform of the autocorrelation of the noise. But this latter feature is just a consequence of the white noise assumption for $\eta(t)$. If the noise itself has a finite correlation time, the second FD theorem *will* involve the Fourier-Laplace transform of the autocorrelation of the noise. We shall see how this comes about in Ch. 17. The first and second FD theorems would then be much more alike in structure.

[14] A proof of this assertion involves LRT, which we have skirted around. (However, see Appendix I.) Also, the particular numerical constant (π^{-1}) appearing in Eq. (16.20) depends on the conventions we have adopted, as explained in the footnote following Eq. (16.10).

16.3 Analyticity of the mobility

As we have already stated, the dynamic mobility $\mu(\omega)$ is a particular example of what is known as a generalized susceptibility in LRT. The results to be derived now are applicable to all such generalized susceptibilities, although we shall continue to work with the mobility in order to be specific.

An important property of the dynamic mobility can be extracted from the integral representation in Eq. (16.19) : since the autocorrelation function $\langle v(0)\, v(t)\rangle_{\text{eq}}$ is a real quantity, we find that

$$\mu(-\omega) = \mu^*(\omega) \quad \text{for all } \textit{real} \quad \omega, \tag{16.22}$$

where the asterisk denotes the complex conjugate[15]. It follows at once that, for all real ω,

$$\text{Re } \mu(-\omega) = \text{Re } \mu(\omega) \quad \text{and} \quad \text{Im } \mu(-\omega) = -\text{Im } \mu(\omega). \tag{16.23}$$

That is,

- the *real* part of the dynamic mobility is an *even* function of the frequency, while the *imaginary* part is an *odd* function of the frequency.

These properties, of course, may also be inferred directly upon writing $e^{i\omega t} = \cos \omega t + i \sin \omega t$ in the Kubo-Green formula (16.19).

Why have we added the phrase, "for all real ω" in Eq. (16.22)? The reason is that it makes sense to speak of $\mu(\omega)$ for complex values of ω as well. In fact,

- a generalized susceptibility can be *analytically continued* to complex values of ω. It has specific analyticity properties that arise from physical requirements, and that have important consequences.

The foremost of these properties is readily deduced from the functional form in Eq. (16.19). The fact that the lower limit of integration is 0 rather than $-\infty$ is traceable to *causality*: namely, the requirement that the effect (the response) cannot precede the cause (the stimulus or the force). Now, if the integral exists for real values of ω, it does so quite obviously even if ω is complex, provided Im ω is *positive*. This is because the factor $\exp{(i\omega t)}$ then leads to an extra damping factor $\exp{[-(\text{Im } \omega)\, t]}$. Such a factor can only *improve* the convergence of the integral, since the integration over t is restricted to non-negative values. We may therefore conclude that:

[15]Once again, this general property is trivially verified for the explicit form obtained in the Langevin model, namely, $\mu(\omega) = 1/[m\,(\gamma - i\omega)]$.

- The expression in Eq. (16.19) is in fact an analytic[16] function of ω in the *upper* half of the complex ω-plane[17].

Now that we know that a generalized susceptibility is analytic in the upper half-plane in ω, we can write down the generalization of the symmetry property implied by Eq. (16.22). It is

$$\mu(-\omega^*) = \mu^*(\omega), \quad \text{Im } \omega \geq 0. \tag{16.24}$$

Note that if ω lies in the upper half of the complex plane, so does $-\omega^*$. Hence the arguments of the functions on both sides of Eq. (16.24) lie in the region in which we are guaranteed that a generalized susceptibility is analytic. On general grounds, and without further input, we really cannot say very much about its possible behavior in the lower half of the complex ω-plane. But we know that an analytic function of a complex variable cannot be holomorphic at *all* points of the extended complex plane[18]. In general, therefore, a generalized susceptibility will have singularities in the lower half-plane in ω. Once again, the explicit form $\mu(\omega) = 1/[m\,(\gamma - i\omega)]$ satisfies this general property. It is certainly analytic for all Im $\omega \geq 0$. In fact, it happens to be analytic for Im $\omega < 0$ as well, except for a simple pole at $\omega = -i\gamma$.

16.4 Dispersion relations

The analyticity of a generalized susceptibility in the upper half-plane in ω enables us to write down **dispersion relations** for its real and imaginary parts. These relations express the fact that the real and imaginary parts form a **Hilbert transform pair**.

The derivation of the dispersions relations for a generalized susceptibility such as $\mu(\omega)$ is straightforward. We begin with the fact that $\mu(\omega)$ is analytic in the upper half-plane in ω. Let us assume, further, that $|\mu(\omega)| \to 0$ as $\omega \to \infty$ along

[16]The more precise term is 'holomorphic', i. e., it satisfies the Cauchy-Riemann conditions.

[17]Causality ensures that the Fourier-Laplace transform of a causal correlation function is analytic in a half-plane in ω. Whether this is the upper or lower half-plane depends on the Fourier transform convention chosen. With our particular convention (see Eqs. (15.27)), in which a function $f(t)$ of the time is expanded as $\int_{-\infty}^{\infty} d\omega\, \widetilde{f}(\omega)\, \exp(-i\omega t)$, it is the *upper* half-plane in the frequency ω in which the susceptibility is analytic. Had we chosen the opposite sign convention and used the factor $\exp(+i\omega t)$ in the expansion above, the region of analyticity would have been the *lower* half-plane in ω. The convention we have adopted is the more commonly used one, at least in physics.

[18]Unless it is just a constant, as we know from Liouville's Theorem in complex analysis.

any direction in the upper half-plane. This is a physically plausible assumption to make in most circumstances, for the following reason. Even for *real* ω, we expect the mobility to vanish as the frequency becomes very large, because the inertia present in any system will not permit it to respond to a sinusoidal applied force oscillating at a frequency much higher than all the natural frequencies present in the system. And when ω is a complex number with a positive imaginary part, the factor $e^{i\omega t}$ in the formula for the mobility lends an extra damping factor $e^{-(\text{Im}\,\omega)\,t}$. The assumption is thus a reasonable one to make[19].

Let ω be a fixed, real, positive frequency. Consider the quantity

$$f(\omega') = \frac{\mu(\omega')}{\omega' - \omega} \tag{16.25}$$

as a function of the complex variable ω'. This function is analytic everywhere in the upper half-plane in ω', as well on the real axis in that variable, except for a simple pole at $\omega' = \omega$ located on the real axis. By Cauchy's integral theorem, its integral over any closed contour C' lying entirely in the upper half-plane is identically equal to zero (see Fig. 16.2). The contour C' can be expanded to the contour C without changing the value of the integral (namely, zero). Thus

$$\oint_{C'} d\omega'\, f(\omega') = \oint_{C} d\omega'\, \frac{\mu(\omega')}{\omega' - \omega} = 0. \tag{16.26}$$

The closed contour C comprises the following:

(i) a large semicircle of radius R in the upper half-plane,

(ii) a line integral on the real axis running from $-R$ to $\omega - \epsilon$,

(iii) a small semicircle, from $\omega - \epsilon$ to $\omega + \epsilon$, lying in the upper half-plane so as to avoid the simple pole of the integrand, and, finally,

(iv) a line integral from $\omega + \epsilon$ to R.

In the limit $R \to \infty$, the contribution from the large semicircle vanishes, because $f(\omega')$ vanishes faster than $1/\omega'$ as $\omega' \to \infty$ along all directions in the upper half-plane—this is ensured by the fact that $\mu(\omega') \to 0$ as $|\omega'| \to \infty$. On the small semicircle, we have $\omega' = \omega + \epsilon\, e^{i\theta}$, where θ runs from π to 0. Therefore, in the limit $\epsilon \to 0$, the contribution from the small semicircle tends to $-i\pi\mu(\omega)$. Hence we get

$$\lim_{\epsilon \to 0} \left\{ \int_{-\infty}^{\omega - \epsilon} d\omega'\, \frac{\mu(\omega')}{\omega' - \omega} + \int_{\omega + \epsilon}^{\infty} d\omega'\, \frac{\mu(\omega')}{\omega' - \omega} \right\} - i\pi\mu(\omega) = 0. \tag{16.27}$$

[19]However, it can be relaxed. See the exercises at the end of this chapter.

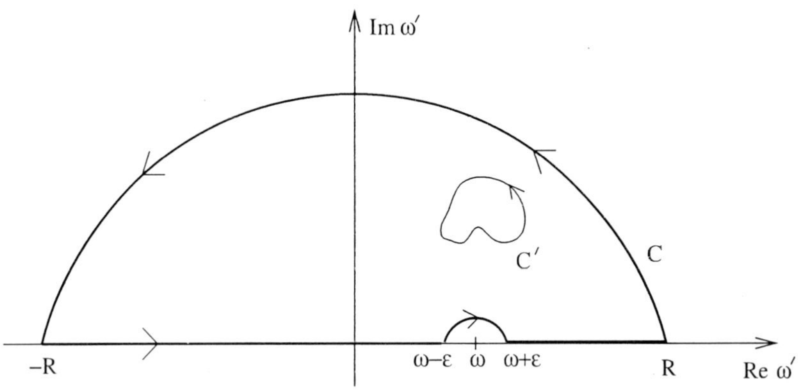

Figure 16.2: The closed contour C' in the ω'-plane over which the line integral $\oint_{C'} d\omega' \, \mu(\omega')/(\omega' - \omega)$ vanishes, and the closed contour C into which C' can be distorted without changing the value (zero) of the integral. Note that all parts of C remain restricted to the upper half-plane and the real axis in ω', the region in which the function $\mu(\omega')$ can have no singularity (as a consequence of causality). On passing to the limits $R \to \infty$ and $\epsilon \to 0$, this leads to the conclusion that the real and imaginary parts of the dynamic mobility at any physical frequency ω constitute a Hilbert transform pair. This relationship results in dispersion relations for the dynamic mobility.

But the limit on the left-hand side of this equation is just the **Cauchy principal value** integral of the integrand $\mu(\omega')/(\omega' - \omega)$. It is a specific prescription for avoiding the divergence or infinity that would otherwise arise, owing to the singularity (a simple pole) of the integrand at $\omega' = \omega$. Denoting the principal value integral by $\mathsf{P}\int(\cdots)$, we get

$$\mu(\omega) = -\frac{i}{\pi}\,\mathsf{P}\int_{-\infty}^{\infty} d\omega'\,\frac{\mu(\omega')}{\omega' - \omega}. \tag{16.28}$$

Note that Eq. (16.28) expresses the mobility at a *real* frequency as a certain weighted sum of the mobility over all other *real* frequencies. No complex frequencies appear anywhere in this formula. An excursion was made into the complex frequency plane. This was made possible by the analyticity properties of the mobility. But we have returned to the real axis, bringing back Eq. (16.28) with us.

Equating the respective real and imaginary parts of the two sides of Eq. (16.28), we get

$$\begin{aligned} \text{Re}\,\mu(\omega) &= \frac{1}{\pi}\,\mathsf{P}\int_{-\infty}^{\infty} d\omega'\,\frac{\text{Im}\,\mu(\omega')}{\omega' - \omega}, \\ \text{Im}\,\mu(\omega) &= -\frac{1}{\pi}\,\mathsf{P}\int_{-\infty}^{\infty} d\omega'\,\frac{\text{Re}\,\mu(\omega')}{\omega' - \omega}. \end{aligned} \tag{16.29}$$

These are the dispersion relations sought. They imply that $\text{Re}\,\mu(\omega)$ and $\text{Im}\,\mu(\omega)$ are Hilbert transforms of each other. These relations can be simplified further by using the symmetry properties expressed in Eqs. (16.23). Doing so, we get

$$\begin{aligned} \text{Re}\,\mu(\omega) &= \frac{2}{\pi}\,\mathsf{P}\int_{0}^{\infty} d\omega'\,\frac{\omega'\,\text{Im}\,\mu(\omega')}{\omega'^2 - \omega^2}, \\ \text{Im}\,\mu(\omega) &= -\frac{2\omega}{\pi}\,\mathsf{P}\int_{0}^{\infty} d\omega'\,\frac{\text{Re}\,\mu(\omega')}{\omega'^2 - \omega^2}. \end{aligned} \tag{16.30}$$

The advantage of this form of the dispersion relations (also called **Kramers-Kronig relations**) is that the range of integration is restricted to physically accessible frequencies, namely, $0 \le \omega' < \infty$.

We emphasize once again:

- It is the physical feature of *causal response* that leads to the analyticity of a generalized susceptibility in a half-plane in the frequency.
- *Thus, dispersion relations are a direct consequence of causality.*

16.5 Exercises

16.5.1 Position autocorrelation of the Brownian oscillator

In Ch. 15, Sec. 15.4.1, we have obtained the velocity autocorrelation function of a Brownian oscillator (a harmonically bound tagged particle). The purpose of the present exercise is to show how the relation between the power spectra of the velocity and the position can be used to find the autocorrelation of the position of the oscillator[20].

Let $t_1 < t_2$. Equation (16.6) for the autocorrelation function of a stationary random process $\xi(t)$ in terms of its power spectrum may be written in the form

$$\langle \xi(t_1)\, \xi(t_2) \rangle = \int_{-\infty}^{\infty} d\omega\, e^{-i\omega(t_2 - t_1)}\, S_\xi(\omega), \tag{16.31}$$

where $S_\xi(\omega)$ is the PSD of ξ. Differentiate both sides of this equation with respect to t_1 and t_2, in succession, and set $t_1 = 0$, $t_2 = t$. Then

$$\left\langle \dot{\xi}(0)\, \dot{\xi}(t) \right\rangle = \int_{-\infty}^{\infty} d\omega\, e^{-i\omega t}\, \omega^2\, S_\xi(\omega). \tag{16.32}$$

But we also have, analogous to the corresponding relationship for the process ξ itself,

$$\left\langle \dot{\xi}(0)\, \dot{\xi}(t) \right\rangle = \int_{-\infty}^{\infty} d\omega\, e^{-i\omega t}\, S_{\dot{\xi}}(\omega). \tag{16.33}$$

Hence the power spectra of the stationary processes[21] $\xi(t)$ and $\dot{\xi}(t)$ are related by

$$S_{\dot{\xi}}(\omega) = \omega^2\, S_\xi(\omega). \tag{16.34}$$

We shall apply this equation to the Brownian oscillator, taking $\xi(t) = x(t)$ and hence $\dot{\xi}(t) = v(t)$.

[20] There are other ways of computing this autocorrelation function, but the method used here is instructive.

[21] We take this opportunity to state explicitly what should be obvious by now. If $\xi(t)$ is a stationary random process, so is the process $\dot{\xi}(t)$. The converse is *not* necessarily true. The obvious example is the velocity $v(t)$ of a freely diffusing tagged particle in thermal equilibrium. As we have pointed out several times, this process is stationary, but the position $x(t)$ is not. In the case of the Brownian oscillator, on the other hand, $x(t)$ *is* a stationary process in thermal equilibrium, and of course so is $v(t)$.

(a) Use Eq. (15.59) for the velocity autocorrelation function of the oscillator in the formula of Eq. (16.4) to show that its PSD is given by

$$S_v(\omega) = \frac{k_B T\gamma}{m\pi}\left\{\frac{\omega^2}{(\omega^2-\omega_0^2)^2+\omega^2\gamma^2}\right\}. \tag{16.35}$$

(b) It follows from Eqs. (16.34) and (16.35) that the PSD of the position of the oscillator is given by

$$S_x(\omega) = \frac{S_v(\omega)}{\omega^2} = \frac{k_B T\gamma}{m\pi}\left\{\frac{1}{(\omega^2-\omega_0^2)^2+\omega^2\gamma^2}\right\}. \tag{16.36}$$

Observe that the singular factor $1/\omega^2$ cancels out neatly, to yield an expression for $S_x(\omega)$ that has no singularity for any real value of ω. Using the inversion formula in Eq. (16.6), we have

$$\langle x(0)\,x(t)\rangle_{\text{eq}} = \int_{-\infty}^{\infty} d\omega\, e^{-i\omega t}\, S_x(\omega) = \frac{k_B T\gamma}{m\pi}\int_{-\infty}^{\infty}\frac{e^{-i\omega t}\,d\omega}{(\omega^2-\omega_0^2)^2+\omega^2\gamma^2}. \tag{16.37}$$

Evaluate this integral for $t > 0$ (it is a simple exercise in contour integration) to obtain the result

$$\langle x(0)\,x(t)\rangle_{\text{eq}} = \frac{k_B T}{m\omega_0^2}\, e^{-\gamma t/2}\left(\cos\,\omega_s t + \frac{\gamma}{2\omega_s}\,\sin\,\omega_s t\right),\quad t \geq 0. \tag{16.38}$$

(c) Now evaluate the integral above for $t < 0$, and show that

$$\langle x(0)\,x(t)\rangle_{\text{eq}} = \frac{k_B T}{m\omega_0^2}\, e^{\gamma t/2}\left(\cos\,\omega_s t - \frac{\gamma}{2\omega_s}\,\sin\,\omega_s t\right),\quad t \leq 0. \tag{16.39}$$

Hence

$$\langle x(0)\,x(t)\rangle_{\text{eq}} = \frac{k_B T}{m\omega_0^2}\, e^{-\gamma|t|/2}\left(\cos\,\omega_s|t| + \frac{\gamma}{2\omega_s}\,\sin\,\omega_s|t|\right),\quad -\infty < t < \infty. \tag{16.40}$$

Observe that $\langle x(0)\,x(t)\rangle_{\text{eq}}$ is indeed an even function of t, as required of the autocorrelation function of a (one-component) stationary random process. Further, its slope at $t = 0$ vanishes, i. e., $\langle x(0)\,\dot{x}(0)\rangle_{\text{eq}} = 0$, as required by stationarity.

16.5.2 Power spectrum of a multi-component process

The extension of the concept of the power spectrum to the case of a stationary multi-component, or vector, random process is straightforward. For our purposes, it is convenient to use the velocity of the tagged particle in the full three-dimensional case as a concrete example. The power spectrum is now a matrix.

Using the Wiener-Khinchin Theorem, the elements of this matrix are given by

$$S_{ij}(\omega) = \frac{1}{2\pi}\int_{-\infty}^{\infty} dt\, e^{i\omega t}\,\phi_{ij}(t), \quad \text{where} \quad \phi_{ij}(t) = \langle v_i(0)\, v_j(t)\rangle_{\text{eq}}\,. \tag{16.41}$$

This is the multi-component version of Eq. (16.2). However, since $\phi_{ij}(t)$ is neither an even function nor an odd function of t, we cannot immediately re-cast $S_{ij}(\omega)$ in a form analogous to that in Eq. (16.4). $S_{ij}(\omega)$ need not even be real for $i \neq j$ (even though all the components of the random process itself are real-valued.)

(a) Use the general symmetry property of the correlation matrix, $\phi_{ij}(-t) = \phi_{ji}(t)$ (recall Eq. (4.36)), to show that the elements of the power spectrum matrix satisfy the property

$$S_{ij}(\omega) = S^{*}_{ji}(\omega), \quad \text{or, in matrix form,} \quad \mathsf{S}(\omega) = \mathsf{S}^{\dagger}(\omega). \tag{16.42}$$

Thus,

- the power spectrum of a multi-component stationary random process is a *hermitian* matrix.

(b) Write the correlation tensor as the sum of its symmetric and antisymmetric parts,

$$\phi_{ij}(t) \equiv \phi^{\text{sym}}_{ij}(t) + \phi^{\text{antisym}}_{ij}(t). \tag{16.43}$$

Recall that this is *also* a break-up of the correlation tensor as the sum of an even function of t and an odd function of t (see Eqs. (4.37)). Hence show that the multi-component counterpart of Eq. (16.4) is

$$S_{ij}(\omega) = \frac{1}{\pi}\int_{0}^{\infty} dt\,\left[\phi^{\text{sym}}_{ij}(t)\,\cos\,\omega t + i\;\phi^{\text{antisym}}_{ij}(t)\,\sin\,\omega t\right]. \tag{16.44}$$

It follows that the symmetric and antisymmetric parts of the tensor $S_{ij}(\omega)$ are directly related to its real and imaginary parts, respectively, according to

$$\begin{aligned} S^{\text{sym}}_{ij}(\omega) &= \text{Re}\; S_{ij}(\omega) = \frac{1}{\pi}\int_{0}^{\infty} dt\;\phi^{\text{sym}}_{ij}(t)\,\cos\,\omega t, \\ S^{\text{antisym}}_{ij}(\omega) &= i\,\text{Im}\; S_{ij}(\omega) = \frac{i}{\pi}\int_{0}^{\infty} dt\;\phi^{\text{antisym}}_{ij}(t)\,\sin\,\omega t\,. \end{aligned} \tag{16.45}$$

We reiterate that these formulas are valid for any stationary vector-valued random process, although our immediate concern is the velocity of the tagged particle.

16.5.3 The mobility tensor

The generalization of the results derived in Sec. 16.2 to the full three-dimensional velocity of the tagged particle is straightforward. The **mobility tensor** $\mu_{ij}(\omega)$ is given by a direct extension of the Kubo-Green formula in the one-dimensional case, Eq. (16.19):

$$\mu_{ij}(\omega) = \frac{1}{k_B T} \int_0^\infty dt \, e^{i\omega t} \, \langle v_i(0) \, v_j(t) \rangle_{\text{eq}} \equiv \frac{1}{k_B T} \int_0^\infty dt \, e^{i\omega t} \, \phi_{ij}(t). \tag{16.46}$$

Since $\phi_{ij}(t)$ is real, it follows immediately that

$$\text{Re } \mu_{ij}(-\omega) = \text{Re } \mu_{ij}(\omega) \quad \text{and} \quad \text{Im } \mu_{ij}(-\omega) = -\text{Im } \mu_{ij}(\omega) \tag{16.47}$$

for all real values of ω, exactly as in the one-dimensional case, Eq. (16.23). Writing $\phi_{ij}(t)$ as the sum of its symmetric and antisymmetric parts, and $e^{i\omega t}$ as the sum of its real and imaginary parts, we see that $\mu_{ij}(\omega)$ is a sum of four integrals.

(a) The multi-component version of the relaxation-response relationship connecting the power spectrum of the velocity and the dynamic mobility, Eq. (16.20), can now be read off easily. Show that

$$\begin{aligned} S_{ij}^{\text{sym}}(\omega) &= \text{Re } S_{ij}(\omega) = \frac{k_B T}{\pi} \text{Re } \mu_{ij}^{\text{sym}}(\omega), \\ S_{ij}^{\text{antisym}}(\omega) &= i \text{ Im } S_{ij}(\omega) = i \, \frac{k_B T}{\pi} \text{Im } \mu_{ij}^{\text{antisym}}(\omega), \end{aligned} \tag{16.48}$$

where μ_{ij}^{sym} and $\mu_{ij}^{\text{antisym}}$ are the symmetric and antisymmetric parts of the mobility tensor. Equations (16.48) can be combined, of course, to give

$$S_{ij}(\omega) = \frac{k_B T}{\pi} \left[\text{Re } \mu_{ij}^{\text{sym}}(\omega) + i \text{ Im } \mu_{ij}^{\text{antisym}}(\omega) \right]. \tag{16.49}$$

16.5.4 Particle in a magnetic field: Hall mobility

We turn (once again) to an application of the formalism to the case of a tagged particle of charge q and mass m in a magnetic field $\mathbf{B} = B\,\mathbf{n}$. In Ch. 4, we have shown (see Eq. (4.34)) that the elements $\phi_{ij}(t) = \langle v_i(0) \, v_j(t) \rangle_{\text{eq}}$ of the velocity correlation matrix (or tensor) are given by

$$\phi_{ij}(t) = \frac{k_B T}{m} \, e^{-\gamma|t|} \, [n_i \, n_j + (\delta_{ij} - n_i \, n_j) \cos \omega_c t - \epsilon_{ijk} \, n_k \sin \omega_c t] \, , \tag{16.50}$$

where $\omega_c = qB/m$ is the cyclotron frequency.

(a) Show that the symmetric and antisymmetric parts of the mobility tensor are given by

$$\mu_{ij}^{\text{sym}}(\omega) = \frac{1}{m}\left\{\frac{n_i\, n_j}{\gamma - i\omega} + \frac{(\delta_{ij} - n_i\, n_j)\,(\gamma - i\omega)}{(\gamma - i\omega)^2 + \omega_c^2}\right\} \tag{16.51}$$

and

$$\mu_{ij}^{\text{antisym}}(\omega) = -\frac{\epsilon_{ijk}\, n_k\, \omega_c}{m[(\gamma - i\omega)^2 + \omega_c^2]}\,. \tag{16.52}$$

(b) Using Eq. (16.44), show that the symmetric and antisymmetric parts of the corresponding power spectrum matrix are given by

$$S_{ij}^{\text{sym}}(\omega) = \frac{\gamma k_B T}{m\pi}\left\{\frac{n_i\, n_j}{\gamma^2 + \omega^2} + \frac{(\delta_{ij} - n_i\, n_j)(\gamma^2 + \omega^2 + \omega_c^2)}{(\gamma^2 + \omega^2 + \omega_c^2)^2 - 4\omega^2\,\omega_c^2}\right\} \tag{16.53}$$

and

$$S_{ij}^{\text{antisym}}(\omega) = -\frac{\gamma k_B T}{m\pi}\left\{\frac{2i\,\epsilon_{ijk}\, n_k\, \omega_c\, \omega}{(\gamma^2 + \omega_c^2 + \omega^2)^2 - 4\omega^2\,\omega_c^2}\right\}. \tag{16.54}$$

(c) Hence verify that the response-relaxation relationships of Eqs. (16.48) are satisfied by the foregoing expressions for the mobility and the power spectrum.

We have seen in Sec. 15.4.2 that the time integral of the symmetric part of the velocity correlation tensor yields the diffusion tensor D_{ij} for the charged tagged particle in a magnetic field. What does the antisymmetric part of the correlation tensor (or the dynamic mobility tensor) signify? In order to be specific, let us choose the magnetic field to be directed along the z-axis. Hence $\mathbf{n} = (0, 0, 1)$. Further, suppose an electric field is also applied along the x-axis, causing an acceleration of the charged particle in the x-direction. We now recognize the familiar Hall geometry: owing to the Lorentz force $q(\mathbf{v} \times \mathbf{B})$, a Hall current is set up in the y-direction. The component $\mu_{12}(\omega)$ of the mobility tensor is a measure of this response. In the case at hand, we find

$$\mu_{12}(\omega) = -\mu_{21}(\omega) = -\frac{\omega_c}{m[(\gamma - i\omega)^2 + \omega_c^2]}\,. \tag{16.55}$$

16.5.5 Simplifying the dispersion relations

Use the symmetry properties of the real and imaginary parts of $\mu(\omega)$ given by Eqs. (16.23) to reduce Eqs. (16.29) to Eqs. (16.30).

16.5.6 Subtracted dispersion relations

It may so happen that a generalized susceptibility (which we shall denote by $\chi(\omega)$) does *not* vanish as $\omega \to \infty$. Instead, it may tend to a constant as $\omega \to \infty$ along some direction or directions in the region Im $\omega \geq 0$. (It is clear that this is most likely to happen as $\omega \to \infty$ along the real axis itself.) When we try to derive dispersion relations for the real and imaginary parts of $\chi(\omega)$, we find that the contribution from the large semicircle of radius R no longer vanishes as $R \to \infty$. If $\chi(\omega) \to$ a constant, say χ_∞, *uniformly* as $|\omega| \to \infty$ along the real axis and along all directions in the upper half-plane, we could go ahead by including $i\pi\chi_\infty$ as the extra contribution from the large semicircle. But there is no guarantee that this will be the case. In fact, it does not happen, in general.

To get over the problem, we assume that the value of $\chi(\omega)$ at some particular real value of ω_0 of the frequency is known. Then, instead of the function in (16.25), consider the quantity

$$f(\omega') = \frac{\chi(\omega') - \chi(\omega_0)}{(\omega' - \omega_0)(\omega' - \omega)} \tag{16.56}$$

as a function of ω'. It is evident that *this* function does vanish faster than $1/\omega'$ as $|\omega'| \to \infty$, owing to the extra factor $(\omega' - \omega_0)^{-1}$. Moreover, it is analytic everywhere in the upper half ω'-plane and on the real axis, except for a simple pole at $\omega' = \omega$, as before[22]. Repeat the arguments used in Sec. 16.4 to show that

$$\chi(\omega) = \chi(\omega_0) - \frac{i}{\pi}(\omega - \omega_0)\, \mathsf{P}\int_{-\infty}^{\infty} d\omega' \, \frac{\chi(\omega') - \chi(\omega_0)}{(\omega' - \omega_0)(\omega' - \omega)}. \tag{16.57}$$

Hence obtain the dispersion relations satisfied by the real and imaginary parts of $\chi(\omega)$ in this case. These are called *once-subtracted* dispersion relations, and ω_0 is the *point of subtraction*[23].

[22]Observe that it does *not* have any singularity at $\omega' = \omega_0$. This is the reason for subtracting $\chi(\omega_0)$ from $\chi(\omega')$ in the numerator.

[23]From a mathematical point of view, it should be evident that the procedure given above can be extended to cover situations when the analytic function $\chi(\omega)$ actually diverges as $\omega \to \infty$ in the upper half-plane. For instance, if $\chi(\omega) \sim \omega^{n-1}$ asymptotically, where n is a positive integer, we can write down an n-fold subtracted dispersion relation for it. The latter will require n constants as inputs. These could be, for instance, the values of $\chi(\omega)$ at n different frequencies or points of subtraction.

Chapter 17

The generalized Langevin equation

The shortcomings of the Langevin model, involving the simultaneous requirements of causality and stationarity, have already been discussed in the preceding chapters. In this final chapter, we return to these considerations. A generalization of the Langevin equation, involving a memory kernel for the frictional force on the tagged particle and, correspondingly, a driving noise with a nonzero correlation time, is shown to resolve these difficulties. The generalized Langevin equation offers a phenomenological approach to the motion of the tagged particle that is fully consistent with Linear Response Theory. We show how fluctuation-dissipation theorems may be derived directly from this equation.

17.1 Shortcomings of the Langevin model

The Langevin model we have discussed at some length in the preceding chapters is, of course, an approximation to the physical problem of the motion of a tagged particle in a fluid. But the model itself has certain intrinsic deficiencies, as we have pointed out at various junctures. We now resume the discussion of this aspect, from where it was left off in Ch. 15, Sec. 15.5.

17.1.1 Short-time behavior

To recapitulate in brief: The difficulty lies in the extreme short-time (or $t \to 0$) behavior of the velocity process as predicted by the Langevin equation. We have seen that the velocity is a stationary random process in equilibrium, in the absence of an external force. The equilibrium autocorrelation function $\langle v(t)\, v(t+t')\rangle_{\text{eq}}$ is therefore a function of the time difference t' alone, and is independent of t. Hence its derivative with respect to t must vanish, i. e.,

$$\frac{d}{dt}\big\langle v(t)\, v(t+t')\big\rangle_{\text{eq}} = \big\langle \dot{v}(t)\, v(t+t')\big\rangle_{\text{eq}} + \big\langle v(t)\, \dot{v}(t+t')\big\rangle_{\text{eq}} = 0. \tag{17.1}$$

Passing to the limit $t' \to 0$ from above in this equation, we must have[1]

$$\langle v(t)\, \dot{v}(t)\rangle_{\text{eq}} = 0 \tag{17.2}$$

at any instant of time t. In other words, the instantaneous velocity and the instantaneous acceleration must be uncorrelated with each other when the system is in thermal equilibrium. But this leads to an inconsistency. We have shown that the equilibrium velocity autocorrelation is a decaying exponential, $\langle v(t)\, v(t+t')\rangle_{\text{eq}} = (k_BT/m)\exp(-\gamma t')$, $t' \geq 0$. Differentiating both sides with respect to t' and taking the limit $t' \to 0$ from above, we find that $\langle v(t)\, \dot{v}(t)\rangle_{\text{eq}} = -\gamma k_BT/m$, which is not identically equal to zero. This is in contradiction with Eq. (17.2).[2]

17.1.2 The power spectrum at high frequencies

The inconsistency may be brought out in another (related) way as well. If $v(t)$ is a stationary random process in a state of thermal equilibrium, Eqs. (17.1) apply. The second of these equations can be written as

$$\begin{aligned}\big\langle \dot{v}(t)\, v(t+t')\big\rangle_{\text{eq}} &= -\big\langle v(t)\, \dot{v}(t+t')\big\rangle_{\text{eq}} = -\langle v(t)\, \frac{d}{dt'}v(t+t')\rangle_{\text{eq}} \\ &= -\frac{d}{dt'}\big\langle v(t)\, v(t+t')\big\rangle_{\text{eq}}.\end{aligned} \tag{17.3}$$

Differentiating both sides with respect to t', we have

$$\big\langle \dot{v}(t)\, \dot{v}(t+t')\big\rangle_{\text{eq}} = -\frac{d^2}{dt'^2}\big\langle v(t)\, v(t+t')\big\rangle_{\text{eq}}. \tag{17.4}$$

Now, the definition of the power spectrum of the velocity as the Fourier transform of the velocity autocorrelation implies the inverse relation given in Eq. (16.6), namely,

$$\big\langle v(t)\, v(t+t')\big\rangle_{\text{eq}} = \int_{-\infty}^{\infty} d\omega\, S_v(\omega)\, e^{-i\omega t'}. \tag{17.5}$$

---

[1] v and $\dot{v}$ are classical variables; hence they commute with other.

[2] Recall that in Ch. 15, Sec. 15.5, we tried to fix this problem by 'deriving' a nonzero value for the equal-time correlator $\langle v(t)\, \eta(t)\rangle_{\text{eq}}$, even though causality would suggest that it is equal to zero. However, this did not resolve the problem unequivocally. In fact, it led to incorrect results.

Therefore Eq. (17.4) becomes

$$\big\langle \dot{v}(t)\, \dot{v}(t+t\,')\big\rangle_{\text{eq}} = \int_{-\infty}^{\infty} d\omega\, \omega^2\, S_v(\omega)\, e^{-i\omega t\,'}. \tag{17.6}$$

Setting $t\,' = 0$, we get a formula for the equilibrium value of the mean squared acceleration in terms of the second moment of the power spectrum of the velocity:

$$\langle \dot{v}(t)\, \dot{v}(t)\rangle_{\text{eq}} = \big\langle \dot{v}^2\big\rangle_{\text{eq}} = \int_{-\infty}^{\infty} d\omega\, \omega^2\, S_v(\omega). \tag{17.7}$$

We certainly expect the mean squared acceleration of the tagged particle to be finite in equilibrium. Now, according to Eq. (16.11), the power spectrum $S_v(\omega)$ of the velocity of the tagged particle (in a state of thermal equilibrium) is given by the Lorentzian

$$S_v(\omega) = \frac{\gamma k_B T}{m\pi\,(\gamma^2 + \omega^2)}\,. \tag{17.8}$$

But the integral on the right-hand side of Eq. (17.7) diverges when the Lorentzian form in Eq. (17.8) is substituted for $S_v(\omega)$, because $\int_{-\infty}^{\infty} d\omega\, \omega^2/(\gamma^2+\omega^2) = \infty$. Similarly, the higher moments of $S_v(\omega)$ can be related to the equilibrium averages of the higher derivatives of x and their powers. These relations are model-independent. As physical quantities, these mean values should be finite; but the higher moments of the Lorentzian diverge, leading to inconsistencies. As the form (17.8) for the power spectrum follows directly from the Langevin equation, it is to this equation that we must trace the problem.

17.2 The memory kernel and the GLE

It is now clear that the Langevin equation cannot be a rigorously consistent way of specifying the velocity process at extremely (and arbitrarily) short intervals of time after any given initial instant. In the frequency domain, this shortcoming shows up in the high-frequency behavior of the velocity power spectrum, which does not vanish sufficiently rapidly as $|\omega| \to \infty$. What, precisely, is the source of the problem with the Langevin model?

We can trace the flaw right back to the modeling of the friction (or the systematic part of the random force on the tagged particle) as $F_{\text{sys}}(t) = -m\gamma v(t)$, in Eq. (2.17) of Ch. 2, Sec. 2.3. This form implies an *instantaneous* reaction on the part of the fluid (or heat bath) in producing the retarding or dissipative force on the tagged particle. In principle, however, a truly instantaneous reaction is impossible. As we have already pointed out (or implied) at several junctures (in particular, in Ch. 2, Sec. 2.1; Ch. 4, Sec. 4.1; and Ch. 11, Sec. 11.2), such an assumption is a justifiable *approximation* if the time scale on which the tagged particle moves is much greater than that on which the fluid molecules themselves move. In that case, provided we restrict ourselves to times of the order of, or greater than, the longer time scale, the Langevin model may be expected to be satisfactory. How, then, should the frictional force $F_{\text{sys}}(t)$ be modeled in order to be applicable on smaller time scales as well? We need to keep in mind three physical requirements:

(i) $F_{\text{sys}}(t)$ should be *causal*: the force at time t can only depend on the velocity history (from $-\infty$, in general) up to time t, and not upon velocities at times later than t.

(ii) $F_{\text{sys}}(t)$ should represent a *retarded* response. The effect of the velocity at an earlier instant t' should be weighted by a function of the elapsed time $(t-t')$.

(iii) $F_{\text{sys}}(t)$ should be *linearly* dependent on the velocity, if we continue to remain in the linear response regime.

All these requirements are met if we model the systematic part of the random force by the following linear functional of the velocity:

$$F_{\text{sys}}(t) = -m \int_{-\infty}^{t} dt'\, \gamma(t-t')\, v(t'). \tag{17.9}$$

Thus, the friction constant γ of the original Langevin equation has been replaced by a **memory kernel** $\gamma(t-t')$, defined for non-negative values of its argument. (Formally, we may define $\gamma(t-t')$ to be identically equal to zero for $t < t'$.) On physical grounds, we expect this kernel to be a non-negative, decreasing function of its argument. Note that the memory kernel has the physical dimensions of $(\text{time})^{-2}$. The original Langevin equation is recovered in the limit of an instantaneous response by the medium, i. e., by setting $\gamma(t) = \gamma\,\delta(t)$ where γ is a positive constant.

With the expression in Eq. (17.9) for $F_{\text{sys}}(t)$, the Langevin equation (2.18) of Sec. 2.3 for the velocity of the tagged particle is replaced by the **generalized Langevin equation**Langevin equation!generalized (GLE)

$$m\dot{v}(t) = -m \int_{-\infty}^{t} dt'\, \gamma(t-t')\, v(t') + \eta(t) + F_{\text{ext}}(t). \tag{17.10}$$

The GLE represents a phenomenological approach that retains the basic idea of modeling the motion of the tagged particle by a stochastic evolution equation for its velocity. This equation is no longer an SDE, but an *integro-differential* equation. We have continued to use the symbol $\eta(t)$ for the noise component of the random force due to the collisions of the tagged particle with the particles of the fluid. However, as shall see in the sequel, the *white* noise assumption for $\eta(t)$ is no longer tenable, once $F_{\text{sys}}(t)$ involves a memory kernel. Let us proceed (for the moment) without specifying the properties of the noise.

We state (without proof) the following consequences of the GLE in the absence of an external applied force: Since the equation is not an SDE of the Itô type, the SDE$\leftrightarrow$FPE correspondence is no longer valid. The PDF of the velocity does not satisfy the Fokker-Planck equation, and $v(t)$ is no longer an Ornstein-Uhlenbeck process, i. e., a stationary, Gaussian, Markov process with an exponentially decaying autocorrelation function. But $v(t)$ continues to be a stationary random process in the state of thermal equilibrium, although it becomes non-Markovian. Its conditional PDF asymptotically approaches the equilibrium Maxwellian distribution[3]. We might therefore expect an analog of the FD relation to hold good in this case as well. The elucidation of this relationship will be the focus of our discussion of the GLE.

17.3 Frequency-dependent friction

The formal solution of the linear equation (17.10) may be obtained in a straightforward manner. The friction term is in the form of a convolution integral. Therefore an obvious method of solution is via Fourier transforms. In conformity with the Fourier transform conventions we have used throughout (recall Eq. (15.27)), we have

$$v(t) = \int_{-\infty}^{\infty} d\omega\, e^{-i\omega t}\, \widetilde{v}(\omega) \quad \text{and} \quad \widetilde{v}(\omega) = \frac{1}{2\pi}\int_{-\infty}^{\infty} dt\, e^{i\omega t}\, v(t), \tag{17.11}$$

with similar definitions of the Fourier transforms $\widetilde{F}_{\text{ext}}(\omega)$ and $\widetilde{\eta}(\omega)$ of $F_{\text{ext}}(t)$ and $\eta(t)$, respectively. Further, let

$$\overline{\gamma}(\omega) = \int_{0}^{\infty} dt\, \gamma(t)\, e^{i\omega t} \tag{17.12}$$

[3]In principle, the presence of the memory kernel in the GLE would imply a partial differential equation of infinite order in the time derivatives for the PDF of v. However, it turns out that if the noise $\eta(t)$ is a stationary Gaussian process, then the PDF of v satisfies a Fokker-Planck-*like* equation.

denote the Fourier-Laplace transform of the memory kernel[4]. (Recall that we may define $\gamma(t) \equiv 0$ for $t < 0$.) $\overline{\gamma}(\omega)$ may be termed the **frequency-dependent friction**. The GLE (17.10) then yields the formal solution

$$\widetilde{v}(\omega) = \frac{1}{m[\overline{\gamma}(\omega) - i\omega]} \left(\widetilde{F}_{\text{ext}}(\omega) + \widetilde{\eta}(\omega)\right) \tag{17.13}$$

for the Fourier transform of the velocity. As before, we may assume that the mean value of the noise $\eta(t)$ vanishes at all times. This gives

$$\langle \widetilde{v}(\omega) \rangle = \frac{1}{m[\overline{\gamma}(\omega) - i\omega]} \widetilde{F}_{\text{ext}}(\omega). \tag{17.14}$$

It follows at once from Eq. (16.16) that the dynamic mobility in the model represented by the GLE is

$$\mu(\omega) = \frac{1}{m[\overline{\gamma}(\omega) - i\omega]}. \tag{17.15}$$

Thus, the friction constant γ in the expression for the dynamic mobility obtained in the case of the ordinary Langevin equation (Eq. (15.32)) is now replaced by the frequency-dependent friction[5] $\overline{\gamma}(\omega)$.

The conclusions drawn in Ch. 16, Sec. 16.3 and Sec. 16.4 regarding the analyticity of the dynamic mobility in the upper half-plane in ω were based on causality. They should therefore remain true for $\mu(\omega)$ as derived from the GLE. Based on the general properties we have assumed for the memory kernel $\gamma(t)$, it can be shown that the expression in Eq. (17.15) has no singularities in the upper half-plane in ω. The dispersion relations deduced as a consequence of this analyticity remain valid.

17.4 Fluctuation-dissipation theorems for the GLE

17.4.1 The first FD theorem from the GLE

We have explained in Ch. 16, Sec. 16.2, that (i) in the context of the motion of a tagged particle in a fluid, the relation between the dynamic mobility and the equilibrium velocity autocorrelation function is the (first) fluctuation-dissipation

---

[4] On physical grounds, we expect $\gamma(t)$ to drop off with increasing t sufficiently rapidly to ensure the convergence of the integral in Eq. (17.12). However, if this integral does not exist as it stands, owing to too slow a decay of $\gamma(t)$, $\overline{\gamma}(\omega)$ may be defined as the analytic continuation to $s = -i\omega$ of the Laplace transform of $\gamma(t)$.

[5] The theory of liquids makes considerable use of this sort of memory kernel formalism.

theorem, and (ii) this relation is derivable from LRT (see Appendix I). Now, the assumptions we have made in Sec. 17.2 regarding the nature of F_{sys} are such that the GLE remains consistent with the general framework of linear response theory. We should therefore expect the first FD theorem to be derivable directly from the GLE, without any further inputs from LRT. Let us now turn to this derivation.

The first step is to compute the equilibrium velocity autocorrelation function $\langle v(t_0)\, v(t_0+t)\rangle_{\text{eq}}$, where t_0 is an arbitrary fiducial instant of time, and $t \geq 0$. In the absence of an external force, we write the GLE (17.10) at time $t_0 + t$ in the form

$$m\,\dot{v}(t_0+t) + m\int_{t_0}^{t_0+t} dt'\,\gamma(t_0+t-t')\,v(t') = h(t_0+t\,;\,t_0), \tag{17.16}$$

where the nonstationary force $h(t_0+t\,;\,t_0)$ is defined as

$$h(t_0+t\,;\,t_0) = \eta(t_0+t) - m\int_{-\infty}^{t_0} dt'\,\gamma(t_0+t-t')\,v(t'). \tag{17.17}$$

The systematic frictional force has been split up into two parts for the following reason. The earlier time argument in the correlation function we seek to compute is t_0. We would like to impose the condition of causality, which requires the velocity at time t_0 to be uncorrelated with the force at subsequent instants of time. It turns out that this force must include, in addition to the noise $\eta(t+t_0)$, a portion of the systematic component F_{sys}. At the instant of time $t_0 + t$, the portion of F_{sys} that depends on the velocity history of the particle up to time t_0 is

$$-m\int_{-\infty}^{t_0} dt'\,\gamma(t_0+t-t')\,v(t'). \tag{17.18}$$

It is this quantity that is retained on the right-hand side of Eq. (17.17) along with the noise term, and used in the identification of the effective force $h(t_0+t\,;\,t_0)$ at time $t_0 + t$. Then, the consistent way to impose the causality condition is to require that $v(t_0)$ be uncorrelated to $h(t_0+t\,;\,t_0)$ for all $t > 0$, i. e.,

$$\langle v(t_0)\,h(t_0+t\,;\,t_0)\rangle_{\text{eq}} = 0, \quad t > 0. \tag{17.19}$$

Multiplying both sides of Eq. (17.16) by $v(t_0)$ and taking complete (i. e, equilibrium) averages on both sides, we therefore have

$$\langle v(t_0)\,\dot{v}(t_0+t)\rangle_{\text{eq}} + \int_{t_0}^{t_0+t} dt'\,\gamma(t_0+t-t')\,\langle v(t_0)\,v(t')\rangle_{\text{eq}} = 0. \tag{17.20}$$

Now multiply both sides of this equation by $e^{i\omega t}$ and integrate over t from 0 to ∞. The first term on the left-hand side is integrated by parts, and use is made of the fact that

$$\lim_{t\to\infty} \langle v(t_0)\, v(t_0+t)\rangle_{\rm eq} = 0. \qquad (17.21)$$

After some algebra[6], we get

$$\int_0^\infty dt\, e^{i\omega t}\, \langle v(t_0)\, v(t_0+t)\rangle_{\rm eq} = \frac{\langle v^2(t_0)\rangle_{\rm eq}}{\widetilde{\gamma}(\omega) - i\omega}. \qquad (17.22)$$

Using the fact that $\langle v^2(t_0)\rangle_{\rm eq} = k_B T/m$, and identifying the dynamic mobility in the GLE according to Eq. (17.15), we have

$$\mu(\omega) = \frac{1}{k_B T}\int_0^\infty dt\, e^{i\omega t}\, \langle v(t_0)\, v(t_0+t)\rangle_{\rm eq}. \qquad (17.23)$$

This is precisely the first FD theorem, Eq. (16.19).

Going back to Eq. (17.20), we may let $t \to 0$ from above. Note how this immediately yields

$$\langle v(t_0)\, \dot{v}(t_0)\rangle_{\rm eq} = 0 \qquad (17.24)$$

for the equal-time correlator of v with $\dot{v}$, as required of a stationary random process. The difficulty we faced with the Langevin equation in this respect thus disappears in the case of the generalized Langevin equation.

Observe, too, that the causality condition (17.19) implies that there *is* a non-vanishing correlation between $v(t_0)$ and the noise $\eta(t_0+t)$. Using the definition of $h(t_0+t\,;\,t_0)$ in Eq. (17.17) and the stationarity of the velocity in equilibrium, we find after some simplification that

$$\langle v(t_0)\, \eta(t_0+t)\rangle_{\rm eq} = \langle v(0)\, \eta(t)\rangle_{\rm eq} = m\int_0^\infty dt'\, \gamma(t+t')\, \langle v(0)\, v(t')\rangle_{\rm eq}, \quad t \geq 0. \qquad (17.25)$$

In particular, letting $t \to 0$ from above, we find the equal-time correlation

$$\langle v(t_0)\, \eta(t_0)\rangle_{\rm eq} = m\int_0^\infty dt'\, \gamma(t')\, \langle v(0)\, v(t')\rangle_{\rm eq}. \qquad (17.26)$$

[6]The steps involve changes of variables and interchanges of the order of integration. They are quite similar to those we have encountered several times in the preceding chapters. See the exercises at the end of this chapter.

Passing now to the limit of an instantaneous response $\gamma(t') = \gamma\,\delta(t')$, i. e., to the case of the usual LE, we find, using the fact that $\langle v^2 \rangle_{\text{eq}} = kBT/m$,

$$\langle v(t_0)\,\eta(t_0)\rangle_{\text{eq}} = \gamma k_B T. \tag{17.27}$$

A comparison of Eqs. (17.24) and (17.27) with Eq. (15.76) in Ch. 15, Sec. 15.5 shows how the introduction of a noise that is not δ-correlated, together with a correct formulation of the causality condition, solves the consistency problems that arose in the case of the LE.

17.4.2 The second FD theorem from the GLE

We have mentioned that $\eta(t)$ can no longer be assumed to be a pure white noise. This means that the noise term in the GLE must have a finite correlation time, i. e., it must necessarily be a 'colored' noise. We may anticipate this, because the second fluctuation-dissipation theorem would relate the power spectrum of $\eta(t)$ to the friction—and in the GLE, the latter is not a constant, γ, but rather a *frequency-dependent* quantity $\overline{\gamma}(\omega)$. The second FD theorem is derived directly from the GLE as follows.

We start with the GLE in the absence of an external force, at time t_0:

$$m\dot{v}(t_0) = h(t_0\,;\,t_0) = \eta(t_0) - m\int_{-\infty}^{t_0} dt'\,\gamma(t_0 - t')\,v(t'). \tag{17.28}$$

Multiply both sides of the equation by $h(t_0 + t\,;\,t_0)$. On the left-hand side of the resulting equation, use the GLE (17.16) for $h(t_0 + t\,;\,t_0)$, and take equilibrium averages. Now multiply both sides of the equation by $e^{i\omega t}$ and integrate over t from 0 to ∞. Simplifying the result using the stationarity property of $v(t)$ in equilibrium and its consequences, we get after some algebra

$$\begin{aligned}\int_0^\infty dt\,e^{i\omega t}\,\langle h(t_0\,;\,t_0)\,h(t_0 + t\,;\,t_0)\rangle_{\text{eq}} &= m^2\left(\overline{\gamma}(\omega) - i\omega\right)\Big[\left\langle v^2(t_0)\right\rangle_{\text{eq}} \\ &+ i\omega\int_0^\infty dt\,e^{i\omega t}\,\langle v(t_0)\,v(t_0 + t)\rangle_{\text{eq}}\Big].\end{aligned} \tag{17.29}$$

But the integral on the right-hand side has already been evaluated—see Eq. (17.22). Using this result, we get

$$m\overline{\gamma}(\omega) = \frac{1}{k_B T}\int_0^\infty dt\,e^{i\omega t}\,\langle h(t_0\,;\,t_0)\,h(t_0 + t\,;\,t_0)\rangle_{\text{eq}}\,. \tag{17.30}$$

The final step is to establish that the autocorrelation of the nonstationary noise h is actually equal to that of the stationary noise η. It is an instructive exercise to show that[7]

$$\langle h(t_0\,;\,t_0)\,h(t_0+t\,;\,t_0)\rangle_{\text{eq}} = \langle \eta(t_0)\,\eta(t_0+t)\rangle_{\text{eq}}\,. \tag{17.31}$$

We thus obtain, finally, the second FD theorem in the form

$$m\overline{\gamma}(\omega) = \frac{1}{k_BT}\int_0^\infty dt\, e^{i\omega t}\, \langle \eta(t_0)\,\eta(t_0+t)\rangle_{\text{eq}}\,. \tag{17.32}$$

This is to be compared with the corresponding theorem in the case of the LE, Eq. (16.21). It is at once evident from Eq. (17.32) that the noise $\eta(t)$ in the GLE cannot be δ-correlated, in general. A δ-correlated noise is tenable only when the memory reduces to $\gamma\delta(t)$, as in the case of the original Langevin equation.

17.5 Velocity correlation time

It is evident that the correlation time of the velocity in the generalized Langevin equation is no longer given by γ^{-1}. However, recall that we may define the correlation time in terms of the time integral of the normalized autocorrelation function of the velocity. Repeating Eq. (4.7) (see Ch. 4, Sec. 4.1) for ready reference, we have

$$\tau_{\text{vel}} = \int_0^\infty C(t)\,dt, \quad \text{where} \quad C(t) = \frac{\langle v(0)\,v(t)\rangle_{\text{eq}}}{\langle v^2\rangle_{\text{eq}}}\,. \tag{17.33}$$

Setting $\omega = 0$ in Eq. (17.22), it follows that

$$\tau_{\text{vel}} = \frac{1}{\overline{\gamma}(0)}\,, \tag{17.34}$$

where

$$\overline{\gamma}(0) = \int_0^\infty dt\,\gamma(t) \tag{17.35}$$

is the time integral of the memory kernel. It must be kept in mind that τ_{vel} is merely an *effective* correlation time. In general, the physical origin of a memory kernel lies in the fact that the process of relaxation is governed by a spectrum of time scales, rather than a single characteristic time.

---

[7] See the exercises at the end of this chapter.

17.6 Exercises

17.6.1 Exponentially decaying memory kernel

A standard model for the memory kernel is an exponentially decaying one,

$$\gamma(t) = \gamma_0^2 \, e^{-t/\tau}, \quad (t \geq 0) \tag{17.36}$$

where γ_0 and τ are positive constants with the physical dimensions of $(\text{time})^{-1}$ and time, respectively.

(a) Show that the correlation time of the velocity in this model is $\tau_{\text{vel}} = \gamma_0^2 \, \tau$.

(b) Evaluate the dynamic mobility $\mu(\omega)$ from the formula (17.15) in this case. Show that $\mu(\omega)$ has no singularities in the upper half-plane in ω, and two simple poles that lie in the lower half-plane for all positive values of γ_0 and τ.

17.6.2 Verification of the first FD theorem

Work out the steps leading from Eq. (17.20) to Eq. (17.23).

17.6.3 Equality of noise correlation functions

Establish the equality

$$\langle h(t_0\,;\, t_0)\, h(t_0 + t\,;\, t_0)\rangle_{\text{eq}} = \langle \eta(t_0)\, \eta(t_0 + t)\rangle_{\text{eq}} \,. \tag{17.37}$$

Use the definition of the nonstationary noise h given in Eq. (17.17). You will need to use the result obtained in Eq. (17.25) for the cross-correlation of v and η, with appropriately altered variables.

Epilogue

We have come to the end of our brief account of the elements of nonequilibrium statistical mechanics. Most of the topics discussed in this book should be regarded as the 'classical' ones in this subject. They are the ones in which nonequilibrium statistical mechanics had its origins. They remain both essential and relevant, but a fascinating vista lies ahead, thanks to the enormous number of possibilities opened up by nonlinearity and stochasticity. The variety of phenomena that can occur in systems with a large number of degrees of freedom interacting nonlinearly with each other is truly astounding. Chemical, biochemical and biological systems; nonequilibrium steady states; transient and steady-state fluctuation theorems and work theorems in nonequilibrium situations; bifurcations, chaos and pattern-formation in far-from-equilibrium, spatially-extended systems; the dynamics of networks; these are just a few of the immensely diverse topics that await you. If the introduction offered in this book helps launch you on this journey of exploration, the book would have served its purpose. Let the real journey begin!

Appendix A

Gaussian integrals

A.1 The Gaussian integral

The basic Gaussian integral in one variable is

$$\int_{-\infty}^{\infty} dx\, e^{-ax^2} = 2\int_{0}^{\infty} dx\, e^{-ax^2} = \sqrt{\frac{\pi}{a}}, \quad (a > 0). \tag{A.1}$$

The convergence of the integral is not affected if a becomes complex, as long as Re a remains positive. Equation (A.1) is therefore valid in the region Re $a > 0$. When the range of integration extends from $-\infty$ to ∞, the integral is not affected by a constant shift of the variable of integration. We have

$$\int_{-\infty}^{\infty} dx\, e^{-a(x-x_0)^2} = \sqrt{\frac{\pi}{a}} \quad (\text{Re}\, a > 0), \tag{A.2}$$

where x_0 is an arbitrary real number. Remarkably enough, the result above remains valid[1] even when the variable of integration is shifted by an arbitrary *complex* number z_0. Hence

$$\int_{-\infty}^{\infty} dx\, e^{-a(x-z_0)^2} = \sqrt{\frac{\pi}{a}}, \quad (\text{Re}\, a > 0,\; z_0 \in \mathbb{C}). \tag{A.3}$$

Using Eq. (A.3), the integral $\int_{-\infty}^{\infty} dx\, \exp(-ax^2 + bx)$ can be evaluated. Complete the square in the exponent, and shift the variable of integration by an appropriate constant amount. We find

$$\int_{-\infty}^{\infty} dx\, e^{-ax^2+bx} = \sqrt{\frac{\pi}{a}}\, \exp\left(\frac{b^2}{4a}\right), \; (\text{Re}\, a > 0). \tag{A.4}$$

[1]This assertion, which is not hard to prove, is left as an exercise for the reader.

There are no restrictions on the constant b. Hence Eq. (A.4) is valid for any b—real, pure imaginary, or complex.

A.2 The error function

Closely related to the Gaussian function is the error function, defined as

$$\operatorname{erf}(x) = \frac{2}{\sqrt{\pi}} \int_0^x du\, e^{-u^2}. \tag{A.5}$$

Hence

$$\frac{d}{dx} \operatorname{erf}(x) = \frac{2}{\sqrt{\pi}}\, e^{-x^2}. \tag{A.6}$$

The error function is a monotonically increasing function of its argument. Some of its useful properties are

$$\operatorname{erf}(-\infty) = -1, \quad \operatorname{erf}(0) = 0, \quad \operatorname{erf}(\infty) = 1, \quad \operatorname{erf}(-x) = -\operatorname{erf}(x). \tag{A.7}$$

It is useful to define also the complementary error function $\operatorname{erfc}(x)$, according to

$$\operatorname{erfc}(x) = 1 - \operatorname{erf}(x) = \frac{2}{\sqrt{\pi}} \int_x^\infty du\, e^{-u^2}. \tag{A.8}$$

This function decreases monotonically from the value $\operatorname{erfc}(-\infty) = 2$ to the value $\operatorname{erfc}(\infty) = 0$.

A.3 The multi-dimensional Gaussian integral

There is an important multi-variable generalization of the Gaussian integral (A.4) that is very useful in practice. Let A be a non-singular, symmetric $(n \times n)$ matrix with real elements a_{ij} and positive eigenvalues, and let $\mathbf{b}$ be an $(n \times 1)$ column vector with elements b_i. Then, with $\mathbf{x} \in \mathbb{R}^n$, the integral

$$\int d^n x \,\exp\big(-\mathbf{x}^{\mathrm{T}} \mathsf{A}\, \mathbf{x} + \mathbf{b}^{\mathrm{T}} \mathbf{x}\big) = \sqrt{\frac{\pi^n}{\det \mathsf{A}}}\, \exp\big(\tfrac{1}{4}\mathbf{b}^{\mathrm{T}} \mathsf{A}^{-1} \mathbf{b}\big), \tag{A.9}$$

where the integration is from $-\infty$ to ∞ in each component of $\mathbf{x}$. Written out explicitly,

$$\int_{-\infty}^{\infty} dx_1 \ldots \int_{-\infty}^{\infty} dx_n \,\exp\big(-a_{ij}\, x_i\, x_j + b_i\, x_i\big) = \sqrt{\frac{\pi^n}{\det \mathsf{A}}}\, \exp\big(\tfrac{1}{4}\alpha_{ij}\, b_i\, b_j\big), \tag{A.10}$$

where repeated indices are summed over, as usual, and α_{ij} is the $(ij)^{\text{th}}$ element of A^{-1}.

Further (and profound!) generalizations of this basic result exist, but we do not require these for our present purposes.

A.4 Gaussian approximation for integrals

Let us return to the basic one-dimensional Gaussian integral in Eq. (A.2). This integral provides us with a very useful way of evaluating a class of definite integrals approximately. Several related and far-reaching generalizations of the basic idea exist, such as the *stationary phase approximation* and the *saddle point method* or the *method of steepest descent*. We consider the simplest case here.

Suppose we have an integral of the form

$$I = \int_a^b dx\, e^{-f(x)}, \tag{A.11}$$

where the function $f(x)$ decreases as x increases from a, till it reaches a minimum value at some point x_0, and then increases as x increases to b. We assume that x_0 is a simple minimum of $f(x)$, i. e., $f'(x_0) = 0$ and $f''(x_0) > 0$. Then, expanding $f(x)$ in a Taylor series about the point x_0, we have

$$f(x) = f(x_0) + \tfrac{1}{2} f''(x_0)\,(x - x_0)^2 + \dots. \tag{A.12}$$

Therefore

$$I = e^{-f(x_0)} \int_a^b dx\, e^{-\frac{1}{2} f''(x_0)\,(x-x_0)^2} \left\{1 + \mathcal{O}\left((x - x_0)^3\right)\right\}. \tag{A.13}$$

The Gaussian factor outside the curly brackets in the integrand drops off extremely rapidly as $|x - x_0|$ increases. Hence the predominant contribution to the integral comes from the immediate neighborhood of x_0. Consequently, in the leading approximation we can (i) extend the range of integration to $(-\infty, \infty)$, and (ii) drop all terms in the curly brackets that are proportional to various powers of $(x - x_0)$. The integral I then reduces, to a very good approximation, to the Gaussian integral in Eq. (A.2). We thus obtain the very useful approximation

$$I \simeq \sqrt{\frac{2\pi}{f''(x_0)}}\; e^{-f(x_0)}. \tag{A.14}$$

Appendix B

The gamma function

The Euler gamma function is the generalization of the factorial of a natural number. A convenient starting point is the definite integral

$$\int_0^\infty dx\, x^{r-1}\, e^{-x} = (r-1)!\,, \tag{B.1}$$

which is valid for $r = 1, 2, \ldots$. (Note that the integral may be used to identify $0!$ with unity.) Next, we observe that the convergence of the integral is dependent on the power of x in the integrand, owing to the fact that

$$\int_0^c dx\, x^\alpha$$

converges as long as $\text{Re}\,\alpha > -1$, for any finite $c > 0$. Hence the definite integral

$$\int_0^\infty dx\, x^{z-1}\, e^{-x},$$

regarded as a function of the complex variable z, converges in the region $\text{Re}\, z > 0$ of the complex z-plane. The gamma function $\Gamma(z)$ of the complex variable z is defined in the region $\text{Re}\, z > 0$ as the integral

$$\Gamma(z) = \int_0^\infty dx\, x^{z-1}\, e^{-x}\,. \tag{B.2}$$

When $z = r + 1$ where $r = 0, 1, \ldots$, the integral reduces to $r!$, so that we have the basic relationship

$$\Gamma(r+1) = r!\,, \quad r = 0, 1, 2\ \ldots \tag{B.3}$$

The gamma function thus interpolates between the factorials of non-negative integers. It satisfies the functional relation

$$z\,\Gamma(z) = \Gamma(z+1). \tag{B.4}$$

With the help of this relation, the gamma function may be defined in (or analytically continued to) the whole of the complex z-plane. It is not difficult to show that $\Gamma(z)$ is an analytic function of z with simple poles at $z = 0, -1, -2, \ldots$. Its residue at the pole at $z = -n$ is equal to $(-1)^n/n!$.

The value $z = \frac{1}{2}$ is an important special case. We have

$$\Gamma(\tfrac{1}{2}) = \int_0^\infty dx\, x^{-1/2}\, e^{-x}\,. \tag{B.5}$$

Setting $x = u^2$, we get

$$\Gamma(\tfrac{1}{2}) = 2\int_0^\infty du\; e^{-u^2} = \sqrt{\pi}, \tag{B.6}$$

on putting in the value of the basic Gaussian integral. It follows that

$$\Gamma(r+\tfrac{1}{2}) = \frac{\sqrt{\pi}\,(2r-1)!}{2^{2r-1}\,(r-1)!} \quad \text{for} \quad r = 1, 2, \ldots\,. \tag{B.7}$$

This expression may also be deduced from the very useful *doubling formula* for the gamma function,

$$\Gamma(2z) = \frac{2^{2z-1}}{\sqrt{\pi}}\,\Gamma(z)\,\Gamma\left(z+\tfrac{1}{2}\right). \tag{B.8}$$

The integral

$$\int_0^\infty dx\, x^r\, \exp\left(-ax^2\right)$$

can be evaluated in terms of the gamma function. The integral converges provided $\text{Re}\, a > 0$, since the exponential in the integrand provides a damping factor in that case. Further, we must also have $\text{Re}\, r > -1$, to ensure integrability at the lower limit of integration. Then, changing variables of integration to x^2, we find

$$\int_0^\infty dx\, x^r\, e^{-ax^2} = \Gamma\left(\tfrac{1}{2}(r+1)\right)\Big/ 2a^{(r+1)/2}, \tag{B.9}$$

where $\Gamma(z)$ is the gamma function. This result is used in several places in the text.

Appendix C

Moments, cumulants and characteristic functions

C.1 Moments

The moments $\langle \xi^n \rangle$ of a discrete-valued or continuous-valued random variable[1] ξ are generated by its **moment generating function** $M(u)$, defined as

$$M(u) \equiv \left\langle e^{u\xi} \right\rangle = 1 + \sum_{n=1}^{\infty} \frac{\langle \xi^n \rangle}{n!}\, u^n. \tag{C.1}$$

It follows that $\langle \xi^n \rangle$ is just the nth derivative of $M(u)$ with respect to u, evaluated at $u = 0$:

$$\langle \xi^n \rangle = \left[\frac{d^n M(u)}{du^n} \right]_{u=0}. \tag{C.2}$$

The first moment or mean value $\mu = \langle \xi \rangle$ is the coefficient of u in the power series expansion of $M(u)$.

The **central moments** of the random variable ξ are the moments of its *deviation* from its mean value, namely, $\langle (\delta\xi)^n \rangle = \langle (\xi - \mu)^n \rangle$. The first of these is zero, by definition. The second is the variance, $\left\langle (\xi - \mu)^2 \right\rangle = \left\langle \xi^2 \right\rangle - \mu^2 = \sigma^2$. This shows at once that the variance of a random variable is always positive, because it is the average value of a squared quantity. The variance vanishes if and only if the variable is a 'sure' one, i. e., the sample space of the variable comprises just a single value. The standard deviation is the square root of the variance. The ratio of the standard deviation to the mean (or first moment) is sometimes referred to as the relative fluctuation of the random variable.

[1]We restrict ourselves to real-valued random variables.

C.2 Cumulants

The cumulants κ_n of a probability distribution generalize the concept of the variance to the higher moments of the distribution. They are generated by the **cumulant generating function** $K(u)$, according to

$$K(u) = \sum_{n=1}^{\infty} \frac{\kappa_n}{n!}\, u^n. \tag{C.3}$$

Analogous to Eq. (C.2), we have

$$\kappa_n = \left[\frac{d^n K(u)}{du^n}\right]_{u=0}. \tag{C.4}$$

The relationship between the generating functions $K(u)$ and $M(u)$ is

$$M(u) = \exp K(u)\,, \quad \text{or} \quad K(u) = \ln M(u). \tag{C.5}$$

Hence the cumulants can be expressed in terms of the moments, and *vice versa.* The first three cumulants are, respectively, the first three central moments themselves. That is,

$$\kappa_1 = \mu\,,\ \ \kappa_2 = \sigma^2\,,\ \ \kappa_3 = \big\langle(\delta\xi)^3\big\rangle\,. \tag{C.6}$$

The fourth cumulant is given by

$$\kappa_4 = \big\langle(\delta\xi)^4\big\rangle - 3\,\big\langle(\delta\xi)^2\big\rangle^2\,. \tag{C.7}$$

In general terms, the second cumulant (i. e., the variance) is a measure of the spread or dispersion of the random variable about its mean value. The third cumulant κ_3 measures the asymmetry or **skewness** of the distribution about the mean value. The fourth cumulant κ_4 characterizes the extent to which the distribution deviates from the normal or Gaussian distribution, as mentioned in Ch. 2, Sec. 2.2. For a probability density function that is symmetric about the mean value μ, the excess of kurtosis κ_4/κ_2^2 measures the extent of this deviation. (We shall see in Appendix D that the Gaussian distribution has exactly zero excess of kurtosis.) Taken together, the first four cumulants of a distribution provide a reasonable approximate description of the salient properties of the probability distribution of the random variable concerned.

C.3 The characteristic function

Let us consider, for definiteness, a continuous random variable ξ with PDF $p(\xi)$. The Fourier transform $\widetilde{p}(k)$ of $p(\xi)$, defined as

$$\widetilde{p}(k) = \int_{-\infty}^{\infty} d\xi\, e^{-ik\xi}\, p(\xi), \tag{C.8}$$

is called its characteristic function. It is closely related to the moment generating function of the random variable, as follows. $\widetilde{p}(k)$ can also be regarded as the expectation value of $\exp(-ik\xi)$, so that

$$\widetilde{p}(k) = \left\langle e^{-ik\xi} \right\rangle = M(-ik). \tag{C.9}$$

The characteristic function carries the same information about the random variable as does its PDF. The significance and utility of the characteristic function will become apparent in the next section, and again in Appendix E and Appendix K.

C.4 The additivity of cumulants

An even more important property of cumulants becomes clear when we consider sums of independent random variables. Let ξ_1 and ξ_2 be independent random variables with PDFs $p_1(\xi_1)$ and $p_2(\xi_2)$, respectively, and let $\xi = \xi_1 + \xi_2$ be their sum. The PDF of ξ is given by[2]

$$p(\xi) = \int d\xi_1 \int d\xi_2 \, p_1(\xi_1)\, p_2(\xi_2)\, \delta\big(\xi - (\xi_1 + \xi_2)\big) = \int d\xi_1 \, p_1(\xi_1)\, p_2(\xi - \xi_1). \tag{C.10}$$

This is a convolution of the PDFs of ξ_1 and ξ_2. Hence, by the convolution theorem for Fourier transforms, the Fourier transform of the function p is just the product of the Fourier transforms of the functions p_1 and p_2. In other words, the corresponding characteristic functions are related according to

$$M(-ik) = M_1(-ik)\, M_2(-ik), \tag{C.11}$$

where M_1 and M_2 are the moment generating functions of ξ_1 and ξ_2. Therefore the corresponding cumulant generating functions simply add up:

$$K(-ik) = K_1(-ik) + K_2(-ik). \tag{C.12}$$

As a consequence, the nth cumulant of $\xi_1 + \xi_2$ is the sum of the nth cumulants of ξ_1 and ξ_2. It is evident that this result can be extended immediately to the sum (more generally, to any linear combination) of any number of independent random variables. Unlike the cumulants, none of the moments higher than the first moment (or mean value) has this property of additivity. This is a major advantage that the cumulants of random variables enjoy over the corresponding moments.

---

[2]Recall Eq. (2.30) in Ch. 2 and the comments made there.

Appendix D

The Gaussian or normal distribution

D.1 The probability density function

The Gaussian or normal distribution plays a central role in statistics. We give here a rather pedestrian and informal account of some of its elementary properties, that are of relevance to the discussion in the main text.

The distribution of a continuous random variable ξ (where $-\infty < \xi < \infty$) is said to be Gaussian if its normalized PDF is given by

$$p(\xi) = \frac{1}{\sqrt{2\pi\sigma^2}} \exp\left(-\frac{(\xi-\mu)^2}{2\sigma^2}\right), \tag{D.1}$$

where μ is any real number and σ is a positive number. The parameter μ is the mean value of ξ, σ^2 is its variance, and σ is its standard deviation. That is,

$$\langle \xi \rangle = \mu, \ \left\langle (\xi-\mu)^2 \right\rangle = \sigma^2. \tag{D.2}$$

The PDF is symmetric about its mean value (which is also the peak value or the mode). The variance σ^2 is a direct measure of the width of the Gaussian. The full-width-at-half-maximum (FWHM) is given by

$$2(2\ \ln\ 2)^{1/2}\sigma \simeq 2.355\,\sigma.$$

D.2 The moments and cumulants

Owing to the symmetry of the Gaussian PDF about the mean value, all the odd moments of the deviation from the mean value (that is, all the odd central moments) vanish identically:

$$\left\langle (\xi - \mu)^{2l+1} \right\rangle = 0, \tag{D.3}$$

where $l = 0, 1, \ldots$. On the other hand, all the even central moments of a Gaussian random variable are determined completely in terms of σ^2, i. e., in terms of the variance of the distribution. We have

$$\left\langle (\xi - \mu)^{2l} \right\rangle = \frac{(2\sigma^2)^l}{\sqrt{\pi}}\, \Gamma(l + \tfrac{1}{2}) = \frac{(2l-1)\,!}{2^{l-1}\,(l-1)!}\, \sigma^{2l}, \tag{D.4}$$

where $l = 1, 2, \ldots$. The moment generating function for the Gaussian distribution is

$$M(u) = \left\langle e^{u\,\xi} \right\rangle = \frac{1}{\sqrt{2\pi\sigma^2}} \int_{-\infty}^{\infty} d\xi\, \exp\left(- \frac{(\xi-\mu)^2}{2\sigma^2} + u\,\xi \right). \tag{D.5}$$

Evaluating the integral using Eq. (A.4), we get

$$M(u) = \exp\left(\mu\, u + \tfrac{1}{2}\sigma^2\, u^2\right), \quad \widetilde{p}(k) = \exp\left(-i\mu\, k - \tfrac{1}{2}\sigma^2\, k^2\right). \tag{D.6}$$

Hence the corresponding cumulant generating function is just

$$K(u) = \mu\, u + \tfrac{1}{2}\sigma^2\, u^2. \tag{D.7}$$

Thus,

- *all the cumulants of a Gaussian distribution higher than the second cumulant vanish identically.*

In particular, the excess of kurtosis is exactly zero, as we have already stated[1].

D.3 The cumulative distribution function

The cumulative distribution function of a Gaussian random variable ξ is given by

$$\mathcal{P}(\xi) = \frac{1}{\sqrt{2\pi\sigma^2}} \int_{-\infty}^{\xi} d\xi\,'\, \exp\left(- \frac{(\xi\,'-\mu)^2}{2\sigma^2} \right). \tag{D.8}$$

[1]For a general distribution that is symmetric about its mean value, a positive excess of kurtosis implies that the distribution is 'fatter' than a Gaussian ('platykurtic'), because larger values of $|\xi - \mu|$ contribute more significantly than they do in a Gaussian. On the other hand, a negative excess of kurtosis implies that the distribution is 'leaner' than a Gaussian ('leptokurtic'): the weight of large values of $|\xi - \mu|$ is less than what it would be in the case of a Gaussian.

$\mathcal{P}(\xi)$ is the total probability that the random variable has a value less than or equal to any prescribed value ξ. It starts at the value $\mathcal{P}(-\infty) = 0$ and increases monotonically with ξ, with $\mathcal{P}(\mu) = \frac{1}{2}$ and finally $\mathcal{P}(\infty) = 1$. By definition, $d\mathcal{P}(\xi)/d\xi = p(\xi)$. The cumulative distribution function can be expressed in terms of an error function as

$$\mathcal{P}(\xi) = \frac{1}{2}\left[1 + \text{erf}\left(\frac{\xi - \mu}{\sqrt{2\sigma^2}}\right)\right]. \tag{D.9}$$

D.4 Linear combinations of Gaussian random variables

The statistical properties of *sums* (more generally, of linear combinations) of independent Gaussian random variables $\xi_1, \xi_2, \ldots, \xi_n$ are of considerable interest. Let the mean and variance of ξ_i be μ_i and σ_i^2, respectively. Then the linear combination $\xi = \sum_i a_i\, \xi_i$ is also a Gaussian random variable, with mean and variance given by

$$\langle \xi \rangle = \sum_{i=1}^{n} a_i\, \mu_i \tag{D.10}$$

and

$$\big\langle (\xi - \langle \xi \rangle)^2 \big\rangle = \sum_{i=1}^{n} a_i\, \sigma_i^2. \tag{D.11}$$

Such properties are characteristic of a family of probability distributions of which the Gaussian is a member, and which will be described in Appendix K.

D.5 The Central Limit Theorem

One of the most important theorems of statistics is the Central Limit Theorem[2]. From the results quoted in Eqs. (D.10) and (D.11) above, it is clear that, if $\xi_1, \xi_2, \ldots, \xi_n$ are independent, identically distributed Gaussian random variables with mean μ and variance σ^2, then the random variable

$$z_n = \frac{\xi_1 + \ldots + \xi_n - n\mu}{\sigma\sqrt{n}} \tag{D.12}$$

is also a Gaussian random variable with zero mean and unit variance. What is remarkable is the following:

[2]This is actually a generic name for a class of convergence theorems in statistics. What we refer to here is the most common of these results.

- Even if the ξ_i are not necessarily Gaussian random variables, but are independent, identically distributed random variables with mean μ and *finite* variance σ^2, the distribution of z_n tends to a Gaussian distribution with zero mean and unit variance in the limit $n \to \infty$.

This is the Central Limit Theorem[3]. Several of the conditions stated above can be relaxed without affecting the validity of the theorem, but we do not go into these details here. The crucial requirement is the finiteness of the mean and variance of each of the random variables making up the sum in z_n.

- The Central Limit Theorem helps us understand why the Gaussian distribution occurs so frequently in all physical applications.

D.6 Gaussian distribution in several variables

In the text, we have also used multi-dimensional or **multivariate Gaussian distributions**—specifically, in connection with distributions in phase space, involving both the position and velocity variables. A general multivariate Gaussian PDF in $\boldsymbol{\xi} = (\xi_1, \dots, \xi_n)$ is of the form

$$p(\boldsymbol{\xi}) = \frac{1}{\sqrt{(2\pi)^n \det \boldsymbol{\sigma}}} \exp\big[-\tfrac{1}{2}(\boldsymbol{\xi} - \boldsymbol{\mu})^{\mathrm{T}} \boldsymbol{\sigma}^{-1} (\boldsymbol{\xi} - \boldsymbol{\mu})\big], \tag{D.13}$$

where $\boldsymbol{\mu}$ is the column vector of the mean values, i. e., $\mu_i = \langle \xi_i \rangle$. The symmetric non-singular matrix $\boldsymbol{\sigma}$, whose elements are non-negative real numbers, is called the covariance matrix.

- Note that the different components of $\boldsymbol{\xi}$ in a multivariate Gaussian distribution are not necessarily statistically independent random variables. They *are* correlated with each other, in general.

- The covariance matrix is a direct measure of the degree of this correlation—it is, in fact, precisely the matrix of correlation functions:

$$\boldsymbol{\sigma} = \big\langle (\boldsymbol{\xi} - \boldsymbol{\mu})(\boldsymbol{\xi} - \boldsymbol{\mu})^{\mathrm{T}} \big\rangle . \tag{D.14}$$

This relationship is perhaps more transparent when written out in terms of the elements of the covariance matrix: we have

$$\sigma_{ij} = \langle (\xi_i - \mu_i)(\xi_j - \mu_j) \rangle . \tag{D.15}$$

[3]Widely regarded as the 'crown jewel' of the subject of probability and statistics.

Thus the diagonal elements of $\boldsymbol{\sigma}$ are the variances of the components ξ_i of the multivariate Gaussian process, while the off-diagonal elements are the corresponding cross-correlations between the different components.

- It is important to note that, in the case when $\boldsymbol{\xi}(t)$ is a multivariate Gaussian random *process* (in time), the elements of the covariance matrix are the corresponding *equal-time* correlations.

- If (one or more components of) $\boldsymbol{\xi}$ is a nonstationary random process, the elements of $\boldsymbol{\mu}$ and $\boldsymbol{\sigma}$ will, in general, be time-dependent[4].

D.7 The two-dimensional case

The two-dimensional case, $\boldsymbol{\xi} = (\xi_1\,,\,\xi_2)$, is of particular interest[5]. Let us write, as usual, $\xi_i - \mu_i = \delta\xi_i$. Then the general expression in Eq. (D.13) becomes, in this case,

$$p(\xi_1\,,\,\xi_2) = \frac{1}{2\pi\,\sqrt{\Delta}}\,\exp\left\{-\,\frac{\sigma_{22}\,(\delta\xi_1)^2 - 2\sigma_{12}\,\delta\xi_1\,\delta\xi_2 + \sigma_{11}\,(\delta\xi_2)^2}{2\Delta}\right\}, \tag{D.16}$$

where

$$\Delta = \det\boldsymbol{\sigma} = \sigma_{11}\,\sigma_{22} - \sigma_{12}^2\,. \tag{D.17}$$

Equation (D.16) gives the PDF corresponding to a **bivariate Gaussian distribution**. It is instructive to re-write Eq. (D.16) in a form that shows precisely how $p(\xi_1\,,\,\xi_2)$ differs from the product of the PDFs of two *independent* Gaussian random variables. This form brings out the role of the correlation between the two components of the process. As already stated, the diagonal elements of the covariance matrix are the variances of ξ_1 and ξ_2, respectively. Let us denote these by $\sigma_{11} \equiv \sigma_1^2$ and $\sigma_{22} \equiv \sigma_2^2$, so that σ_1 and σ_2 are the respective standard deviations of ξ_1 and ξ_2. It is customary to define the **correlation** between ξ_1 and ξ_2 as the normalized cross-average

$$\rho_{12} = \frac{\sigma_{12}}{\sqrt{\sigma_{11}\,\sigma_{22}}} = \frac{\sigma_{12}}{\sigma_1\,\sigma_2}\,. \tag{D.18}$$

---

[4]This is precisely what happens in the case of the $(x\,,\,v)$ process discussed in Ch. 12. The solution to the Fokker-Planck equation in phase space is a two-dimensional Gaussian distribution. As we have stated several times, the velocity by itself is a stationary random process, while the position is not.

[5]We have encountered this case several times in the text.

In terms of these quantities, we have

$$
\begin{aligned}
p(\xi_1\,,\,\xi_2) &= \frac{1}{\sqrt{(2\pi\,\sigma_1^2)(2\pi\sigma_2^2)(1-\rho_{12}^2)}}\times \\
&\times \;\exp\left\{-\frac{1}{(1-\rho_{12}^2)}\left(\frac{(\delta\xi_1)^2}{2\sigma_1^2}-\frac{\rho_{12}\,\delta\xi_1\,\delta\xi_2}{\sigma_1\,\sigma_2}+\frac{(\delta\xi_2)^2}{2\sigma_2^2}\right)\right\}.
\end{aligned}
\tag{D.19}
$$

Appendix E

From random walks to diffusion

We have mentioned in Ch. 9 that diffusion, as governed by the diffusion equation, is the continuum limit of a simple random walk. In this Appendix, we see how this comes about.

E.1 A simple random walk

The random walk we consider is as follows. The walker starts at an arbitrary origin in three-dimensional space, and at the end of each time step τ, takes a step of fixed length l in an *arbitrary* direction. Each successive step is taken independently, and is uncorrelated with the steps preceding it. Figure E.1 depicts a realization of such a random walk. We ask for the normalized probability density function $\mathsf{p}(\mathbf{r}, n\tau)$ of the position vector $\mathbf{r}$ of the walker at the end of n time steps, i. e., at time $n\tau$. We shall show that this PDF reduces to the fundamental Gaussian solution (7.20) of the diffusion equation when the limits $n \to \infty$, $l \to 0$, $\tau \to 0$ are taken appropriately.

Let the successive steps of the walker be given by the vectors $\mathbf{s}_1, \mathbf{s}_2, \ldots, \mathbf{s}_n$, so that

$$\mathbf{r} = \sum_{j=1}^{n} \mathbf{s}_j \,. \tag{E.1}$$

Let $\langle \cdots \rangle$ denote the statistical average over all realizations of the random walk (i. e., over all possible configurations of the steps). As each step is equally likely to be in any direction, it is obvious that $\langle \mathbf{s}_j \rangle = 0$. Further, using the fact that the

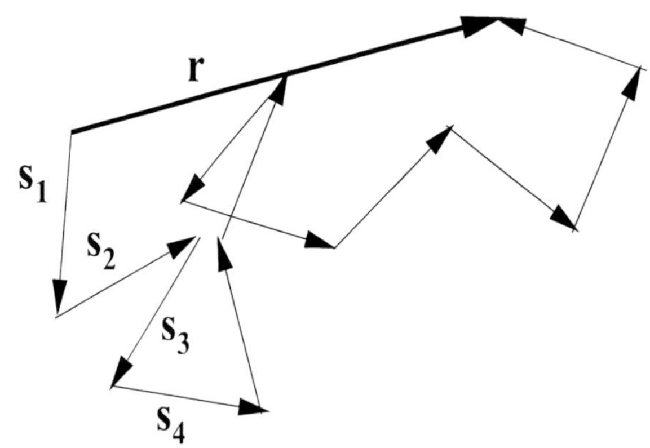

Figure E.1: A simple n-step random walk. The individual steps are independent and uncorrelated with each other. Each step is equally likely to be taken in any direction in space, although the length l of each step is the same. The quantity of interest is the PDF of the end-to-end vector $\mathbf{r}$. The root mean squared value of the end-to-end distance is proportional to $n^{1/2}$. When the number n of steps tends to infinity and the step length l tends to zero such that nl^2 tends to a finite limit, the random walk goes over into diffusive motion. The PDF of $\mathbf{r}$ tends to the fundamental Gaussian solution of the diffusion equation.

magnitude of every step is equal to l,

$$\langle r^2 \rangle = \langle \mathbf{r} \cdot \mathbf{r} \rangle = nl^2 + l^2 \sum_{\substack{i,j=1 \\ i \neq j}}^{n} \langle \cos \theta_{ij} \rangle , \tag{E.2}$$

where θ_{ij} is the angle between the vectors $\mathbf{s}_i$ and $\mathbf{s}_j$. But $\langle \cos \theta_{ij} \rangle = 0$, because the angle between the two vectors is as likely to have a value θ as it is to have a value $(\pi - \theta)$, and $\cos(\pi - \theta) = -\cos\theta$. It follows that $\langle r^2 \rangle = nl^2$. Since n is proportional to the duration t of the walk, we see already the emergence of the $t^{1/2}$ behavior of the root mean squared displacement that is characteristic of diffusive motion.

E.2 The characteristic function

In order to determine the PDF $\mathsf{p}(\mathbf{r}, n\tau)$, it is convenient to begin with its Fourier transform or characteristic function. In keeping with our Fourier transform conventions (see, for instance, Eqs. (7.7) and (7.9)), let us define the Fourier transform pair

$$\widetilde{\mathsf{p}}(\mathbf{k}, n\tau) = \int d^3r \; e^{-i\mathbf{k}\cdot\mathbf{r}} \; \mathsf{p}(\mathbf{r}, n\tau), \quad \mathsf{p}(\mathbf{r}, n\tau) = \frac{1}{(2\pi)^3} \int d^3k \; e^{i\mathbf{k}\cdot\mathbf{r}} \; \widetilde{\mathsf{p}}(\mathbf{k}, n\tau). \tag{E.3}$$

The idea is to compute $\widetilde{\mathsf{p}}(\mathbf{k}, n\tau)$, and then invert the transform to obtain $\mathsf{p}(\mathbf{r}, n\tau)$. Now, $\widetilde{\mathsf{p}}(\mathbf{k}, n\tau)$ can be interpreted as the average value of the quantity $\exp(-i\mathbf{k}\cdot\mathbf{r})$ over all possible n-step walks. Therefore we have

$$\widetilde{\mathsf{p}}(\mathbf{k}, n\tau) = \left\langle e^{-i\mathbf{k}\cdot\mathbf{r}} \right\rangle = \left\langle e^{-i\sum_{j=1}^{n} \mathbf{k}\cdot\mathbf{s}_j} \right\rangle = \left\langle e^{-i\mathbf{k}\cdot\mathbf{s}_1} \ldots e^{-i\mathbf{k}\cdot\mathbf{s}_n} \right\rangle . \tag{E.4}$$

A great simplification occurs now, because *the individual steps are completely independent of each other.* Hence the average value of the product in the final step above becomes equal to the product of average values, and we get

$$\widetilde{\mathsf{p}}(\mathbf{k}, n\tau) = \prod_{j=1}^{n} \left\langle e^{-i\mathbf{k}\cdot\mathbf{s}_j} \right\rangle . \tag{E.5}$$

Let $\pi(\mathbf{r}) \, (\equiv \mathsf{p}(\mathbf{r}, \tau))$ denote the PDF of a *single* step. Since this PDF is the same[1] for each of the steps, we have

$$\left\langle e^{-i\mathbf{k}\cdot\mathbf{s}_j} \right\rangle = \int d^3\mathbf{s}_j \, e^{-i\mathbf{k}\cdot\mathbf{s}_j} \, \pi(\mathbf{s}_j) = \widetilde{\pi}(\mathbf{k}), \tag{E.6}$$

[1]The derivation can be extended to the case in which the PDF of each step has a different functional form. See the remarks made in Sec. E.4.

the Fourier transform of the single-step PDF. Thus

$$\widetilde{\mathsf{p}}(\mathbf{k}, n\tau) = [\widetilde{\pi}(\mathbf{k})]^n \,. \tag{E.7}$$

We turn now to the single-step PDF $\pi(\mathbf{r})$. The only condition imposed on a step is that its magnitude be equal to l. Hence $\pi(\mathbf{r})$ must be proportional to $\delta(r-l)$. The constant of proportionality is fixed by the normalization condition $\int d^3r\ \pi(\mathbf{r}) = 1$. The result is

$$\pi(\mathbf{r}) = \frac{1}{4\pi l^2}\,\delta(r-l). \tag{E.8}$$

Therefore[2]

$$\widetilde{\pi}(\mathbf{k}) = \int d^3r\, e^{-i\mathbf{k}\cdot\mathbf{r}}\,\pi(\mathbf{r}) = \frac{\sin kl}{kl}\,, \tag{E.9}$$

where $k = |\mathbf{k}|$.

E.3 The diffusion limit

Inserting Eq. (E.9) in Eq. (E.7) and inverting the Fourier transform, we get

$$\mathsf{p}(\mathbf{r}, n\tau) = \frac{1}{(2\pi)^3}\int d^3k\ e^{i\mathbf{k}\cdot\mathbf{r}}\left(\frac{\sin kl}{kl}\right)^n. \tag{E.10}$$

The evaluation of the integral in Eq. (E.10) for successive small values of n is an interesting exercise in its own right. It enables us to see how the PDF $\mathsf{p}(\mathbf{r}, n\tau)$, which starts out for $n=1$ as $\pi(\mathbf{r})$, i. e., a singular density proportional to $\delta(r-l)$, gets 'smoother' as n increases. Our immediate purpose, however, is to examine $\mathsf{p}(\mathbf{r}, n\tau)$ in the limit $n\to\infty\,, \tau\to 0$ such that $\lim n\tau = t$. Now, $|(\sin kl)/(kl)| < 1$ for all $kl \neq 0$. Hence the factor $[(\sin kl)/(kl)]^n$ in the integrand in Eq. (E.10) causes the integral to vanish as $n\to\infty$, *unless* l^2 *also* tends to zero simultaneously, like n^{-1}, so that

$$\left(\frac{\sin kl}{kl}\right)^n \simeq \left(1 - \frac{k^2\,l^2}{6}\right)^n \tag{E.11}$$

has a finite limit[3]. But $n^{-1} \sim \tau$, so that we must let l^2 tend to zero like τ. Therefore let $\tau \to 0$ and $l \to 0$, such that

$$\lim \frac{l^2}{6\tau} = D, \tag{E.12}$$

[2]The evaluation of the integral in Eq. (E.9) is left as a simple exercise for the reader. Hint: It is obvious that the result is a scalar, and hence independent of the direction of the vector **k**. We may therefore work in spherical polar coordinates, and choose the direction of **k** as the polar axis.

[3]A little thought shows that this is the *only* possible way in which a nontrivial limiting PDF can arise.

a finite quantity[4]. Then

$$\left(1 - \frac{k^2\, l^2}{6}\right)^n = \left(1 - Dk^2\tau\right)^n = \left(1 - \frac{Dk^2 t}{n}\right)^n \longrightarrow e^{-Dk^2 t} \tag{E.13}$$

in the limit $n \to \infty$. We thus obtain

$$\mathsf{p}(\mathbf{r}, t) = \frac{1}{(2\pi)^3} \int d^3k\; e^{-Dk^2 t + i\mathbf{k}\cdot\mathbf{r}}\,. \tag{E.14}$$

It is easy to carry out the integral above in Cartesian coordinates (!) Put $\mathbf{r} = (x, y, z)$, $\mathbf{k} = (k_1\,, k_2\,, k_3)$, and use the Gaussian integral in Eq. (A.4). Recombining $x^2 + y^2 + z^2$ into r^2, we arrive finally at

$$\mathsf{p}(\mathbf{r}, t) = \frac{1}{(4\pi D t)^{3/2}}\, \exp\left(-\frac{r^2}{4Dt}\right). \tag{E.15}$$

This is precisely the fundamental Gaussian solution (7.20) to the three-dimensional diffusion equation (7.15).

E.4 Important generalizations

The result just derived is an illustration of the Central Limit Theorem. We started with n independent, identically-distributed random variables $\mathbf{s}_j\,,\ 1 \le j \le n$. Each of these has a PDF given by Eq. (E.8). Note that this PDF is *not* a Gaussian. However, it has finite first and second moments. The mean is zero and the variance is l^2. These properties suffice to yield a Gaussian PDF for the sum $\sum_1^n \mathbf{s}_j$ in the limit $n \to \infty$. This conclusion is a remarkably robust one. It remains true under very general conditions, as will be seen from the remarks that follow.

The dimensionality of the space in which the random walk takes place does not appear anywhere in the derivation of the Gaussian limiting distribution. Hence the result is valid in an arbitrary number of dimensions. We already know this to be so for the fundamental solution of the diffusion equation in d dimensions, Eq. (9.22). In d dimensions, we would merely have to define the diffusion constant as $D = \lim l^2/(2d\,\tau)$. Note that $l^2/d \equiv a^2$ is just the mean value of the square of the projection of a step onto any of the Cartesian axes. Hence the diffusion

[4]The choice of the precise numerical factor (= 6) in Eq. (E.12) is unimportant. We have simply tailored it to obtain the same numerical factors in the limiting expression for $\mathsf{p}(\mathbf{r}, n\tau)$ as the fundamental Gaussian solution of the diffusion equation. See the remarks made in the next section.

constant may be defined uniformly in all dimensions as simply $\lim a^2/(2\tau)$, in the limit $a \to 0\,,\ \tau \to 0$.

The Gaussian form for the limiting distribution remains valid even when the step length distribution is a general one, and not restricted to a single value l. But the mean and variance of the step length must be finite. In fact, the result remains valid even if the individual steps are not drawn from the same distribution, but from different distributions. Again, the mean and variance must be finite for each of the steps.

The steps need not be taken at regular intervals in time. They could occur at random instants of time. These instants must not be correlated with each other. Let the mean time between successive steps be τ (i. e., the mean rate at which steps are taken is τ^{-1}). It is then easily shown that the probability that exactly n steps are taken in a time interval t is given by a Poisson distribution in n, with a mean value equal to t/τ. In particular, the normalized **waiting-time distribution** $\psi(t)$ between successive steps is a decaying exponential in time, $e^{-t/\tau}$. The exponential form of the waiting-time distribution ensures that the random walk is a Markov process, and that its continuum limit satisfies a Markovian master equation, namely, the diffusion equation (which is of first order in the time derivative).

Finally, we remark that relaxing the two conditions above (finite mean and variance of the step length distribution, and an exponential waiting time distribution) leads to various kinds of anomalous diffusion—among which are **fractional Brownian motion**, **Lévy flights** and non-Markovian random walks or **continuous-time random walks**[5].

---

[5]I apologize regretfully to the reader for the omission from this book of these very interesting and important topics. The reason for this omission is simply that they form a subject in their own right, and a separate book would be required to do them justice.

Appendix F

The exponential of a (2×2) matrix

The well-known Euler formula

$$e^{iz} = \cos z + i \sin z \tag{F.1}$$

is valid for every complex number z. Remarkably enough, it turns out that an analogous formula holds good for (2×2) matrices as well. As the exponential of a (2×2) matrix is required in numerous applications, this formula is of considerable practical use.

Let $M = \begin{pmatrix} a & b \\ c & d \end{pmatrix}$ be an arbitrary (2×2) matrix. This representation of M is actually short-hand for the expansion

$$M = a \begin{pmatrix} 1 & 0 \\ 0 & 0 \end{pmatrix} + b \begin{pmatrix} 0 & 1 \\ 0 & 0 \end{pmatrix} + c \begin{pmatrix} 0 & 0 \\ 1 & 0 \end{pmatrix} + d \begin{pmatrix} 0 & 0 \\ 0 & 1 \end{pmatrix}, \tag{F.2}$$

in terms of the so-called 'natural basis' that consists of the four matrices on the right-hand side of Eq. (F.2). While this basis has many advantages, it is not very convenient for certain operations like exponentiation. The reason lies in the commutation properties of the matrices comprising the basis. Instead of the natural basis, we can equivalently write M in the basis constituted by the unit matrix I and the Pauli matrices σ_1, σ_2 and σ_3, according to

$$M = \alpha_0 \begin{pmatrix} 1 & 0 \\ 0 & 1 \end{pmatrix} + \alpha_1 \begin{pmatrix} 0 & 1 \\ 1 & 0 \end{pmatrix} + \alpha_2 \begin{pmatrix} 0 & -i \\ i & 0 \end{pmatrix} + \alpha_3 \begin{pmatrix} 1 & 0 \\ 0 & -1 \end{pmatrix}, \tag{F.3}$$

or

$$M = \alpha_0 \, I + \sum_{i=1}^{3} \alpha_i \, \sigma_i \,. \tag{F.4}$$

The coefficients $\{a, b, c, d\}$ and the coefficients $\{\alpha_0, \alpha_1, \alpha_2, \alpha_3\}$ are linear combinations of each other, and determine each other uniquely. For instance,

$$\alpha_0 = \frac{a+d}{2}\,,\; \alpha_1 = \frac{c+b}{2}\,,\; \alpha_2 = \frac{c-b}{2i}\,,\; \alpha_3 = \frac{a-d}{2}\,. \tag{F.5}$$

We then have

$$e^M = e^{\alpha_0 \, I + \alpha_1 \, \sigma_1 + \alpha_2 \, \sigma_2 + \alpha_3 \, \sigma_3} = e^{\alpha_0} \, e^{\boldsymbol{\alpha} \cdot \boldsymbol{\sigma}} = e^{\alpha_0} \sum_{n=0}^{\infty} \frac{(\boldsymbol{\alpha} \cdot \boldsymbol{\sigma})^n}{n!}\,, \tag{F.6}$$

where, formally, $\boldsymbol{\alpha} = (\alpha_1 \,,\, \alpha_2 \,,\, \alpha_3)$, $\boldsymbol{\sigma} = (\sigma_1 \,,\, \sigma_2 \,,\, \sigma_3)$, and

$$\boldsymbol{\alpha} \cdot \boldsymbol{\sigma} \equiv \alpha_1 \, \sigma_1 + \alpha_2 \, \sigma_2 + \alpha_3 \, \sigma_3 \,. \tag{F.7}$$

Note that the components of the 'vector' $\boldsymbol{\alpha}$ need not be real numbers! We now make use of two important properties of the Pauli matrices:

(i) $\sigma_i^2 = I$ for each i: The square of each of the Pauli matrices is the unit matrix.

(ii) $\sigma_i \, \sigma_j + \sigma_j \, \sigma_i = 0$ when $i \neq j$: the Pauli matrices *anti*commute with each other.

These properties enable us to sum the infinite series representing the exponential on the right-hand side of Eq. (F.6) in closed form, even though the Pauli matrices do not commute with each other. The result is[1]

$$e^M = e^{\alpha_0} \left(I \, \cosh \alpha + \frac{\boldsymbol{\alpha} \cdot \boldsymbol{\sigma}}{\alpha} \, \sinh \alpha \right), \tag{F.8}$$

where $\alpha \equiv (\alpha_1^2 + \alpha_2^2 + \alpha_3^2)^{1/2}$. In writing down Eq. (F.8), we have used the formulas

$$\sum_{n=0}^{\infty} \frac{\alpha^{2n}}{(2n)!} = \cosh \alpha \quad \text{and} \quad \sum_{n=0}^{\infty} \frac{\alpha^{2n+1}}{(2n+1)!} = \sinh \alpha \tag{F.9}$$

that are valid for every complex number α with $|\alpha| < \infty$. Equation (F.8) is thus formally valid for all complex numbers α of finite magnitude.

[1] As always, you must work this out and verify the result quoted.

In the special case when α is real, and t is any real number, we have the following useful formulas:

$$e^{Mt} = e^{\alpha_0 t}\left(I \cosh \alpha t + \frac{\boldsymbol{\alpha}\cdot\boldsymbol{\sigma}}{\alpha}\sinh \alpha t\right) \tag{F.10}$$

and

$$e^{iMt} = e^{i\alpha_0 t}\left(I \cos \alpha t + i\,\frac{\boldsymbol{\alpha}\cdot\boldsymbol{\sigma}}{\alpha}\sin \alpha t\right). \tag{F.11}$$

All other cases may be obtained by appropriate analytic continuation from the general formula (F.8).

One might wonder whether a similar 'Euler formula' exists for $(n\times n)$ matrices with $n \geq 3$. The answer is that there is no such formula for a *general* $(n\times n)$ matrix. In special cases, however, it may be possible to obtain a compact expression for the exponential of a matrix. An example is the (3×3) matrix M first encountered in Ch. 4, Sec. 4.3.2 in the problem of a charged particle in a magnetic field. In that instance, a simple expression (see Eq. (4.27)) could be found for the exponential of M, because of the relation $\mathsf{M}^3 = -\mathsf{M}$.

Appendix G

Velocity distribution in a magnetic field

In this Appendix, I give a simple method of solving the Fokker-Planck equation (14.2) of Ch. 14 for the conditional PDF of the velocity of the tagged particle in the presence of a constant magnetic field. The solution is given by Eq. (14.9).

Let us go back to the corresponding Langevin equation(14.1), and write it as in Eq. (4.20), namely,

$$\begin{aligned} \dot{v}_j(t) &= -\gamma v_j(t) - \omega_c \, v_k(t) M_{kj} + \frac{1}{m}\, \eta_j(t) \\ &= -\gamma v_j(t) + \omega_c \, M_{jk} \, v_k(t) + \frac{1}{m}\, \eta_j(t), \end{aligned} \tag{G.1}$$

in terms of the antisymmetric matrix M. Recall that ω_c is the cyclotron frequency qB/m. In vector form, the Langevin equation is

$$\dot{\mathbf{v}}(t) = (-\gamma I + \mathsf{M}\, \omega_c)\; \mathbf{v}(t) + \frac{1}{m}\, \boldsymbol{\eta}(t). \tag{G.2}$$

Now, we know that the deterministic evolution of any initial velocity under the magnetic field alone is given by $\exp(\mathsf{M}\omega_c t)\, \mathbf{v}_0$, as in Eq. (14.8). We also know that the effect of the friction is to introduce the damping factor $\exp(-\gamma t)$. This suggests that we make a change of independent variables from $\mathbf{v}$ to $\mathbf{w}$ by taking out these factors. That is, set

$$\mathbf{v} = e^{-\gamma t}\, e^{\mathsf{M}\omega_c t}\, \mathbf{w}, \quad \text{or} \quad \mathbf{w} = e^{(\gamma I - \mathsf{M}\, \omega_c)t}\, \mathbf{v}. \tag{G.3}$$

Equation (G.2) is then transformed to the stochastic differential equation

$$\dot{\mathbf{w}}(t) = \boldsymbol{\nu}(t), \tag{G.4}$$

in terms of the vector noise process

$$\boldsymbol{\nu}(t) = \frac{1}{m}\, e^{(\gamma I - \omega_c \mathsf{M})\, t}\, \boldsymbol{\eta}(t). \tag{G.5}$$

The statistical properties of $\boldsymbol{\nu}(t)$ follow readily from those of the white noise $\boldsymbol{\eta}(t)$, namely, Eqs. (4.10) and (4.11). A short calculation gives

$$\overline{\nu_i(t)} = 0, \quad \overline{\nu_i(t_1)\, \nu_j(t_2)} = \frac{\Gamma}{m^2}\, \delta_{ij}\, e^{2\gamma t_1}\, \delta(t_1 - t_2). \tag{G.6}$$

Thus each Cartesian component of $\boldsymbol{\nu}(t)$ is also a δ-correlated noise, uncorrelated with the other components. But it is a *nonstationary* noise, owing to the presence of the explicitly time-dependent factor $\exp(2\gamma t_1)$ in its autocorrelation[1]. The general SDE$\leftrightarrow$FPE correspondence may now be used to write down the Fokker-Planck equation satisfied by the conditional PDF of $\mathbf{w}$. Denote this PDF by $F(\mathbf{w}, t)$. Then the FPE corresponding to the SDE (G.4) is simply

$$\frac{\partial F}{\partial t} = \frac{\Gamma}{2m^2}\, e^{2\gamma t}\, \nabla_{\mathbf{w}}^2\, F, \tag{G.7}$$

where $\nabla_{\mathbf{w}}$ is the del operator with respect to the components of $\mathbf{w}$. The initial condition on F is

$$F(\mathbf{w}, 0) = \delta^{(3)}(\mathbf{w} - \mathbf{w}_0), \tag{G.8}$$

where $\mathbf{w}_0 = \mathbf{v}_0$, from the definition of $\mathbf{w}$ in Eq. (G.3). Equation (G.7) has the form of an ordinary diffusion equation, but with a time-dependent 'diffusion coefficient'. In order to remove this explicit time-dependence, we change *independent* variables from t to the monotonically increasing function of t given by

$$\widetilde{t} = \frac{(e^{2\gamma t} - 1)}{2\gamma}. \tag{G.9}$$

Thus $\widetilde{t} = 0$ when $t = 0$, and $\widetilde{t} \to \infty$ as $t \to \infty$. For simplicity of notation, let us retain the symbol F for the PDF of $\mathbf{w}$ even after the change of variables from t to $\widetilde{t}$. Then Eq. (G.7) is transformed to

$$\frac{\partial F}{\partial \widetilde{t}} = \frac{\Gamma}{2m^2}\, \nabla_{\mathbf{w}}^2\, F = \frac{\gamma k_B T}{m}\, \nabla_{\mathbf{w}}^2\, F, \tag{G.10}$$

on putting in the FD relation $\Gamma = 2m\gamma k_B T$. The fundamental solution to this ordinary diffusion equation is of course the normalized three-dimensional Gaussian

[1] The factor $\exp(2\gamma t_1)$ could also be replaced with $\exp(2\gamma t_2)$, because of the presence of the δ-function.

in the Cartesian components of $\mathbf{w}$. It is given by[2]

$$F(\mathbf{w},\widetilde{t}\,|\,\mathbf{w}_0)=\left(\frac{m}{4\pi\gamma k_B T\,\widetilde{t}}\right)^{3/2}\exp\left[-\frac{m(\mathbf{w}-\mathbf{w}_0)^{\mathrm{T}}(\mathbf{w}-\mathbf{w}_0)}{4\gamma k_B T\,\widetilde{t}}\right]. \tag{G.11}$$

Substituting for $\widetilde{t}$ from its definition in Eq. (G.9), we get the conditional PDF of $\mathbf{w}$ as a function of $\mathbf{w}$ and t. To obtain the PDF of the velocity $\mathbf{v}$ itself, we need to multiply F by the appropriate Jacobian of the transformation from $\mathbf{v}$ to $\mathbf{w}$, according to

$$p(\mathbf{v},t\,|\,\mathbf{v}_0)=\left|\frac{\partial(w_1\,,\,w_2\,,\,w_3)}{\partial(v_1\,,\,v_2\,,\,v_3)}\right|\,F(\mathbf{w},t\,|\,\mathbf{w}_0). \tag{G.12}$$

Here $|\cdots|$ stands for the determinant of the corresponding Jacobian matrix. This quantity is (elegantly!) found by noting that

$$\left|\frac{\partial(w_1\,,\,w_2\,,\,w_3)}{\partial(v_1\,,\,v_2\,,\,v_3)}\right|=\det\exp\left[(\gamma\,I-\mathsf{M}\,\omega_c)t\right]=\exp\operatorname{trace}\left[(\gamma\,I-\mathsf{M}\,\omega_c)t\right]. \tag{G.13}$$

In the second equation, we have used the famous identity between the determinant of the exponential of a matrix and the exponential of its trace[3]. Since trace $\mathsf{M}=0$ and trace $I=3$, we get

$$\left|\frac{\partial(w_1\,,\,w_2\,,\,w_3)}{\partial(v_1\,,\,v_2\,,\,v_3)}\right|=e^{3\gamma t}. \tag{G.14}$$

From Eqs. (G.11), (G.12) and (G.14), we get

$$\begin{aligned} p(\mathbf{v},t\,|\,\mathbf{v}_0) &= \left[\frac{m}{2\pi k_B T(1-e^{-2\gamma t})}\right]^{3/2}\times \\ &\quad\times\ \exp\left[-\frac{m\big(e^{-\mathsf{M}\omega_c t}\mathbf{v}-\mathbf{v}_0\big)^{\mathrm{T}}\big(e^{-\mathsf{M}\omega_c t}\mathbf{v}-\mathbf{v}_0\big)}{2k_B T\,(1-e^{-2\gamma t})}\right], \end{aligned} \tag{G.15}$$

where we have used the fact that $\mathbf{w}_0=\mathbf{v}_0$. Finally, we note that M is an anti-symmetric matrix with real elements. Hence $e^{\mathsf{M}\omega_c t}$ is an orthogonal matrix, i. e., the transpose of $e^{-\mathsf{M}\omega_c t}$ is $e^{\mathsf{M}\omega_c t}$. It then follows in a couple of steps[4] that

$$\big(e^{-\mathsf{M}\omega_c t}\mathbf{v}-\mathbf{v}_0\big)^{\mathrm{T}}\big(e^{-\mathsf{M}\omega_c t}\mathbf{v}-\mathbf{v}_0\big)=\big(\mathbf{v}-e^{\mathsf{M}\omega_c t}\mathbf{v}_0\big)^{\mathrm{T}}\big(\mathbf{v}-e^{\mathsf{M}\omega_c t}\mathbf{v}_0\big). \tag{G.16}$$

Insert Eq. (G.16) in Eq. (G.15), and revert to the usual notation $\mathbf{v}^2$ for the scalar product of a vector $\mathbf{v}$ with itself. We obtain the expression quoted in Eq. (14.9) for $p(\mathbf{v},t\,|\,\mathbf{v}_0)$.

---

[2]As we have represented vectors by column matrices for the moment, we need to be consistent in the notation. We must therefore write $\mathbf{w}^{\mathrm{T}}\mathbf{w}$ for the dot product of $\mathbf{w}$ with itself.

[3]We can't possibly end this book without using this marvellous identity at least once!

[4]Fill them in.

Appendix H

The Wiener-Khinchin Theorem

The power spectral density of a stationary random process $\xi(t)$ is defined (see Eq. (16.1)) as[1]

$$S_\xi(\omega) = \lim_{T\to\infty} \frac{1}{2\pi T} \left| \int_0^T dt\, e^{i\omega t}\, \xi(t) \right|^2 . \tag{H.1}$$

The Wiener-Khinchin theorem (Eq. (16.2)) states that $S_\xi(\omega)$ is equal to the Fourier transform of the autocorrelation function of $\xi(t)$. The steps leading from Eq. (16.1) to Eq. (16.2) are quite elementary (except for a crucial one), and are detailed below. The 'derivation' might give the erroneous impression that the theorem is a trivial one. But this is not so. A crucial step in the derivation (which we slur over, but point out) requires careful treatment, and this is what makes the theorem nontrivial[2].

We have

$$\begin{aligned} \left| \int_0^T dt\, e^{i\omega t}\, \xi(t) \right|^2 &= \int_0^T dt_1 \int_0^T dt_2\, \xi(t_1)\, \xi(t_2)\, \cos\omega(t_1 - t_2) \\ &= 2\int_0^T dt_1 \int_0^{t_1} dt_2\, \xi(t_1)\, \xi(t_2)\, \cos\omega(t_1 - t_2). \end{aligned} \tag{H.2}$$

The first equation follows from the fact that the left-hand side is a real quantity, so that its imaginary part must vanish identically. (It is easy to see that it does

[1] In fact, one should start a step earlier than Eq. (H.1), and replace the integral on the right-hand side of this equation by the sum of a very long time series made up of the values of the random variable at the instants at which it is sampled.

[2] We make this point because some of the literature extant might give the (wrong) impression that the theorem is nothing more than a *definition* of the power spectrum as the Fourier transform of the autocorrelation function!

so, because $\sin(t_1 - t_2)$ is an odd function of $(t_1 - t_2)$.) The second equation follows on using the symmetry of the integrand under the interchange $t_1 \leftrightarrow t_2$ of the dummy variables of integration. Next, change variables of integration from t_2 to $t = t_1 - t_2$. This gives

$$\begin{aligned}\left|\int_0^T dt\, e^{i\omega t}\, \xi(t)\right|^2 &= 2\int_0^T dt_1 \int_0^{t_1} dt\, \xi(t_1)\, \xi(t_1 - t)\, \cos\omega t \\ &= 2\int_0^T dt \int_t^T dt_1\, \xi(t_1)\, \xi(t_1 - t)\, \cos\omega t, \qquad \text{(H.3)}\end{aligned}$$

where the second line is obtained by interchanging the order of integration. Now change variables of integration once again, from t_1 to $t' = t_1 - t$. Thus

$$\left|\int_0^T dt\, e^{i\omega t}\, \xi(t)\right|^2 = 2\int_0^T dt\, \cos\omega t \int_0^{T-t} dt'\, \xi(t')\, \xi(t + t'). \qquad \text{(H.4)}$$

Putting this into Eq. (H.1),

$$S_\xi(\omega) = \lim_{T\to\infty} \frac{1}{\pi T}\int_0^T dt\, \cos\omega t \int_0^{T-t} dt'\, \xi(t')\, \xi(t + t'). \qquad \text{(H.5)}$$

It is the next step that must be justified rigorously. As we have already stated, we shall not do so. We use the ergodicity of the process $\xi(t)$ to write

$$\lim_{T\to\infty} \frac{1}{T}\int_0^{T-t} dt'\, \xi(t')\, \xi(t + t') = \langle \xi(t')\, \xi(t + t')\rangle = \phi_\xi(t). \qquad \text{(H.6)}$$

An interchange of limits[3] then leads to

$$S_\xi(\omega) = \frac{1}{\pi}\int_0^\infty dt\, \phi_\xi(t)\, \cos\omega t. \qquad \text{(H.7)}$$

We now use the fact that $\phi_\xi(t) = \phi_\xi(-t)$, which is a general property of the autocorrelation function of a stationary process (as we have shown in Eq. (4.8)). Hence

$$S_\xi(\omega) = \frac{1}{2\pi}\int_{-\infty}^\infty dt\, \phi_\xi(t)\, \cos\omega t = \frac{1}{2\pi}\int_{-\infty}^\infty dt\, e^{i\omega t}\, \phi_\xi(t). \qquad \text{(H.8)}$$

That is, the power spectrum of $\xi(t)$ is just the Fourier transform of its autocorrelation function.

[3]This shuffling requires proper justification.

Appendix I

Classical linear response theory

In several chapters of this book, we have been concerned with velocity correlation functions and the dynamic mobility. As already mentioned in the text, these are specific examples of response functions and generalized susceptibilities, respectively. These quantities play a key role in Linear Response Theory. In view of their importance, it is essential to understand how they arise in LRT, and how formulas of the Kubo-Green type for generalized susceptibilities originate. This Appendix is devoted to a brief account of these aspects. In keeping with the rest of this book, we restrict our considerations to the classical (rather than quantum mechanical) case.

I.1 Mean values

The system (more precisely, the subsystem) of interest, described by an unperturbed Hamiltonian $H_0(q,p)$, is assumed to be in thermal equilibrium in contact with a heat bath at a temperature T. Here q denotes the set $\{q_k\}$ of generalized coordinates of the system, and p denotes the set $\{p_k\}$ of the conjugate generalized momenta[1]. The equilibrium phase space density (or density matrix) is proportional to $\rho^{\text{eq}} \equiv \exp\left(-\beta H_0(q,p)\right)$, where $\beta = (k_B T)^{-1}$ as usual. The equilibrium average of any observable $Y(q,p)$ pertaining to the system is given by

$$\langle Y \rangle_{\text{eq}} = \frac{\text{Tr}\left(\rho^{\text{eq}}\, Y\right)}{\text{Tr}\,\rho^{\text{eq}}} \equiv \frac{\int dq\, dp\, Y(q,p)\, \exp\left[-\beta H_0(q,p)\right]}{\int dq\, dp\, \exp\left[-\beta H_0(q,p)\right]}\,. \tag{I.1}$$

The denominator in Eq. (I.1) is of course the canonical partition function Z_{can}. It is convenient to choose a normalization of the equilibrium density matrix such

[1]In this book, we have mostly dealt with the case of a tagged particle in a fluid as the subsystem of interest. But the formalism of LRT is of course applicable in much greater generality.

that $\mathrm{Tr}\,\rho^{\,\mathrm{eq}} = 1$. We note that the equilibrium average value $\langle Y \rangle_{\mathrm{eq}}$ of any function $Y(q,p)$ (that has no *explicit* time dependence) is independent of time. Indeed, this is the very import of a stationary or **invariant probability density**, of which the Boltzmann-Gibbs equilibrium measure $\exp(-\beta H_0)$ is a pre-eminent example[2].

We consider the perturbation of the system by an external or applied 'force' that is switched on at $t = -\infty$, and is time-dependent, in order to be as general as possible. The perturbed Hamiltonian has the form

$$H(q,p,t) = H_0 + H'(t) = H_0(q,p) - X(q,p)\, F_{\mathrm{ext}}(t), \tag{I.2}$$

where $X(q,p)$ is the system variable to which the applied 'force' couples[3]. The objective of LRT is to derive an expression for the mean value of any observable $Y(q,p)$ in the presence of the perturbation, correct to first order in $F_{\mathrm{ext}}(t)$. This expression is of the form

$$\langle Y \rangle = \langle Y \rangle_{\mathrm{eq}} + \langle \delta Y \rangle\,, \tag{I.3}$$

where δY is a linear functional of $F_{\mathrm{ext}}(t)$. Thus LRT is essentially first-order time-dependent perturbation theory at a finite (nonzero) temperature T, as already mentioned in Sec. 1.1. The task before us is the determination of the quantity $\langle \delta Y \rangle$.

I.2 The Liouville equation

Consider, first, free evolution under the unperturbed Hamiltonian H_0, and in the absence of the heat bath. From Hamilton's equations $\dot{q}_k = \partial H_0/\partial p_k$, $\dot{p}_k = -\partial H_0/\partial q_k$, we know that the time evolution of any function $Y(q,p)$ of the dy-

[2]Recall also the remarks made in Sec. 6.5.2.

[3]The minus sign preceding the perturbation is purely a matter of convention. If the perturbation is a spatially uniform physical force $F_{\mathrm{ext}}(t)$ acting on a particle moving in one dimension (the x-axis, say), it is clear that the corresponding potential energy is simply $-x\, F_{\mathrm{ext}}(t)$. The perturbation term in Eq. (I.2) is just a generalization of this form. X need not necessarily have the physical dimensions of length, nor F_{ext} that of force, but their product has the physical dimensions of energy.

We must also note that Eq. (I.2) does not cover all possible perturbations. The case of a particle of charge e in external electric and magnetic fields is a notable instance. As we know, this case involves a velocity-dependent force. The free-particle Hamiltonian $\mathbf{p}^2/(2m)$ is replaced by $(\mathbf{p} - e\mathbf{A})^2/(2m) + e\phi$, where $\mathbf{A}$ and ϕ are the vector and scalar potentials corresponding to the fields, and $\mathbf{p}$ is the momentum canonically conjugate to the position $\mathbf{r}$ of the particle. The formalism of LRT can be extended to cover such cases. We have, of course, used the more direct Langevin equation approach in this book.

namical variables is given by the equation of motion

$$\frac{dY}{dt} = \{Y, H_0\} \equiv i\mathcal{L}_0 Y\,. \tag{I.4}$$

Here $\{\,,\,\}$ denotes the Poisson bracket, and

$$\mathcal{L}_0 = i\sum_k \left(\frac{\partial H_0}{\partial q_k}\frac{\partial}{\partial p_k} - \frac{\partial H_0}{\partial p_k}\frac{\partial}{\partial q_k}\right) \tag{I.5}$$

is the **Liouville operator** corresponding to the Hamiltonian H_0. The *formal* solution to Eq. (I.4) is

$$Y\big(q(t), p(t)\big) = e^{i\mathcal{L}_0 t}\, Y\big(q(0), p(0)\big). \tag{I.6}$$

We shall keep track of time arguments by writing just $Y(t)$ for $Y\big(q(t), p(t)\big)$ (and similarly for other functions of the dynamical variables), for brevity[4].

Let us now return to the problem at hand, namely, the determination of the mean value of any physical quantity Y in the presence of the heat bath and the applied force. LRT proceeds to calculate δY by deducing, first, the correction to the phase space density arising from the presence of the perturbation $H'(t)$ in the Hamiltonian. The exact density $\rho(t)$ satisfies the **Liouville equation**

$$\frac{\partial \rho(t)}{\partial t} = \{H(t)\,, \rho(t)\} \equiv -\,i\,\mathcal{L}(t)\,\rho(t), \tag{I.7}$$

where

$$\mathcal{L}(t) = \mathcal{L}_0 + \mathcal{L}'(t) \tag{I.8}$$

is the 'total' Liouville operator[5]. The initial condition on $\rho(t)$ is $\rho(-\infty) = \rho^{\,\mathrm{eq}}$. Setting $\rho(t) = \rho^{\,\mathrm{eq}} + \delta\rho(t)$, we get the following inhomogeneous equation for $\delta\rho(t)$:

$$\frac{\partial\,\delta\rho}{\partial t} + i\,\mathcal{L}_0\,\delta\rho = -\,i\,\mathcal{L}'\,\rho^{\,\mathrm{eq}}. \tag{I.9}$$

[4]This is more than just a matter of notation. We retain this notation even when the Hamiltonian is perturbed: the time argument will stand for the dynamical variables $q(t), p(t)$ at time t, *as if* the evolution has been governed by the *unperturbed* Hamiltonian H_0 (or the unperturbed Liouville operator $\mathcal{L}_0$). The quantum mechanical counterpart is perhaps more familiar to you: Essentially, we want to work in the interaction picture. For this purpose it is convenient to first express all operators in the Heisenberg picture as generated by the *unperturbed* Hamiltonian.

[5]The extra minus sign on the right-hand side of Eq. (I.7), in marked contrast to the right-hand side of the equation of motion (I.4) for any dynamical observable, should not escape your notice! *A probability density (or distribution) by itself is not an observable.*

Here we have used the fact that $\mathcal{L}_0\,\rho^{\mathrm{eq}} \equiv 0$, and also *dropped the higher-order term* $\mathcal{L}'\,\delta\rho$. The solution to Eq. (I.9) satisfying the initial condition $\delta\rho = 0$ as $t \to -\infty$ is

$$\delta\rho(t) = \int_{-\infty}^{t} dt'\, e^{-i(t-t')\mathcal{L}_0}\,\big(-i\mathcal{L}'\,\rho^{\mathrm{eq}}\big)\,F_{\mathrm{ext}}(t'). \tag{I.10}$$

- The great simplification that obtains in LRT is, of course, the fact that only the time-*independent* operator $\mathcal{L}_0$ has to be exponentiated.

I.3 The response function

The 'correction' $\langle \delta Y\rangle$ we seek is now obtained by an argument that is perhaps more familiar to you in the context of quantum mechanics—namely, we equate expectation values calculated in the Schrödinger and Heisenberg pictures, respectively. After some simplification, the following result is obtained:

$$\langle \delta Y\rangle = \int_{-\infty}^{t} dt'\,\Phi_{XY}(t-t')\,F_{\mathrm{ext}}(t'), \tag{I.11}$$

where

$$\Phi_{XY}(t-t') = \big\langle\{X(t')\,,\,Y(t)\}\big\rangle_{\mathrm{eq}} = \big\langle\{X(0)\,,\,Y(t-t')\}\big\rangle_{\mathrm{eq}}\,, \quad (t \ge t') \tag{I.12}$$

is the **response function** connecting the stimulus (represented by the system variable X) to the response (represented by the measured system variable Y). The integral on the right-hand side in Eq. (I.11) represents a response that is

- *linear*, because the integral is a linear functional of the applied force F_{ext};
- *causal*, because the integration over the force history cuts off at the upper limit t;
- *retarded*, because the response function (or weight factor) depends only on the time *interval* between the occurrence of the stimulus at time t' and the measurement of the response at any later time t.

Some re-arrangement (using the cyclic property of the trace) enables us to express the response function in the alternative form

$$\Phi_{XY}(t-t') = \mathrm{Tr}\Big(\{\rho^{\mathrm{eq}}\,,\,X(0)\}\;Y(t-t')\Big). \tag{I.13}$$

But

$$
\begin{aligned}
\{\rho^{\rm eq}\,,\,X(0)\} &= \sum_k \left(\frac{\partial\, e^{-\beta H_0}}{\partial q_k}\frac{\partial X(0)}{\partial p_k} - \frac{\partial\, e^{-\beta H_0}}{\partial p_k}\frac{\partial X(0)}{\partial q_k} \right) \\
&= -\beta\, e^{-\beta H_0} \sum_k \left(\frac{\partial H_0}{\partial q_k}\frac{\partial X(0)}{\partial p_k} - \frac{\partial H_0}{\partial p_k}\frac{\partial X(0)}{\partial q_k} \right) \\
&= \beta\, \rho^{\rm eq}\, \dot{X}(0)\,. \qquad \text{(I.14)}
\end{aligned}
$$

Therefore the response function (which originally involved a Poisson bracket of two functions of the dynamical variables at different instants of time) essentially reduces to the equilibrium average value of a *product* of functions, namely, the cross-correlation function of the quantities $\dot{X}$ and Y, according to

$$\Phi_{XY}(t-t') = \beta \left\langle \dot{X}(0)\, Y(t-t') \right\rangle_{\rm eq}, \quad t \geq t'. \qquad \text{(I.15)}$$

In passing, we note that the quantum mechanical counterpart of this result is somewhat more involved, owing to the noncommutativity of operators in general[6].

We can read off the *time reversal property*, if any, of the response function $\Phi(t)$ from the foregoing formula. Let us consider situations in which there are no complications such as external magnetic fields (which must also be reversed under time reversal, as we have seen in Ch. 4). Further, suppose the dynamical quantities X and Y have definite 'time-parities' $\epsilon_X (= \pm 1)$ and $\epsilon_Y (= \pm 1)$ under time-reversal. We then see from Eq. (I.15) that

$$\Phi_{XY}(-t) = -\,\epsilon_X\, \epsilon_Y\, \Phi(t), \qquad \text{(I.16)}$$

using the fact that $\dot{X}$ and X must have opposite time parities. Thus $\Phi_{XY}(t)$ must be either an even function of t, or an odd function of t, in such cases. In general, however, there is no such requirement[7].

[6]For completeness, let us write down the actual quantum mechanical expression. Dynamical variables are replaced by operators, of course, and Poisson brackets are replaced by commutators multiplied by $(i\hbar)^{-1}$. The cross-correlation in Eq. (I.15) is replaced by a so-called **Kubo canonical correlation function**, and the response function has the form

$$\Phi_{XY}(t-t') = (i\hbar)^{-1} \left\langle [X(0)\,,\, Y(t-t')] \right\rangle_{\rm eq} = \int_0^\beta d\lambda \left\langle e^{\lambda H_0}\, \dot{X}(0)\, e^{-\lambda H_0}\, Y(t-t') \right\rangle_{\rm eq}.$$

It is trivially checked that Eq. (I.15) is recovered if $[H_0\,,\, \dot{X}(0)] = 0$.

[7]This can happen even if X and Y have definite time-parities. The velocity of a particle has negative time-parity (it changes sign upon time reversal). In spite of this, any off-diagonal element of the velocity correlation tensor for a particle moving in a magnetic field is neither an even function of t, nor an odd function of t, as is clear from Eq. (4.34).

I.4 The generalized susceptibility

The generalized susceptibility that is a measure of the response is defined, as we already know, by the Fourier-Laplace transform of the response function. We have

$$\chi_{XY}(\omega) = \int_0^\infty dt\, e^{i\omega t}\, \Phi_{XY}(t). \tag{I.17}$$

Thus, we arrive at the compact formula

$$\chi_{XY}(\omega) = \beta \int_0^\infty dt\, e^{i\omega t}\, \left\langle \dot{X}(0)\, Y(t) \right\rangle_{\text{eq}}. \tag{I.18}$$

If X and Y are real-valued, then $\text{Re}\, \chi_{XY}(\omega)$ and $\text{Im}\, \chi_{XY}(\omega)$ are, respectively, even and odd functions of ω. We have already seen that these properties hold good for the mobility $\mu(\omega)$ (recall Eqs. (16.23)). The argument given in Sec. 16.3 to establish the analyticity of $\mu(\omega)$ in the upper half of the ω-plane are applicable to $\chi_{XY}(\omega)$ as well, based on the representation of this function provided by Eq. (I.18). The counterpart of Eq. (16.24) also holds good, namely,

$$\chi_{XY}(-\omega^*) = \chi^*_{XY}(\omega), \quad \text{Im}\, \omega \geq 0. \tag{I.19}$$

Moreover, the real and imaginary parts of the generalized susceptibility constitute a Hilbert transform pair, satisfying the dispersion relations we have derived in the case of $\mu(\omega)$.

Finally, which generalized susceptibility does the dynamic mobility $\mu(\omega)$ of the tagged particle represent? This susceptibility is readily identified. The function $X(q,p)$ is, in this case, simply the x-coordinate of the particle, as the perturbation is just $-x\, F_{\text{ext}}(t)$. The function $Y(q,p)$ (the response that is measured) is its velocity v. Therefore, since $\dot{x} \equiv v$, we have

$$\mu(\omega) \equiv \chi_{xv}(\omega) = \beta \int_0^\infty dt\, e^{i\omega t}\, \langle v(0)\, v(t) \rangle_{\text{eq}}. \tag{I.20}$$

from Eq. (I.18). This is precisely the Kubo-Green formula, Eq. (15.43).

Appendix J

Power spectrum of a random pulse sequence

In Ch. 16, we have defined and discussed the power spectral density (or power spectrum) of a stationary random process. We have also considered the power spectra of the random processes that occur in the Langevin equation: the driving white noise, $\eta(t)$, and the driven process, the velocity $v(t)$ of the tagged particle.

The following fact is of very great generality:

- Experimental information regarding dynamics is usually obtained in the form of time series. The power spectrum, which is directly deduced from a time series, is a fundamental characterizer of the underlying dynamics.

These statements are applicable to stochastic processes, deterministic processes, and combinations of the two. In view of this practical importance, it is useful to consider some basic aspects of power spectra that are not restricted to the immediate context of this book, but which are relevant to any discussion of noise and fluctuations.

J.1 The transfer function

The results obtained in Sec. 16.1.3 also enable us to write down an important relationship between the 'input' and 'output' power spectra in the Langevin model: that is, a relation between $S_\eta(\omega)$ and $S_v(\omega)$. For ready reference, let us repeat Eq. (16.8) for $S_\eta(\omega)$ and Eq. (16.11) for $S_v(\omega)$:

$$S_\eta(\omega) = \frac{m\gamma k_B T}{\pi}, \quad S_v(\omega) = \frac{\gamma k_B T}{m\pi\,(\gamma^2 + \omega^2)}. \tag{J.1}$$

We have already seen that the dynamic mobility in the Langevin model is given by

$$\mu(\omega) = \frac{1}{[m\,(\gamma - i\omega)]}\,. \tag{J.2}$$

This expression satisfies the general property $\mu(-\omega) = \mu^*(\omega)$ for real values of ω (Eq. (16.22)). It is therefore evident that

$$S_v(\omega) = |\mu(\omega)|^2\, S_\eta(\omega). \tag{J.3}$$

In the analogous case of thermal noise in a resistor, we have Eq. (16.10) for the power spectrum $S_V(\omega)$ of the noise voltage, and Eq. (16.12) for the power spectrum $S_I(\omega)$ of the current. The corresponding susceptibility is the complex admittance $Y(\omega)$, given by Eq. (16.18). Once again, we see that

$$S_I(\omega) = |Y(\omega)|^2\, S_V(\omega). \tag{J.4}$$

Equations (J.3) and (J.4) are special cases of a general relation. Suppose the stationary random process $\xi(t)$ is driven by another stationary random process $\eta(t)$ by means of a linear equation (such as a linear stochastic differential equation, for instance). Note that $\eta(t)$ need not necessarily be a white noise. However, for consistency of notation, we continue to use the symbol η for the driving or input process, and ξ for the driven or output variable. Then:

- The output power spectrum is the product of the input power spectrum and a function of the frequency called the **transfer function**, denoted by $H(\omega)$. The transfer function is the square of the modulus of the corresponding generalized susceptibility $\chi(\omega)$ connecting the input and output processes[1] :

$$S_\xi(\omega) = H(\omega)\, S_\eta(\omega) = |\chi(\omega)|^2\, S_\eta(\omega). \tag{J.5}$$

The transfer function is a very important property of a system. It finds much use in engineering applications, in particular. Note that $H(\omega)$ depends only on the *magnitude* of $\chi(\omega)$. To measure the *phase* of $\chi(\omega)$, in practice we must use two or more driven variables coupled to each other, and consider the set of cross-susceptibilities[2].

[1]Sometimes the generalized susceptibility $\chi(\omega)$ itself is called the transfer function, but this terminology is likely to cause confusion, and is best avoided.

[2]The foregoing statements are quite generally applicable. In those instances when LRT is applicable, we have an additional special feature: namely, susceptibilities can be computed in terms of *equilibrium* correlation functions.

Throughout this book, we have been interested in obtaining information about the driven or output process (such as the velocity of the tagged particle), given the statistical properties of the driving noise. However, the concept of the transfer function has another practical use. The relation between input and output power spectra also provides a means for the determination of a noise spectrum. Suppose we require the power spectrum of the random variable $\eta(t)$ characterizing the noise in a system[3]. The noise is generally a weak signal in practical applications. When it is passed through a sharply tunable amplifier with transfer function $H(\omega) = |\chi(\omega)|^2$, the power spectrum of the output signal $\xi(t)$ is given by Eq. (J.1). Hence the formula of Eq. (16.7) for the mean squared value of ξ gives

$$\langle \xi^2 \rangle = 2 \int_0^\infty d\omega \, S_\xi(\omega) = 2 \int_0^\infty d\omega \, |\chi(\omega)|^2 \, S_\eta(\omega). \tag{J.6}$$

Now suppose the amplifier is very sharply tuned to a frequency ω_0, with a mid-band response $|\chi(\omega_0)|^2$ and an effective band-width B_{eff} defined by

$$|\chi(\omega_0)|^2 \, B_{\text{eff}} = \int_0^\infty d\omega \, |\chi(\omega)|^2. \tag{J.7}$$

The quantities $|\chi(\omega_0)|^2$ and B_{eff} are essentially amplifier parameters. They may be determined with the help of a signal generator. Then, assuming that $S_\eta(\omega)$ does not also resonate or vary rapidly near $\omega = \omega_0$, we have

$$S_\eta(\omega_0) \simeq \frac{\langle \xi^2 \rangle}{2|\chi(\omega_0)|^2 \, B_{\text{eff}}}. \tag{J.8}$$

Thus, a measurement of the mean squared *output* with a quadratic detector enables us to determine S_η at the frequency $\omega = \omega_0$. Repeating this procedure at other frequencies (by tuning the amplifier appropriately), one may, in principle, map the input spectrum $S_\eta(\omega)$.

J.2 Random pulse sequences

We turn to another important question that occurs very frequently in noise studies. Suppose there is a noise source that emits a sharp pulsed signal at random instants of time, t_i. The individual pulses are uncorrelated with each other, and occur independently, in a statistically stationary manner. This kind of **pulse process** is a generalized form of what is called **shot noise**. Given the shape of the pulse, can one predict the form of the power spectrum of the pulse sequence? In

[3]Remember that $\eta(t)$ need not be a white noise.

the more general case, the shapes of successive pulses may themselves be different: the pulse shape may vary at random, although its *functional* form is expected to remain the same in a given problem. For example, if the pulses are rectangular ones, the height and duration of the pulses may vary randomly. Results applicable to these and several other generalizations are available. We discuss some of the simplest cases.

On the time axis, the instants $\{t_i\}$ form an uncorrelated random sequence of points. More precisely, they represent a stationary Poisson pulse process, exactly as described in Ch. 5, Sec. 5.6.1, in our discussion of the symmetric dichotomous Markov process. Let ν be the average rate at which the pulses occur[4]: that is, the mean interval between consecutive members of the set of instants $\{t_i\}$ is ν^{-1}. The probability that r of these points lie in any interval of time t is $e^{-\nu t}(\nu t)^r/r!$, where $r = 0, 1, 2, \ldots$. Let us consider the case in which all the pulses are identical in shape, and given by a functional form $f(t - t_i)$, where t_i is the time at which the ith pulse begins[5]. For instance, in the case of a rectangular pulse of unit height and duration τ, we have

$$f(t - t_i) = \theta(t - t_i) - \theta(t - \tau - t_i), \tag{J.9}$$

where $\theta(t)$ is the usual unit step function. While we are usually concerned with pulses of finite duration, the results to be discussed below are also applicable, under certain conditions, to 'noncompact' pulse shapes—for example, an exponentially decaying pulse. In this case we have

$$f(t - t_i) = e^{-\lambda(t-t_i)}\,\theta(t - t_i), \tag{J.10}$$

where λ^{-1} is the time constant associated with the pulse. The output or resultant signal at time t is given by

$$\xi(t) = \sum_{i=-\infty}^{\infty} h_i\, f(t - t_i), \tag{J.11}$$

where h_i is the amplitude (or peak height) of the ith pulse. This height is also permitted to vary randomly: $\{h_i\}$ constitutes a set of independent random variables assumed to be drawn from some prescribed distribution. We are interested in the statistical properties of the stationary random process represented by $\xi(t)$.

[4]We have used the symbol λ for this mean rate in Sec. 5.6.1. But in the present context, ν is the symbol customarily used in the literature.

[5]Thus $f(t)$ is defined for $t \geq 0$, and may be defined to be identically zero for $t < 0$.

Campbell's Theorem states that the mean and variance of $\xi(t)$ are given by

$$\langle\xi\rangle = \nu\,\langle h\rangle \int_{-\infty}^{\infty} dt\, f(t), \quad \langle\xi^2\rangle - \langle\xi\rangle^2 = \nu\,\langle h^2\rangle \int_{-\infty}^{\infty} dt\, f^2(t). \tag{J.12}$$

More generally, the theorem states that the nth cumulant[6] of $\xi(t)$ is given by

$$\kappa_n = \nu\,\langle h^n\rangle \int_{-\infty}^{\infty} dt\,[f(t)]^n \tag{J.13}$$

where $\langle h^n\rangle$ is the nth moment of the distribution of the pulse height h. Further generalizations of Campbell's Theorem also exist.

The power spectrum of $\xi(t)$ is given by another theorem. Let $\widetilde{f}(\omega)$ denote the Fourier transform of the pulse shape, i. e.,

$$\widetilde{f}(\omega) = \frac{1}{2\pi}\int_{-\infty}^{\infty} dt\, e^{i\omega t}\, f(t). \tag{J.14}$$

Carson's Theorem then asserts that[7]

$$S_\xi(\omega) = (2\pi)^2\,|\tilde{f}(\omega)|^2\left\{2\nu\,\langle h^2\rangle + 4\pi\nu^2\,\langle h\rangle^2\,\delta(\omega)\right\}. \tag{J.15}$$

But $|\tilde{f}(\omega)|^2\,\delta(\omega)$ can be written as $\big(\tilde{f}(0)\big)^2\,\delta(\omega)$. Using Campbell's Theorem for the mean value of ξ, we therefore get

$$S_\xi(\omega) = 8\pi^2\,\nu\,\langle h^2\rangle\,|\tilde{f}(\omega)|^2 + 4\pi\,\langle\xi\rangle^2\,\delta(\omega). \tag{J.16}$$

This is the form in which Carson's Theorem is usually stated[8]. Note the presence of a zero-frequency δ-function peak in the power spectrum. This 'DC term' occurs whenever the pulse sequence has a nonzero mean value.

For the rectangular pulse of Eq. (J.9), we find that the mean value and the variance are given by

$$\langle\xi\rangle = \nu\tau\,\langle h\rangle\ , \ \langle\xi^2\rangle - \langle\xi\rangle^2 = \nu\tau^2\,\langle h^2\rangle\,. \tag{J.17}$$

---

[6]Recall the definition of the cumulants of a random variable given in Appendix C.

[7]The numerical factor in Eq. (J.15) depends on the Fourier transform convention we have adopted, given in Eqs. (15.27). Had we used the alternative convention in which the factor $(2\pi)^{-1}$ is not in the integral over t, but rather in the integral over ω, then the overall factor $(2\pi)^2$ outside the curly brackets in Eq. (J.15) would have been absent.

[8]Once again, with the alternative Fourier transform convention, Carson's Theorem reads

$$S_\xi(\omega) = 2\,\nu\,\langle h^2\rangle\,|\tilde{f}(\omega)|^2 + 4\pi\,\langle\xi\rangle^2\,\delta(\omega).$$

The power spectrum is given by

$$S_\xi(\omega) = \frac{8\nu \langle h^2 \rangle}{\omega^2} \sin^2\left(\tfrac{1}{2}\omega\tau\right) + 4\pi\nu^2\tau^2 \langle h \rangle^2 \delta(\omega). \tag{J.18}$$

Similarly, for a sequence of exponential pulses given by Eq. (J.10), we find

$$\langle \xi \rangle = \frac{\nu \langle h \rangle}{\lambda}, \ \langle \xi^2 \rangle - \langle \xi \rangle^2 = \frac{\nu \langle h^2 \rangle}{2\lambda}. \tag{J.19}$$

The corresponding power spectrum is

$$S_\xi = \frac{2\nu \langle h^2 \rangle}{\omega^2 + \lambda^2} + \frac{4\pi^2\nu^2 \langle h \rangle^2}{\lambda^2} \delta(\omega). \tag{J.20}$$

If the pulse *shape* itself varies randomly, a further averaging process must be carried out to obtain the mean, variance, power spectrum, etc. For example, if the duration τ of each rectangular pulse varies independently and randomly with a specified PDF, then

$$S_\xi(\omega) = \frac{8\nu \langle h^2 \rangle}{\omega^2} \left\langle \sin^2\left(\tfrac{1}{2}\omega\tau\right)\right\rangle, \tag{J.21}$$

where $\left\langle \sin^2\left(\tfrac{1}{2}\omega\tau\right)\right\rangle$ denotes the average value of the function concerned over the distribution of τ. If several parameters specifying the pulse shape vary randomly, more complicated extensions of the theorem must be used. Finally, the parameters of the pulse shape (e. g., τ and h) may be *correlated* random variables, and not independent ones. Some results exist in this considerably more complicated case as well.

J.3 Shot noise

These results may be applied immediately to a simple case, namely, the shot noise in a diode arising from the spontaneous fluctuations in the number n of electrons emitted *per unit time* by the cathode[9]. Since the electrons are emitted independently and at random. The total number of electrons emitted up to time t is evidently

$$N(t) = \int_{-\infty}^{t} dt' \sum_{i=-\infty}^{\infty} \delta(t' - t_i), \tag{J.22}$$

[9] It is assumed that saturation conditions have been reached, so that the random process is stationary.

where $\{t_i\}$ is an uncorrelated Poisson-distributed sequence of instants, as before. Therefore the number of electrons emitted per unit time at time t is formally given by

$$n(t) = \frac{dN(t)}{dt} = \sum_{i=-\infty}^{\infty} \delta(t - t_i). \tag{J.23}$$

The pulse shape in this case is therefore just a δ-function of unit height. Its Fourier transform is written down trivially. But we must take care to include the whole pulse, and not just half of it: the Fourier transform of the pulse shape is given by

$$\widetilde{f}(\omega) = \frac{1}{2\pi}\int_{-\infty}^{\infty} dt\, e^{i\omega t}\,\delta(t) = \frac{1}{2\pi}\,, \tag{J.24}$$

and *not* by $(2\pi)^{-1}\int_{-\infty}^{\infty} dt\, e^{i\omega t}\,\delta(t)\,\theta(t) = \pi^{-1}$. Another way to see this is as follows. Let us go back to the case of a rectangular pulse, set each pulse height equal to h such that the product $h\tau = 1$. Now take the limit $\tau \to 0$ in the expression derived for the power spectrum of a rectangular pulse sequence, Eq. (J.18). Since $(\sin\omega\tau)/(\omega\tau) \to 1$ as $\tau \to 0$, we

$$S_n(\omega) = 2\nu + 4\pi\nu^2\delta(\omega) \tag{J.25}$$

for the power spectrum of the process $n(t)$. The fluctuating current at the anode is given by $I(t) = en(t)$, where e is the charge on an electron. Therefore the power spectrum of the current is, for $\omega > 0$,

$$S_I(\omega) = e^2\, S_n(\omega) = 2e^2\,\nu = 2e\,\langle I\rangle\,, \tag{J.26}$$

where $\langle I\rangle = e\nu$ is the average current. This basic result is known as **Schottky's Theorem**. A similar result holds for charge carriers crossing potential barriers, as in *p-n* junction diodes. Note that the power spectrum has a 'strength' dependent on the electronic *charge*, and not on the *temperature* T as in the case of thermal noise. This feature is typical of cases in which an external force causes the system to depart from thermal equilibrium, making the original form of the Nyquist theorem inapplicable.

J.4 Barkhausen noise

As a second example, consider the discontinuous changes in the magnetization of a ferromagnetic material under the action of an external magnetic field, owing to the dynamics of the domains in the specimen (primarily, the irregular motion of domain walls). This effect is known as **Barkhausen noise**. If a ferromagnet is placed in a slowly varying magnetic field, these jumps in the magnetization induce random voltage pulses in a coil wound around the specimen. We may obtain an expression for the power spectrum of the noise voltage as follows. Suppose the saturation magnetization of an elementary domain changes, on the average, from $-\mathsf{m}$ to $+\mathsf{m}$ in each jump. Then the flux change per jump decays with time according to

$$\Phi(t) = 2\,\mathsf{m}\,e^{-t/\tau}, \tag{J.27}$$

where τ is the corresponding relaxation time. The induced voltage pulse in an N-turn coil is therefore

$$f(t) = -N\,\frac{d\Phi}{dt} = \frac{2N\,\mathsf{m}}{\tau}\,e^{-t/\tau}, \quad t > 0. \tag{J.28}$$

We thus have a sequence of exponential pulses. Let the average rate of occurrence of the pulses be ν, as usual. Then the power spectrum of the noise voltage is given, for $\omega > 0$, by

$$S_V(\omega) = 8\pi^2\,\nu\,|\tilde{f}(\omega)|^2 = \frac{8\nu\,N^2\,\mathsf{m}^2}{1+\omega^2\tau^2}\,. \tag{J.29}$$

This is a Lorentzian spectrum. It is in good agreement with experimental data on Barkhausen noise. The time constant τ is typically of the order of 10^{-4} s, while ν is proportional to the frequency of the applied magnetic field.

Appendix K

Stable distributions

K.1 The family of stable distributions

A topic of great importance in statistics is that of the probability distribution of the sum of independent random variables. In particular, a crucial question is the existence of a limiting distribution (or **limit law**) when the number of random variables becomes infinite. We have already seen a fundamental example of such a limit law in Appendix D, Secs. D.4 and D.5, and again in Appendix E on the passage from random walks to diffusion. The stable distributions are intimately connected with such limit laws.

Suppose we have n independent, identically distributed (or 'i.i.d.') random variables $\xi_1, \xi_2, \ldots, \xi_n$, each with a cumulative probability distribution $\mathcal{P}$. Are there forms of $\mathcal{P}$ such that the sum $\sum_{i=0}^{n} \xi_i$, possibly shifted by an n-dependent constant and re-scaled by another n-dependent constant, also has the *same* distribution function? In other words, when is it possible to find a positive constant a_n and a real constant b_n such that the probability distribution of the quantity

$$\xi^{(n)} = \frac{1}{a_n}\left\{\sum_{i=0}^{n} \xi_i - b_n\right\} \tag{K.1}$$

is also given by $\mathcal{P}$ itself, for every positive integer $n \geq 2$? The complete answer to this question is one of the key results in statistics. There is a whole family of distributions with the property required, the stable distributions—the full name being the Lévy skew alpha-stable distributions. When the property is satisfied even without the shift b_n in Eq. (K.1), the corresponding distributions are said to be **strictly stable**. These comprise a subset of the class of stable distributions.

Before we describe the stable distributions themselves, we note that there are several alternative (and equivalent) ways of stating the defining property of stable distributions. The statement made above, in terms of the sum $\xi^{(n)}$, is just one among these. An alternative statement is as follows: Let ξ_1 and ξ_2 be two i.i.d. random variables, with distribution function $\mathcal{P}$. Given any two arbitrary positive numbers a_1 and a_2, if and only if[1] we can find a positive number a and a real number b such that $(a_1 \xi_1 + a_2 \xi_2 - b)/a$ has the same distribution $\mathcal{P}$, then $\mathcal{P}$ is a stable distribution. Again, if this can be done without a shift constant b, the distribution is strictly stable.

This last formulation can be re-expressed as a property of the distribution function itself. Given any two positive numbers a_1 and a_2, if (and only if) we can find a positive number a and a real number b such that the distribution function $\mathcal{P}$ satisfies

$$\mathcal{P}(\xi/a_1) * \mathcal{P}(\xi/a_2) = \mathcal{P}\left((\xi - b)/a\right), \tag{K.2}$$

where the symbol $*$ denotes the convolution of the two distributions, then $\mathcal{P}(\xi)$ is a stable distribution. If this relation is satisfied with $b = 0$ in all cases, we have a strictly stable distribution.

K.2 The characteristic function

Not surprisingly, the most explicit way of specifying the stable distributions is in terms of their characteristic functions. In fact, Eq. (K.2) already suggests that the characteristic function of a stable distribution must satisfy some sort of 'multiplication property', and must involve exponential functions. This intuitive guess is indeed borne out.

The stable distributions are characterized by four parameters, as we shall see below. It turns out that the scaling constant a_n in Eq. (K.1) must have the power-law form $a_n = n^{1/\alpha}$, where $0 < \alpha \leq 2$. The exponent or index α is the primary characterizer of the members of the family. The stable distributions have continuous PDFs that are unimodal[2]. Let ξ denote the random variable, as usual, and $p(\xi)$ its PDF. In general (that is, for an arbitrary value of α), $p(\xi)$ cannot be

[1] Hence the conditions that follow are both necessary and sufficient.

[2] A unimodal distribution essentially means that there is a single peak in the PDF, or a unique most probable value for the random variable. We shall not get into the technicalities of a more precise definition of unimodality here.

written down in explicit closed form[3]. However, the characteristic function *can* be so expressed.

- In essence, the modulus of the characteristic function $\widetilde{p}(k)$ for a stable distribution with exponent α behaves like $\exp(-|k|^{\alpha})$.

More precisely: in its most general form, the characteristic function of a stable distribution involves four real parameters, α, β, c and μ, and is given by[4]

$$\widetilde{p}(k) = \exp\left[-ik\mu - c|k|^{\alpha}\left\{1 + i\beta\tan\left(\tfrac{1}{2}\pi\alpha\right)\operatorname{sign}k\right\}\right] \quad \text{for } \alpha \neq 1, \tag{K.3}$$

and by

$$\widetilde{p}(k) = \exp\left[-ik\mu - c|k|\left\{1 - \frac{2i\beta}{\pi}(\ln|k|)\operatorname{sign}k\right\}\right] \quad \text{for } \alpha = 1. \tag{K.4}$$

The parameter $\mu \in (-\infty, \infty)$ is related to the overall 'location' of the distribution (it is the mean value when the latter exists), $c \in (0, \infty)$ is a scale factor, and $\beta \in [-1, 1]$ is a measure of the asymmetry of the PDF about its center. When $\alpha < 1$ and $\beta = \pm 1$, the distribution is one-sided: the support of the distribution in ξ is $(-\infty, 0]$ if $\beta = -1$, while it is $[0, \infty)$ for $\beta = +1$. The symmetric stable distributions correspond to $\beta = 0$. These constitute the family of **Lévy symmetric alpha-stable distributions**.

K.3 The three important special cases

There are three notable and important cases in which the formula for $\widetilde{p}(k)$ can be inverted to yield explicit expressions for the PDF $p(\xi)$ in terms of *elementary* functions[5]. These are:

(i) $\alpha = 2$: the distribution is a Gaussian with mean μ and variance $2c$, the PDF being

$$p(\xi) = (4\pi c)^{-1/2}\exp\left[-(\xi - \mu)^2/(4c)\right]. \tag{K.5}$$

[3]For certain special values of the parameters, it *can* be written down in such a form. See Sec. K.3 below.

[4]Equations (K.3) and (K.4) represent one out of a number of alternative ways of writing $\widetilde{p}(k)$ for stable distributions.

[5]There do exist a couple of other cases, corresponding to $\alpha = \frac{2}{3}$ and $\alpha = \frac{3}{2}$, when the PDF can be obtained explicitly in terms of known functions. But these do not occur as frequently in physical applications. Moreover, in these cases $p(\xi)$ is only expressible in terms of certain hypergeometric functions (to be precise, Whittaker functions), and does not reduce to any elementary function.

Since $\tan \pi = 0$, the parameter β plays no role in this case. The Gaussian distribution occurs very frequently in this book. The first such example is the Maxwellian shown in Fig. 2.1, Ch. 2.

(ii) $\alpha = 1$, $\beta = 0$: this is the **Cauchy distribution**. The PDF has a Lorentzian shape, symmetric about its center μ, and is given by

$$p(\xi) = \frac{c}{\pi[(x-\mu)^2 + c^2]}\,. \tag{K.6}$$

Figure 16.1 in Ch. 16 is an example of a Lorentzian.

(iii) $\alpha = \frac{1}{2}$, $\beta = +1$: this case by itself is often called the Lévy distribution[6]. When $\mu = 0$, the PDF is restricted to positive values of ξ, and is given by[7]

$$p(\xi) = \begin{cases} \left[c/(2\pi\xi^3)\right]^{1/2} \exp\left[-c/(2\xi)\right], & \xi > 0 \\ 0, & \xi < 0. \end{cases} \tag{K.7}$$

The first-passage time density $q(t\,,\,x\,|\,x_0)$ for diffusion in one dimension, shown in Fig. 10.1, Ch. 10 , is an example of such a PDF. As emphasized in the text, $q(t\,,\,x\,|\,x_0)$ is a stable density with exponent $\frac{1}{2}$ in the random variable t—the random variable here is the *time* of first passage from x_0 to x.

There are close relationships among the three special distributions listed above. In Ch. 7, Sec. 7.5.6, two such connections have been pointed out, in the context of diffusion in one dimension. Again, in Ch. 10, Sec. 10.3.3, a similar relationship has been obtained—in this case, in the context of first-passage times for one-dimensional diffusion. All these are special cases of certain remarkable connections between different members of the family of stable distributions.

For example: If ξ has a Gaussian distribution ($\alpha = 2$), then ξ^{-2} has a Lévy distribution ($\alpha = \frac{1}{2}, \beta = 1$). This is a special case of a **duality** that exists between different stable distributions: A stable distribution with index α (where $1 \le \alpha \le 2$) for ξ is essentially equivalent to a stable distribution with index α^{-1} (so that $\frac{1}{2} \le \alpha^{-1} \le 1$) for $\xi^{-\alpha}$.

[6] This can be a little confusing: as we have stated, the complete name for the whole family of stable distributions is 'Lévy alpha-stable distributions'.

[7] Essentially the same distribution is obtained by reflection about the ordinate axis, with support restricted to *negative* values of ξ, when $\alpha = \frac{1}{2}$, $\beta = -1$.

K.4 Asymptotic behavior: 'heavy-tailed' distributions

Although the Fourier transform $\widetilde{p}(k)$ cannot be inverted explicitly in general, the asymptotic or large-$|\xi|$ behavior of $p(\xi)$, or that of the cumulative distribution function $\mathcal{P}(\xi)$, can be deduced for any allowed value of α.

- *Except for the Gaussian case* (for which α has the limiting value 2), all the stable distributions are 'heavy-tailed' in the sense that they fall off like an inverse power of $|\xi|$ as $|\xi| \to \infty$: the cumulative distribution function $\mathcal{P}(\xi) \sim |\xi|^{-\alpha}$, to leading order in this limit. Correspondingly, the PDF $p(\xi) \sim |\xi|^{-\alpha-1}$.

This power-law behavior leads to a **self-similarity** property that has important consequences in the applications of these distributions.

- As a direct result of the fact that PDF $p(\xi) \sim |\xi|^{-\alpha-1}$, we find that *none of the stable distributions except the limiting case of the Gaussian has a finite mean squared value, and hence a finite variance.* In fact, for $\alpha < 1$, even the mean value is infinite[8] .

K.5 Generalized Central Limit Theorem

We have seen that the Central Limit Theorem (Appendix D, Sec. D.5) essentially says that the limit law for appropriate linear combinations of independent random variables, each of which has a distribution with finite mean and variance, is the Gaussian distribution. A notable example is that of the passage from random walks to diffusion, detailed in Appendix E. Recall the remarks made there regarding the generality of the result.

Similar limit-law results obtain for sums of independent random variables, each of which has a distribution with a power-law tail characterized by an exponent $-\alpha$. (The distribution of each random variable in the sum could actually be different in detailed form from that of the others.) The limit law in such a case is precisely a stable distribution with index α.

- In this sense, the stable distributions act as attractors in the space of distributions. Under fairly general conditions, it can be shown that there are no

[8] Even in the marginal case $\alpha = 1$, it is formally 'logarithmically divergent'. But for the (symmetric) Lorentzian distribution, we could argue that the first central moment is zero by invoking symmetry—or, more formally, by agreeing to define the central moment as the principal value of the integral concerned.

other such attractors[9].

The standard Central Limit Theorem is thus a special case (perhaps the most important one) of a more general result.

K.6 Infinite divisibility

We have seen that the re-scaled sum of n i.i.d. random variables, each of which has a stable distribution, is also a stable distribution. This holds good for every positive integer value of n. We can now ask the converse question: given a random variable ξ, can it be written as the sum of n independent, *identically distributed* random variables ξ_i $(1 \le i \le n)$ (which may be termed the 'components' of ξ), for every positive integer value of n? If so, ξ is said to be an **infinitely divisible random variable**. Note that the (common) distribution of each component ξ_i need *not* be the same as that of the sum ξ. Moreover, if the set of possible values of a random variable is *bounded*, it *cannot* be infinitely divisible.

If a random variable is infinitely divisible, its probability distribution is an **infinitely divisible distribution**. It is clear that this is only possible if the characteristic function of ξ is the nth power of a characteristic function (that of ξ_i) for every n. Now, a function $\widetilde{p}(k)$ has to be quite special in order to be the characteristic function of a probability distribution. For instance, $\widetilde{p}(0)$ must be equal to unity, so that the random variable ξ is a proper random variable, with a normalized probability distribution. Moreover, the inverse Fourier transform of $\widetilde{p}(k)$ must be a real, non-negative function of ξ, in order to be an acceptable PDF. If $\widetilde{p}_{\text{comp}}(k)$ is the (common) characteristic function of any of the n components ξ_i of an infinitely divisible random variable, and $\widetilde{p}(k)$ is that of the sum ξ, we must have a relationship of the form

$$\widetilde{p}(k) = \big[\,\widetilde{p}_{\text{comp}}(k)\,\big]^n \tag{K.8}$$

for every positive integer value of n. It turns out that

- all the Lévy alpha-stable distributions are also infinitely divisible distributions.

It is easy to see how this comes about in an important special case, the standard Gaussian distribution (which is a stable distribution with $\alpha = 2$). If the mean value is μ and the variance is σ^2, characteristic function is given by

$$\widetilde{p}(k) = \exp\left\{-ik\mu - \tfrac{1}{2}\sigma^2 k^2\right\} = \big[\,\widetilde{p}_{\text{comp}}(k)\,\big]^n, \tag{K.9}$$

[9]More precisely: the stable distributions have non-empty basins of attraction in the space of distributions.

where

$$\widetilde{p}_{\text{comp}}(k) = \exp\left\{-ik\,(\mu/n) - \tfrac{1}{2}(\sigma/\sqrt{n})^2\,k^2\right\}. \tag{K.10}$$

It is an instructive exercise (left to the reader) to check out the case of a general stable distribution in similar fashion.

We conclude this brief account of stable distributions with the following remarks.

- The form of the characteristic function $\widetilde{p}_{\text{comp}}(k)$ in Eq. (K.10) shows that each component ξ_i of the infinitely divisible Gaussian random variable ξ is itself a Gaussian random variable. This is a feature of every stable distribution: the corresponding random variable is infinitely divisible into components that have the same stable distribution as the original one.

- While all the stable distributions are infinitely divisible distributions, the converse is not true. *The stable distributions comprise a subset of the set of infinitely divisible distributions.*

A prominent example of this assertion is the Poisson distribution. Recall that a random variable r $(= 0,\,1,\,\ldots)$ is Poisson-distributed with a mean value[10] ν when $P(r) = e^{-\nu}\,\nu^r/r!$. The characteristic function is therefore

$$\widetilde{P}(k) = \left\langle e^{-ikr}\right\rangle = \exp\left\{\nu\,(e^{-ik}-1)\right\} = \left[\exp\left\{(\nu/n)\,(e^{-ik}-1)\right\}\right]^n, \tag{K.11}$$

showing that the Poisson distribution is infinitely divisible. But it is not a member of the family of Lévy alpha-stable distributions[11].

[10]It is well known that the variance is equal to the mean value for a Poisson-distributed random variable. But it is easily seen that, in fact, *every* cumulant is equal to the mean value for this distribution.

[11]Once again, in the interests of accuracy and completeness we need this (last!) footnote. The final equation in Eqs. (K.11) shows that $\widetilde{p}_{\text{comp}}(k)$ is again the characteristic function of a Poisson distribution. This suggests that, although the Poisson distribution is not a stable distribution in the standard sense, it comes close to being one! It turns out that the concept of a stable distribution can be extended in certain respects to distributions on $\mathbb{Z}_+$, the set of non-negative integers. In this modified sense, the Poisson distribution can indeed be regarded as a 'stable' distribution. The analog of the index α is 1 in this case.

Suggested Reading

1. L. Arnold, *Stochastic Differential Equations: Theory and Applications*, Wiley, New York, 1974.

2. V. I. Arnold and A. Avez, *Ergodic Problems of Classical Mechanics*, Benjamin, New York, 1968.

3. R. Balescu, *Equilibrium and Nonequilibrium Statistical Mechanics*, Wiley, New York, 1975.

4. R. Balescu, *Statistical Dynamics: Matter Out of Equilibrium*, Imperial College Press, London, 2000.

5. T. K. Caughey, in *Advances in Applied Mechanics*, Vol. 11, ed. C.-S. Yih, Academic Press, New York, 1971.

6. W. T. Coffey, Yu. P. Kalmykov, and J. T. Waldron, *The Langevin Equation: With Applications to Stochastic Problems in Physics, Chemistry and Electrical Engineering*, World Scientific, Singapore, 2004.

7. D. R. Cox and H. D. Miller, *The Theory of Stochastic Processes*, Chapman & Hall, London, 1994.

8. S. Dattagupta, *Relaxation Phenomena in Condensed Matter Physics*, Academic Press, New York, 1987.

9. S. Dattagupta and S. Puri, *Dissipative Phenomena in Condensed Matter: Some Applications*, Springer, New York, 2004.

10. J. L. Doob, *Stochastic Processes*, Wiley, New York, 1953.

11. J. Dorfman, *An Introduction to Chaos in Nonequilibrium Statistical Mechanics*, Cambridge University Press, Cambridge, 2001.

12. W. Feller, *An Introduction to Probability Theory and Its Applications*, Vols. 1 & 2, Wiley Eastern, New Delhi, 1972.

13. G. Gallavotti, F. Bonetto, and G. Gentile, *Aspects of Ergodic, Qualitative and Statistical Theory of Motion*, Springer-Verlag, Berlin, 2004.

14. C. W. Gardiner, *Handbook of Stochastic Processes for Physics, Chemistry and the Natural Sciences*, Springer-Verlag, Berlin, 1997.

15. P. Gaspard, *Chaotic Scattering and Statistical Mechanics*, Cambridge University Press, Cambridge, 1998.

16. D. T. Gillespie, *Markov Processes: An Introduction for Physical Scientists*, Academic, New York, 1991.

17. W. T. Grandy, Jr., *Foundations of Statistical Physics, Vol. II: Nonequilibrium Phenomena*, Reidel, Dordrecht, 1988.

18. G. Grimmett and D. Stirzaker, *Probability and Random Processes*, 3rd edition, Oxford University Press, Oxford, 2001.

19. K. Itô and H. P. McKean, Jr., *Diffusion Processes and Their Sample Paths*, Springer-Verlag, Berlin, 1965.

20. M. Kac, *Probability and Related Topics in Physical Sciences*, Wiley-Interscience, New York, 1959.

21. J. Keizer, *Statistical Thermodynamics of Nonequilibrium Processes*, Springer-Verlag, Berlin, 1987.

22. R. Kubo, in *Quantum Statistical Mechanics in the Natural Sciences*, eds. B. Kursunoglu, S. L. Mintz, and S. M. Widmeyer, Plenum Press, New York, 1973.

23. R. Kubo, M. Toda, and N. Hashitsume, *Statistical Physics II: Nonequilibrium Statistical Mechanics*, Springer-Verlag, Berlin, 1985.

24. L. D. Landau and E. M. Lifshitz, *Statistical Physics*, Part 1, 3rd edition, Elsevier, New York, 1999.

25. B. H. Lavenda, *Nonequilibrium Statistical Thermodynamics*, Wiley, New York, 1985.

26. G. F. Lawler and L. N. Coyle, *Lectures on Contemporary Probability*, Amer. Math. Soc., Providence, RI, 1999.

27. G. F. Lawler, *Introduction to Stochastic Processes*, 2nd edition, Chapman & Hall, New York, 2006.

28. J. L. Lebowitz and E. W. Montroll, eds., *Studies in Statistical Mechanics, Vol. 11: Nonequilibrium Phenomena from Stochastics to Hydrodynamics*, Elsevier, New York, 1984.

29. D. S. Lemons, *An Introduction to Stochastic Processes*, The Johns Hopkins University Press, Baltimore, 2002.

30. G. F. Mazenko, *Nonequilibrium Statistical Mechanics*, Wiley-VCH, Weinheim, 2006.

31. R. M. Mazo, *Brownian Motion: Fluctuations, Dynamics and Applications*, Clarendon Press, Oxford, 2002.

32. J. L. Mijnheer, *Sample Path Properties of Stable Processes*, Mathematical Centre Tracts No. 59, Mathematics Centrum, Amsterdam, 1975.

33. E. W. Montroll and J. L. Lebowitz, eds., *Fluctuation Phenomena*, North-Holland, Amsterdam, 1987.

34. E. Nelson, *Dynamical Theories of Brownian Motion*, Princeton University Press, Princeton, 1967.

35. G. Nicolis and I. Prigogine, *Self-organisation in Nonequilibrium Systems*, Wiley, New York, 1977.

36. I. Oppenheim, K. Shuler and G. Weiss, *Stochastic Processes in Chemical Physics: The Master Equation*, MIT Press, Cambridge, 1977.

37. S. Redner, *A Guide to First-Passage Processes*, Cambridge University Press, Cambridge, 2001.

38. L. E. Reichl, *A Modern Course in Statistical Physics*, Wiley-Interscience, New York, 1998.

39. F. Reif, *Fundamentals of Statistical and Thermal Physics*, McGraw-Hill, New York, 1965.

40. H. Risken, *The Fokker-Planck Equation: Methods of Solution and Applications*, 2nd edition, Springer-Verlag, Berlin, 1996.

41. B. Simon, *Functional Integration and Quantum Physics*, 2nd edition, AMS Chelsea Publishing, Providence, RI, 2000.

42. R. L. Stratonovich, *Topics in the Theory of Random Noise*, Vol. I, Gordon and Breach, New York, 1981.

43. T. T. Soong, *Random Differential Equations in Science and Engineering*, Academic Press, New York, 1973.

44. N. G. van Kampen, *Stochastic Processes in Physics and Chemistry*, North-Holland, Amsterdam, 1985.

45. N. Wax, ed., *Selected Papers in Noise and Stochastic Processes*, Dover, New York, 1954.

46. G. H. Weiss, *Aspects and Applications of the Random Walk*, North-Holland, Amsterdam, 1994.

47. E. Wong, *An Introduction to Random Processes*, Springer-Verlag, New York, 1983.

48. V. M. Zolotarev, *One-dimensional Stable Distributions*, American Mathematical Society, Providence, R. I., 1986.

49. D. N. Zubarev, *Nonequilibrium Statistical Mechanics*, Consultants Bureau, New York, 1974.

50. R. Zwanzig, *Nonequilibrium Statistical Mechanics*, Oxford University Press, Oxford, 2004.

Index